Evolutionary
Paleoecology

The Ecological Context
of Macroevolutionary Change

Evolutionary Paleoecology

The Ecological Context
of Macroevolutionary Change

EDITED BY

Warren D. Allmon
David J. Bottjer

Columbia University Press

Columbia University Press
New York   Chichester, West Sussex

Copyright © 2001 Columbia University Press
All rights reserved

Library of Congress Cataloging-in-Publication Data

Evolutionary paleoecology : the ecological context of macroevolutionary
change / edited by Warren D. Allmon, David J. Bottjer.
    p. cm.
  Includes bibliographical references and index.
  ISBN 0-231-10994-6 (cloth : alk. paper)—ISBN 0-231-10995-4 (pbk. :
alk. paper)
    1. Evolutionary paleoecology. I. Allmon, Warren D. II. Bottjer, David J.
  QE721.2.E87 E96 2000
  560'.45—dc21

                                                    00-064522

Printed in the United States of America
c 10 9 8 7 6 5 4 3 2 1
p 10 9 8 7 6 5 4 3 2 1

Contents

Dedication

J. JOHN SEPKOSKI JR. stands as one of the preeminent leaders of the late twentieth century in the ongoing effort to synthesize evolutionary paleobiology and paleoecology into the new discipline of evolutionary paleoecology. Many scientific disciplines, born recently, collect data with new technology at enormous rates. The avid practice of paleontology dates back to the nineteenth century, and given the nature of the materials, production of data is time-intensive because it is typically "hand-crafted" by paleontologists. Jack was one of the first paleontologists to recognize the treasure trove of data that existed in the paleontological literature of the past 150 years, which if extracted, could allow paleontologists sufficient quantities of data to allow statistical analysis and modeling of broad trends in the fossil record. And this is where Jack's great success lies. His legacy resides in such fundamental contributions as establishing the broad diversity trend of marine families in the Phanerozoic; the statistical analysis of mass extinctions and their timing, including recognition of the "Big 5"; delineation of the three Great Evolutionary Faunas of the Phanerozoic; and characterization of onshore–offshore trends. On his shoulders he lifted paleontology up, and much of what is evolutionary paleoecology today begins with his accomplishments.

Jack collaborated with many individuals to produce these achievements, and his name will always be linked with the highly productive association he had with Dave Raup. Many of us who worked with Jack were energized by his vision and creativity. Perhaps what was most impressive about this giant in our field was his humility and enormous generosity, particularly to the younger practitioners of paleontology. Jack mixed this all in with a great sense of humor, and evenings with him commonly combined conversations on paleontology with high adventure. In recent years his marriage to Christine Janis seemed the perfect match, and he talked with great excitement on their life together. His premature departure from our lives leaves both a personal and a professional void. His research interests and activities had never been greater, as reflected in his broad involvement with the production of this book. He read and made detailed comments on all the contributions and was preparing to write a final summary chapter when he died on May 1, 1999. Jack Sepkoski set the stage for much of what we do, and it is to his memory that we dedicate this volume.

Contributors

Warren D. Allmon
Paleontological Research Institution
1259 Trumansburg Road
Ithaca, NY 14850

Richard B. Aronson
Dauphin Island Sea Lab
101 Bienville Boulevard, Dauphin Island, AL 36528
Department of Marine Sciences
University of South Alabama
Mobile, AL 36688

Richard M. Bateman
The Natural History Museum
Cromwell Road
London SW7 5BD, UK

David J. Bottjer
Department of Earth Sciences
University of Southern California
Los Angeles, CA 90089-0740

Paul Cutlip
Department of Geology and Geophysics
Texas A&M University
College Station, TX 77843

William A. DiMichele
Department of Paleobiology
Smithsonian Institution
Washington, DC 20560, USA

Mary L. Droser
Department of Earth Sciences
University of California
Riverside, CA 92521

Thor A. Hansen
Department of Geology
Western Washington University
Bellingham, WA 98225

Patricia H. Kelley
Department of Earth Sciences
University of Carolina at Wilmington
Wilmington, NC 28403–3297

Bruce S. Lieberman
Department of Geology
University of Kansas
120 Lindley Hall
Lawrence, KS 66045

George R. McGhee Jr.
Department of Geological Sciences
Rutgers University
New Brunswick, NJ 08903

William Miller III
Department of Geology
Humboldt State University
Arcata, CA 95521-8299

William F. Precht
PBS & J
2001 Northwest 107th Avenue
Miami, FL 33308

Anne Raymond
Department of Geology and Geophysics
Texas A&M University
College Station, TX 77843

Robert M. Ross
Paleontological Research Institution
1259 Trumansburg Road
Ithaca, NY 14850

Peter M. Sheehan
Department of Geology
Milwaukee Public Museum
Milwaukee, WI 53233

William E. Stein
Center for Paleobotany
Binghamton University
Binghamton, NY 13902

Merrill Sweet
Department of Biology
Texas A&M University
College Station, TX 77843

Carol M. Tang
Department of Geology
Arizona State University
Tempe, AZ 85287–1404

James W. Valentine
Museum of Paleontology and Department of
    Integrative Biology
University of California
Berkeley, CA 94720

1

Evolutionary Paleoecology: The Maturation of a Discipline

Warren D. Allmon and David J. Bottjer

EVERY BOOK HAS A HISTORY. In this instance, the history says much about the changes in the discipline of evolutionary paleoecology. Around 1990, one of us proposed the idea for a symposium on evolutionary paleoecology to the Paleontological Society. There was only moderate interest in the topic, however, and it entered the queue of symposium topics to be almost forgotten, even by the proposer. In early 1995 the coordinator for the Paleontological Society reminded the proposer that the symposium was approaching the top of the pile and that he needed to begin to get things organized. This time, interest among potential contributors was much greater and the response to participate was so enthusiastic that when the symposium was finally held in October 1996, in Denver, it had too many speakers, and presentations had to be limited to 15 minutes instead of the usual 20.

Why the difference? We think that something (perhaps several things) has happened in the last few years that has made the topic of evolutionary paleoecology one of the most active and exciting in paleontology.

The taxonomy of disciplines is always subjective. What we call *evolutionary paleoecology* is a loosely connected skein of research programs that focus on the environmental and ecological context for long-term (i.e., macroevolutionary) changes seen in the fossil record. This conceptualization is sufficiently broad to successfully encompass two recent definitions of the term. Valentine (1973:2) defined *evolutionary paleoecology* as "the study of the evolution of biological organization"; Kitchell (1985:91) labeled it the study of "the macroevolutionary consequences of ecological roles and strategies."

These definitions distinguish *evolutionary paleoecology* from what Kitchell called simply *paleoecology*, defined as "studies of past environments that contribute to applied problems and theory in the geological sciences, particularly facies analysis and the reconstruction of past environments" (1985:91). If a more specific term for such studies is required, *descriptive paleoecology* may suffice. Basic references for this field include Ladd (1957), Ager (1963), Imbrie and Newell (1964), Schäfer (1972), Boucot (1981), Gall (1983), Newton and Laporte (1989), and Dodd and Stanton (1991). This definition may also distinguish *evolutionary paleoecology* from what has frequently been called *community paleoecology*, the subfield devoted to describing the diversity, environmental setting, structure, and patterns of change in paleocommunities, and to understanding the factors that affect those features (e.g., Ziegler et al. 1974; Rollins and Donahue 1975; Scott and West 1976; Miller 1990).

Thus defined, *evolutionary paleoecology* has been around for a long time. Almost since the publication of *The Origin of Species* (1859), researchers have attempted to understand how the environment has affected evolutionary history, often using the fossil record as their primary data (e.g., Allmon 1994). So why the evident recent rise in activity and interest?

We detect the beginnings of a fundamental shift in thinking about the way in which ecology affects macroevolutionary patterns and processes. This shift may (or may not) mark the beginnings of a truly adequate understanding of how environment and ecology affect the evolutionary process over long timescales. In any case, it has dramatically affected the problems that many paleontologists find interesting and the methods by which they approach them. We point to five recent developments that may have heralded this shift:

1. Large-scale paleoecological patterns. The last 20 years have seen the documentation of a number of major patterns in the ecological history of life on Earth. Large-scale patterns of Phanerozoic diversity are now fairly well described (e.g., Sepkoski 1993). From these and similar data also came an understanding of patterns of onshore origination of morphological novelties (and so higher taxa) among many marine invertebrates (e.g., Bottjer and Jablonski 1988; Jablonski and Bottjer 1991). Over the course of the entire Phanerozoic Eon, benthic marine faunas show a distinctive pattern of changing position above and below the sediment-water interface (e.g.,

Ausich and Bottjer 1982; Bottjer and Ausich 1986); this pattern of tiering describes much of the overall shape of marine faunas over the last 540 million years. Last but probably not least, the nature of resource utilization over the Phanerozoic appears to include increasing bioturbation (Thayer 1983) and escalation between predators and prey (Vermeij 1977, 1987), and both of these patterns may be part of an overall increase in food supply in the oceans during this time (Bambach 1993; Vermeij 1995).

2. *Rise of the taxic view.* It is now reasonably clear that morphological stasis is a widespread evolutionary phenomenon, at least among some clades (e.g., Gould and Eldredge 1993; Eldredge 1995). To the degree that stasis is dominant in a clade, long-term morphological patterns in that clade must be explained largely through the patterns of origination and extinction of species that do not change significantly during their duration. This taxic view is very different from the transformational view, under which morphological trends within clades are produced largely by gradual changes within species lineages (Eldredge 1979, 1982). The dominance of morphological stasis in a clade calls into question the role of natural selection in producing long-term morphological trends; selection may be responsible for stasis via stabilizing selection (Eldredge 1985), it may act mainly at speciation (Avise 1976; Dobzhansky 1976), or it may not be very important at all at higher hierarchical levels of the evolutionary process (Gould 1985). The taxic view compels us to take morphological stasis seriously in explorations of the large-scale history of life, and in the context of paleoecology, it forces us to be specific about exactly where and how ecology might matter to evolution. The taxic view also has important methodological implications in that we may see much of the history of life as fundamentally a branching process (e.g., Raup 1985).

The pattern of "coordinated stasis" (Brett et al. 1996) and the "turnover-pulse hypothesis" (Vrba 1993) have further highlighted and encouraged the taxic view, particularly around the issue of exactly how (or even whether) the environment may interact with individual lineages to create patterns of origination, stability, and extinction. We have long known that there are "intrinsic" as well as "extrinsic" factors in evolution (Allmon and Ross 1990); we are now beginning to focus on what role particular intrinsic and extrinsic factors may be playing in determining many taxonomic patterns (e.g., Morris et al. 1995).

3. *Appreciation of scale.* Can processes acting at one timescale adequately explain phenomena at all timescales? Are patterns at one timescale reducible or expandable to other timescales? We once thought we knew the answer. Much of the power of Darwinism lies in its purported ability to explain

long-term changes in the history of life via processes visible in the backyard pigeon cage. However, it has become increasingly evident that application of Darwinian natural selection or any other evolutionary process must occur at the appropriate temporal and spatial scale (e.g., Gould 1985; Aronson 1994; Martin 1998). Processes acting at one scale may not apply at another; patterns at one scale may not be recognizable at another. This means that the recognition of large-scale paleoecological patterns such as those described above may or may not be explicable by processes acting at ecological timescales accessible to human investigators today.

4. *Uniformitarianism revisited.* Along with problems of temporal scaling, it has also become increasingly apparent that there are paleoecological questions that do not yield satisfactory solutions through the strict application of uniformitarian approaches. Although the usual approach for reconstructing history in the natural world uses uniformitarianism as a dominant guiding principle, reconstruction of Earth's biological history differs from using immutable physical and chemical axioms. The reason for this difference is that biological and physical features of Earth's environments, by their very nature, have changed through time because of organic evolution. Thus, it is possible for ancient biological attributes of the environment to no longer exist or be predominant in modern settings (e.g., Kauffman 1987; Berner 1991; Sepkoski et al. 1991; Hagadorn and Bottjer 1997). Nonuniformitarian approaches have been most commonly taken by Precambrian paleoecologists. Phanerozoic paleoecologists, however, have begun to adopt some of the healthy skepticism about uniformitarianism that characterizes the methodology of the Precambrian paleoecologist. Much of the growth of the new discipline of evolutionary paleoecology will depend on the insights provided through application of a nonuniformitarian viewpoint (e.g., Bottjer et al. 1995; Vannier, Babin, and Rocheboeuf 1995; Fischer and Bottjer 1995; Bottjer 1998).

5. *Geobiology.* Although we have long known that the earth's physical environment "matters" to evolution, we have struggled to understand exactly how. One common problem is that we have frequently lacked sufficiently detailed data on the nature of the physical environment in the geological past to allow us to compare environmental and evolutionary changes. With the advent of much more precise geochronology and stable isotope biogeochemistry, however, more and more researchers are attempting very precise comparisons between ancient physical environmental changes and evolutionary events, from the Precambrian to the Holocene, from protists to hominids (e.g., Knoll 1992; Feibel 1997). This pursuit is referred to by some as *geobiology.* (This word is also sometimes used as

almost synonymous with paleobiology; see Bottjer 1995b.) As we begin to learn more about the nature of Earth's physical history, we may be able to learn a great deal more about how life has responded to that history.

Prospect

One of the most important questions we can ask about the history of life is, "does ecology matter" (Jackson 1988)? Most biologists and paleontologists were trained to believe that it does, but the exact mechanisms by which ecology matters to patterns that play out over tens or hundreds of millions of years have never been entirely clear. As we learn more about these patterns, the search for their causes becomes even more pressing. Research has refined the questions. As Carl Brett and co-authors have put it in a recent major volume on coordinated stasis: "the most significant goal and challenge of evolutionary paleoecology lies in seeking a new synthetic view of the evolutionary process which integrates the processes of species evolution, ecology, and mass extinction" (Brett, Ivany, and Schopf 1996:17).

This summary is amply borne out in the chapters of this volume. This book is not an encyclopedic synthesis of evolutionary paleoecology, but a benchmark sampler of active research in a very active field. The chapters do not so much answer whether, or the way in which, ecology matters as they explore in fairly explicit directions the ways in which it might. In these directions must lie the solution to the question of how the biotic and abiotic environment affect evolutionary change on this planet.

REFERENCES

Ager, D. 1963. *Principles of Paleoecology.* New York: McGraw-Hill.

Allmon, W. D. 1994. Taxic evolutionary paleoecology and the ecological context of macroevolutionary change. *Evolutionary Ecology* 8:95–112.

Allmon, W. D. and R. M. Ross. 1990. Specifying causal factors in evolution: The paleontological contribution. In R. M. Ross and W. D. Allmon, eds., *Causes of Evolution: A Paleontological Perspective,* pp. 1–17. Chicago: University of Chicago Press.

Aronson, R. 1994. Scale-dependent biological interactions in the marine environment. *Annual Review of Oceanography and Marine Biology* 32:435–460.

Ausich, W. I. and D. J. Bottjer. 1982. Phanerozoic tiering in suspension-feeding communities on soft substrata throughout the Phanerozoic. *Science* 216:173–174.

Avise, J. C. 1976. Genetic differentiation during speciation. In F. J. Ayala, ed., *Molecular Evolution,* pp. 106–122. Sunderland MA: Sinauer Associates.

Bambach, R. K. 1993. Seafood through time: Changes in biomass, energetics and productivity in the marine ecosystem. *Paleobiology* 19:372–397.

Berner, R. A. 1991. A model for atmospheric CO_2 over Phanerozoic time. *American Journal of Science* 291:339–376.

Bottjer, D. J. 1995a. Evolutionary paleoecology: Diverse approaches. *Palaios* 10(1):1–2.

Bottjer, D. J. 1995b. Our unique perspective. *Palaios* 10(6):491–492.

Bottjer, D. J. 1998. Phanerozoic non-actualistic paleoecology. *Geobios* 30:885–893.

Bottjer, D. J. and W. I. Ausich. 1986. Phanerozoic development of tiering in soft sub-strata suspension-feeding communities. *Paleobiology* 12:400–420.

Bottjer, D. J. and D. Jablonski. 1988. Paleoenvironmental patterns in the evolution of post-Paleozoic benthic marine invertebrates. *Palaios* 3:540–560.

Bottjer, D. J., K. A. Campbell, J. K. Schubert, and M. L. Droser. 1995. Palaeoecological models, non-uniformitarianism, and tracking the changing ecology of the past. In D. W. J. Bosence and P. A. Allison, eds., *Marine Palaeoenvironmental Analysis from Fossils*, pp. 7–26. Geological Society Special Publication No. 83. London: The Geological Society.

Boucot, A. J. 1981. *Principles of Benthic Marine Paleoecology.* New York: Academic Press.

Brett, C. E., L. C. Ivany, and K. M. Schopf. 1996. Coordinated stasis: An overview. *Palaeogeography, Palaeoclimatology, Palaeoecology* 127:1–21.

Darwin, C. 1859. *On the Origin of Species.* London: John Murray.

Dobzhansky, T. 1976. Organismic and molecular aspects of species formation. In F. J. Ayala, ed., *Molecular Evolution*, pp. 95–105. Sunderland MA: Sinauer Associates.

Dodd, J. R. and R. J. Stanton Jr. 1991. *Paleoecology: Concepts and Applications*, 2nd ed. New York: John Wiley and Sons.

Eldredge, N. 1979. Alternative approaches to evolutionary theory. *Bulletin of the Carnegie Museum of Natural History* 13:7–19.

Eldredge, N. 1982. Phenomenological levels and evolutionary rates. *Systematic Zoology* 31:338–347.

Eldredge, N. 1985. *Unfinished Synthesis: Biological Hierarchies and Modern Evolutionary Thought.* New York: Oxford University Press.

Eldredge, N. 1995. Species, speciation, and the context of adaptive change in evolution. In D. Erwin and R. Anstey, eds., *New Approaches to Speciation in the Fossil Record*, pp. 39–66. New York: Columbia University Press.

Feibel, C. S. 1997. Debating the environmental factor in hominid evolution. *GSA Today* 7(3):1–7.

Fischer, A. G. and D. J. Bottjer. 1995. Oxygen-depleted waters: A lost biotope and its role in ammonite and bivalve evolution. *Neues Jahrbuch fur Palaontologie Abhandlungen* 19:133–146.

Gall, J.-C. 1983. *Ancient Sedimentary Environments and the Habitats of Living Organisms.* Berlin: Springer-Verlag.

Gould, S. J. 1985. The paradox of the first tier: An agenda for paleobiology. *Paleobiology* 11(1):2–12.

Gould, S. J. and N. Eldredge. 1993. Punctuated equilibrium comes of age. *Nature* 366: 223–227.

Hagadorn, J. W. and D. J. Bottjer. 1997. Wrinkle structures: Microbially mediated sedimentary structures common in subtidal siliciclastic settings at the Proterozoic-Phanerozoic transition. *Geology* 25:1047–1050.

Imbrie, J. and N. Newell, eds. 1964. *Approaches to Paleoecology*. New York: Wiley.

Jablonski, D. and D. J. Bottjer. 1991. Environmental patterns in the origins of higher taxa: The post-Paleozoic fossil record. *Science* 252:1831–1833.

Jackson, J. B. C. 1988. Does ecology matter? *Paleobiology* 14:307–312.

Kauffman, E. G. 1987. The uniformitarian albatross. *Palaios* 2:531.

Kitchell, J. A. 1985. Evolutionary paleoecology: Recent contributions to evolutionary theory. *Paleobiology* 11(1):91–104.

Knoll, A. H. 1992. Biological and biogeochemical preludes to the Ediacaran radiation. In J. Lipps and P. Signor, eds., *The Origin and Early Evolution of the Metazoa*, pp. 53–84. New York: Plenum Press.

Ladd, H. S., ed. 1957. Treatise on marine ecology and paleoecology. Volume 2, *Paleoecology*. Geological Society of America Memoir 67. Boulder CO: The Geological Society of America.

Martin, R. E. 1998. *One Long Experiment: Scale and Process in Earth History*. New York: Columbia University Press.

Miller, W. III, ed. 1990. Paleocommunity temporal dynamics: The long-term development of multispecies assemblies. Special Publication No. 5. Knoxville TN: The Paleontological Society.

Morris, P. J. L. C. Ivany, K. M. Schopf, and C. E. Brett. 1995. The challenge of paleoecological stasis: Reassessing sources of evolutionary stability. *Proceedings of the National Academy of Sciences* 92:11269–11273.

Newton, C. R. and L. Laporte. 1989. *Ancient Environments*, 3rd ed. Englewood Cliffs NJ: Prentice Hall.

Raup, D. M. 1985. Mathematical models of cladogenesis. *Paleobiology* 11(1):42–52.

Rollins, H. B. and J. Donahue. 1975. Towards a theoretical basis of paleoecology: Concepts of community dynamics. *Lethaia* 8:255–270.

Schäfer, W. 1972. *Ecology and Paleoecology of Marine Environments*. Chicago: University of Chicago Press.

Scott, R. W. and R. R. West, eds. 1976. *Structure and Classification of Paleocommunities*. Stroudsburg PA: Dowden, Hutchinson, and Ross.

Sepkoski, J. J. Jr. 1993. Ten years in the library: New data confirm paleontological patterns. *Paleobiology* 19:43–51.

Sepkoski, J. J. Jr., R. K. Bambach, and M. L. Droser. 1991. Secular changes in Phanerozoic event bedding and the biological overprint. In G. Einsele, W. Ricken, and A. Seilacher, eds., *Cycles and Events in Stratigraphy*, pp. 298–312. Berlin: Springer.

Thayer, C. H. 1983. Sediment-mediated biological disturbance and the evolution of marine benthos. In M. J. S. Tevesz and P. L. McCall, eds., *Biotic Interactions in Recent and Fossil Benthic Communities*, pp. 480–626. New York: Plenum Press.

Valentine, J. W. 1973. *Evolutionary Paleoecology of the Marine Biosphere*. Englewood Cliffs NJ: Prentice Hall.

Vannier, J., C. Babin, and P. R. Rocheboeuf. 1995. Le principe d'actualisme applique aux faunes paleozoiques: Un outil or un leurre? *Geobios* 18:395–407.

Vermeij, G. J. 1977. The Mesozoic marine revolution: evidence from snails, predators, and grazers. *Paleobiology* 3:245–258.

Vermeij, G. J. 1987. *Evolution and Escalation.* Princeton NJ: Princeton University Press.

Vermeij, G. J. 1995. Economics, volcanoes, and Phanerozoic revolutions. *Paleobiology* 21:125–252.

Vrba, E. S. 1993. Turnover-pulses, the Red Queen, and related topics. *American Journal of Science* 293a:418–452.

Ziegler, A. M., K. R. Walker, E. J. Anderson, E. G. Kauffman, R. N. Ginsburg, and N. P. James. 1974. Principles of benthic community analysis: Notes for a short course. Sedimenta IV, University of Miami Comparative Sedimentology Laboratory.

2

Scaling Is Everything:
Brief Comments on
Evolutionary Paleoecology

James W. Valentine

BECAUSE I USED THE TERM EVOLUTIONARY PALEOECOLOGY in the title of a book in 1973 when the field was developing (Valentine 1973), the editors of this volume asked me to write briefly about the genesis of this term and to comment on how this field has fared. That book, *Evolutionary Paleo-ecology of the Marine Biosphere*, was indeed part of a broad movement to apply what was known about invertebrate fossils to attempt to answer biological questions. This movement involved a long series of contributions by many workers. My remarks are restricted to marine invertebrate studies.

The title, *Evolutionary Paleoecology of the Marine Biosphere*, was meant to carry two messages. The first was that the subject of the book was biological (or paleobiological) rather than geological. Although there had been many fine pioneering studies in what is now called paleoecology, the term *paleoecology* was being increasingly employed to describe the field of paleoenvironmental reconstruction. Some studies labeled as paleoecology did not involve organisms at all, but were sedimentological or petrographic, and were dedicated to understanding environments of deposition, not of habitation. Still other paleoecological studies that did involve organisms were nevertheless devoted only to reconstructing depositional environments for geological purposes.

Although those research programs were certainly valuable contributions to geology, they did not necessarily yield information on ecological processes of the past, except fortuitously as by-products. In search of an appropriate title for a treatment of paleoecology, I tried to find a phrase that connoted biology rather than geology. Paleobiological paleoecology sounded ridiculous, and even biological paleoecology was much too redundant, so evolutionary paleoecology it became, all 13 syllables. I'm not certain whether this was the first use of the term. Coincidentally, that same year Dobzhansky published his famous dictum that "nothing in biology makes sense except in the light of evolution" (Dobzhansky 1973), which rather nicely supported my choice.

Second, and more important, the title also implied a paleoecology at large scales, studied over evolutionary time rather than case by case. The best parts of the book were concerned with trends through time or with comparisons between conditions at different periods of time. With trivial exceptions, it is clearly not possible to study ecological or evolutionary processes directly from the fossil record. For a given fossil assemblage, about the best that can be done is use ecological theory to frame the various interpretations. What can uniquely be studied, however, are the results of ecological processes as they were worked out by evolution over stretches of time far longer than the life of a single investigator studying living ecosystems, or even than a single stratum bearing a fossil assemblage. A wide variety of ecological processes may be in play within a living community, but in order to determine which are important for biotic history, the fossil record is indispensable. A reasonable, widely followed, research strategy for the paleobiologist is to investigate some aspect of the fossil record to understand which biological questions might profitably be studied; to learn everything that is known of the processes that seem appropriate to the question from biological work; and then to proceed with a formal research project dedicated to testing relevant hypotheses over time and across circumstances in the fossil record. Curiously, not many biologists have reversed this strategy, although many hypotheses that are formulated to account for recent patterns are found to fail in the fossil record, and are thus at least incomplete.

Evolutionary paleoecology, then, would for a start use an ecological theory as a framework within which to examine and evaluate paleoecological processes, which famously form the theater of the evolutionary play, over time. The evolutionary events revealed in such studies are chiefly macroevolutionary, involving scales appropriate to the fossil record. Furthermore, the rising fields of biodiversity and of macroecology, although not strictly paleontological, have strong historical underpinnings, especially involving processes at scales perfectly familiar to investigators in paleoecology and macroevolution.

It is interesting that the literature of these neontological fields tends to be an easy read for paleoecologists, who are accustomed to the scales and even employ similar conceptual tools. Scale seems to be a key feature of evolutionary paleoecology. The fossil and Recent data and the range of hypotheses available to evolutionary paleoecologists are expanding continuously.

It is clearly impossible to evaluate or even mention all the current trends in evolutionary paleoecology; however, this volume provides at least an introduction. One of the stimuli for large-scale studies was the rise of the theory of plate tectonics: if there could be global tectonics, could there not be global paleobiology? Because plate tectonic processes were more or less incessant, they should provide a continuous but ever-changing template of physical environments to which ecological structures might be molded, and within which the evolutionary history of the biota, ever adapting to the new conditions, could be interpreted right across the Phanerozoic Eon. To be sure, for many parameters, the relationships between geological and biological processes are indirect and intricate, and prediction of cause and effect is difficult, especially considering the scale of the data. Nevertheless, after the appearance of global tectonics, Phanerozoic studies began to flourish. These studies present the phenomena not otherwise appreciated and provide a framework for more detailed research at finer scales.

The topics of global Phanerozoic research can be quite varied; Phanerozoic studies that are global for their subjects have been composed of, among other things, ecospace occupation (Bambach 1977), family diversity (Sepkoski 1981; Sepkoski and Hulver 1985); extinction (Raup and Sepkoski 1982; Jablonski 1986); vertical community structure (Ausich and Bottjer 1982); biological disturbance (Thayer 1983); shell-breaking predation (Vermeij 1983); of onshore–offshore origination (Jablonski et al. 1983); morphological patterns in corals (Coates and Jackson 1985); bioclastic accumulation (Kidwell and Brenchly 1994, 1996); and carbonate shell mineralogy (Stanley and Hardie 1998). This is not a scientific sampling of the literature, but it does suggest that there has been a lag and perhaps some revival in broad-scale studies, which is most welcome. The earlier of these studies have come to be regarded as seminal.

When finer-scale studies are made of features for which Phanerozoic data are available, they usually produce different results, and therefore the utility of the larger scales is sometimes questioned. Global diversity profiles of families commonly vary greatly from their orders and of the orders from their phyla, and regional variations exist in essentially all paleoecological parameters, raising questions as to which of the scales provides real results. Of course they all do, but the results do pertain to different questions on different scales. There is a good chance that the interrelationships themselves among data at different scales may prove to be a help to evolutionary paleoecology, but they have not yet been adequately investigated. Raup et al. (1973) modeled small-number samples of clade diversifications, repeated under the same rules but stochastic within certain constraints, and produced great variability in the resulting diversity profiles. However, if large-number samples were run with those rules, the variability between runs would be reduced (see Stanley 1979). But of course as long as there are stochastic elements in such a model, some variability will always remain; the largest of sample sizes is not fixed. The largest sample size of diversity available displays a well-known profile across the Phanerozoic (Sepkoski 1981). It is hard to believe that many of the processes that gave rise to this profile do not have stochastic elements. There must be a potential parental distribution of which our actual diversity history (assuming it is fairly represented by the profile) represents a sample. How much difference, then, would there be in the profile if we re-ran metazoan history? Or Phanerozoic history? I don't think that we know, but it's certainly a problem in evolutionary paleoecology, and one that might be solved, at the appropriate scale.

REFERENCES

Ausich, W. I. and D. J. Bottjer. 1982. Tiering in suspension-feeding communities on soft substrata throughout the Phanerozoic. *Science* 216:173–174.

Bambach, R. K. 1977. Species richness in marine benthic habitats through the Phanerozoic. *Paleobiology* 3:152–167.

Coates, A. G. and J. B. C. Jackson. 1985. Morphological themes in the evolution of clonal and aclonal marine invertebrates. In J. B. C. Jackson, L. W. Buss, and R. E. Cook, eds., *Population Biology and Evolution of Clonal Organisms,* pp. 67–106. New Haven CT: Yale University Press.

Dobzhansky, Th. 1973. Nothing in biology makes sense except in the light of evolution. *American Biology Teacher* 35:125–129.

Jablonski, D. 1986. Background and mass extinctions: the alternation of macroevolutionary regimes. *Science* 231:129–133.

Jablonski, D., J. J. Sepkoski Jr., D. J. Bottjer, and P. M. Sheehan. 1983. Onshore-offshore patterns in the evolution of Phanerozoic shelf communities. *Science* 222:1123–1125.

Kidwell, S. M. and P. J. Brenchley. 1994. Patterns of bioclastic accumulation throughout the Phanerozoic: Changes in input or in destruction? *Geology* 22:1139–1143.

Kidwell, S. M. and P. J. Brenchley. 1996. Evolution of the fossil record: Thickness trends in marine skeletal accumulations and their implications. In D. Jablonski, D. H. Erwin, and J. H. Lipps, eds., *Evolutionary Paleobiology,* pp. 290–336. Chicago: University of Chicago Press.

Raup, D. M. and J. J. Sepkoski Jr. 1982. Mass extinctions in the marine fossil record. *Science* 215:1501–1503.

Raup, D. M., S. J. Gould, T. J. M. Schopf, and D. S. Simberloff. 1973. Stochastic models of phylogeny and the evolution of diversity. *Journal of Geology* 81:525–542.

Sepkoski, J. J. Jr. 1981. A factor analytic description of the Phanerozoic marine fossil record. *Paleobiology* 7:36–53.

Sepkoski, J. J. Jr. and M. L. Hulver. 1985. An atlas of Phanerozoic clade diversity diagrams. In J. W. Valentine, ed., *Phanerozoic Diversity Patterns,* pp. 11–39. Princeton NJ: Princeton University Press.

Stanley, S. M. 1979. *Macroevolution.* San Francisco: W. H. Freeman.

Stanley, S. M. and L. A. Hardie. 1998. Secular oscillations in the carbonate mineralogy of reef-building and sediment-producing organisms driven by tectonically forced shifts in seawater chemistry. *Palaeogeography, Palaeoclimatology, Palaeoecology* 144:3–19.

Thayer, C. W. 1983. Sediment-mediated biological disturbance and the evolution of marine benthos. In M. J. S. Tevesz and P. L. McCall, eds. *Biotic Interactions in Recent and Fossil Benthic Communities,* pp. 479–625. New York: Plenum Press.

Valentine, J. W. 1973. *Evolutionary Paleoecology of the Marine Biosphere.* Englewood Cliffs NJ: Prentice-Hall.

Vermeij, G. J. 1983. Shell-breaking predation through time. In M. J. S. Tevesz and P. L. McCall, eds., *Biotic Interactions in Recent and Fossil Benthic Communities,* pp. 649–669. New York: Plenum Press.

3

What's in a Name?
Ecologic Entities and the Marine
Paleoecologic Record

William Miller III

IMAGINE A COMMUNITY ECOLOGIST venturing into the litera-
ture of marine paleoecology for the first time. Let us say that her first exposure
will be in the reading of a volume of contributed chapters, such as this one. If
our colleague scratches her head each time she is confused over inconsistent
and illogical usage of unit definitions, by the end of the book she might be
bald. This would be no reflection on the quality of data or analytical rigor in
such volumes, but rather a consequence of a prevailing indifference to fundamental properties of the ecologic entities recorded in fossil deposits. Should
paleoecologists do something about the situation or continue to promote
depilation in this way?

Paleoecology is usually considered to be the study of ecologic properties of
fossil organisms or assemblages of organisms. A better definition would state
that paleoecology is more concerned with organisms and assemblages viewed
at larger or more inclusive spatial and temporal scales than those typically considered in neoecology. What paleoecologists do is fairly clear, but why they do
it, the *purpose* of paleoecology, is far from clear. Although this chapter will seem
at first to be a rehashing of terminology, it is really about the issue of purpose,
the approach here being an assessment of the entities, or things, paleoecologists

study. Specifically, the approach will consist of a review of recent systems of paleoecologic unit classification and a proposal of a way to evaluate entities detected in the fossil record that could stabilize terminology and help to settle the ontologic aspect of purpose. I also illustrate some of the consequences of ignoring these issues.

The relationships of paleoecology to evolutionary biology in general and ecology in particular have always been uncertain, and occasionally someone says so unambiguously (Hoffman 1979, 1983; Gould 1980; Kitchell 1985; Allmon 1992; A. I. Miller 1993). One way to see this uncertainty is to notice how liberally paleoecologists have borrowed concepts and techniques from ecology, but how oblivious most ecologists seem to be about what goes on in paleoecology. As ecologists have begun to scale up their observations to encompass large units of biotic organization, large-scale environmental contexts, and climate history, they have started to work at levels familiar to paleoecologists. The ecologists, however, are developing their own brand of *macroecology* (e.g., Turner 1989; Delcourt and Delcourt 1991; Gilpin and Hanski 1991; Brown 1995; Hansson, Fahrig, and Merriam 1995; Wu and Loucks 1995). Perhaps the reason for the continuing separation of disciplines has to do with our attending separate conferences, publishing in different journals, or using different methods, but it might also relate to the fact that paleoecology somehow skipped a crucial stage in its conceptual development that Eldredge (1985:163) has described as ". . . frankly groping for an ontology of ecological entities. . . ." Terms such as *community, paleocommunity, assemblage,* and *biofacies* are used to mean almost any kind of multispecies aggregate. Ecologists are not entirely free from this confusion over terms (McIntosh 1985, 1995; Fauth et al. 1996), but paleoecologists, in terms of words available for use and spatiotemporal scaling dimensions, have more to be confused about.

If we take *deme* and *species-lineage* to be potentially real things whose meaning and significance need to be understood before evolutionary patterns and processes are interpreted satisfactorily (Mayr 1970, 1988; Stanley 1979; Eldredge 1989; Ereshefsky 1992; Gould 1995), why should we be unconcerned about the validity of the terms *community* and *ecosystem?* This is not the same as the debate over whether multispecies assemblies are strongly interacting, stable entities (the Clementsian–Eltonian view) or happenstance aggregations of populations merely tolerating local environmental factors (the Gleasonian view) (DeMichele et al., chapter 11, this volume). Instead, what I attempt to address is the problem, for instance, of letting a community be any of the following: fossils loaded into a sample bag at a particular locality; samples having generally similar fossil content collected at several different localities or stratigraphic levels; or statistically defined clusters of taxa or samples at many scales of resolution.

TABLE 3.1. Kauffman-Scott System of Units[a]

Global biota
   Contemporaneous global biota
      Realm
         Region
            Province
               Subprovince
                  Endemic center
                     Ecosystem
                        Sere
                           Assemblage
                              Community (paleocommunity)
                                 Association (many kinds)
                                   Population
                                     Individual organism

[a] Kauffman 1974; Kauffman and Scott, 1976.

TABLE 3.2. Boucot-Brett System[a]

Ecological-Evolutionary Units
   Ecological-Evolutionary Subunits
      Assemblage
         Biofacies / community group / community type
         Community

[a] Boucot 1975, 1983, 1990a,b,c; Brett, Miller, and Baird 1990; Brett and Baird 1995

TABLE 3.3. Bambach-Bennington System (1996)

Community type (paleocommunity type)
   Community (paleocommunity)
      Local community (local paleocommunity)
         Avatar (no fossil equivalent)

Classifications of Paleoecologic Units

Here I review five essentially hierarchical classification systems for fossil deposits that have a more or less explicit ecologic character (whether or not real ecologic entities or systems are in fact represented) and have been fairly well publicized (tables 3.1–3.4). There are other, mostly older, systems, but these are the ones paleoecologists are likely to think about when they consider units. To build a consensus regarding terminology, the practice of redefining units in every new publication should be discouraged. Parts of the classifications are compared in table 3.5.

TABLE 3.4. Valentine System[a]

Biosphere system
   Historical biotic system
      Province system
         Regional ecosystem
            Biotope system
               Local ecosystem
                  Ecologic association / interaction cell
                     Local population system / avatar
                        Individual organism

[a] Based on Valentine 1968, 1973; Eldredge and Salthe 1984; Eldredge 1985; W. Miller 1990, 1991, 1996

TABLE 3.5. Possible Correlation of Units Employed in Recent Paleoecologic Literature

Kauffman–Scott	*Boucot–Brett*	*Bambach–Bennington*	*Valentine*
Global biota	——	——	Biosphere system
Contemporaneous global biota[a]	Ecological– Evolutionary Unit[a]	——	Historical biotic system[a]
Province	——	——	Province system
Subprovince / endemic center[a]	Ecological– Evolutionary Subunit[a]		Regional ecosystem[a]
Assemblage	Biofacies	Community-type	Biotope system
Community[a]	Community[a]	Community[a]	——
——	——	Local community	Local ecosystem
Association[a]	——	——	Ecologic association[a]
Population[a]	——	Avatar[a]	Population system / avatar[a]
Organism	(Organism)	(Organism)	Organism

[a] Parts of units compared are essentially equivalent; or scale is nearly the same, but criteria vary somewhat.

Kidwell System

Kidwell and co-workers have developed a classification based on the degree of time-averaging of skeletal remains in a particular sample or bedding unit (Kidwell and Bosence 1991; Kidwell 1993; Kidwell and Flessa 1995). The system is not really hierarchical because less time-averaged units do not necessarily form parts of more time-averaged units. I mention it, however, because the extent of blending of original ecologic units is a criterion in the scheme, making it a useful starting place in the ecologic analysis of fossil deposits. The classification includes four categories of assemblage (Kidwell and Flessa 1995:288–289): *ecological snapshots* or *census assemblages*, providing a record of local communities having "zero to minimal time-averaging"; *within-habitat time-averaged assemblages*, recording "temporally persistent" communities over time spans of 1 to 10^3 yr; *environmentally condensed assemblages*, containing ecologic mixtures of skeletons that accumulated "over periods of significant environmental change" in the order of 10^2 to 10^4 yr; and *biostratigraphically condensed assemblages*, encompassing "major environmental changes as well as evolutionary time" and containing a record spanning $10^5–10^6$ yr.

Kauffman–Scott System

An elaborate classification was proposed by Kauffman (1974) and later expanded by Kauffman and Scott (1976). The scheme is in part hierarchical because higher levels may consist of the lower levels of organization, but it includes units that could be viewed as ecologic, developmental patterns, and as biogeographic divisions (table 3.1). The units are defined and compared by Kauffman and Scott (1976:13–21) in one of the only paleoecologic lexicons anyone has ever bothered to compile. The criteria for judging membership in the units are varied. For the multispecies aggregates, spatiotemporal co-occurrence of taxa and "vertical" position in the scheme are the most important characteristics.

Boucot–Brett System

Boucot (1975, 1983, 1986, 1990a,b,c; also Sheehan 1991, 1996) has repeatedly pointed out that extensive, practical biostratigraphic experience is the most reliable approach to ecologic classification of fossil deposits. His Ecological–Evolutionary Units have been adopted in the work on *coordinated stasis* by Brett and Baird (1995) as the most inclusive divisions of Phanerozoic ecologic history. This is a hierarchical classification of descriptive units (table 3.2) in that the more localized, short-lived units are contained in the interregional, long-lived divisions. The main criteria used to identify and organize the units are biostratigraphic position at varied scales of resolution and inclusiveness (based on size and duration). Brett and Baird (1995) recommended dividing the largest units into Ecological–Evolutionary Subunits. Beyond this, Boucot, Sheehan, Brett, and others have used terminology for the divisions of subunits including *assemblages, community groups* and *types*, and *biofacies*. Each major unit is viewed as a record of biotic stability or reorganization following an episode of extinctions; the smaller local units record environmentally controlled variations on the larger regimes. Boucot (1978, 1990c) has discussed the evolutionary dynamics associated with appearance and collapse of the largest divisions; Brett and co-workers (Miller, Brett, and Parson 1988; Brett, Miller, and Baird 1990; Brett and Baird 1995; Morris et al. 1995) have concentrated their attention on the properties of the smaller, more localized subdivisions.

Bambach–Bennington System

Another classification based largely on the criterion of species co-occurrences was proposed by Bambach and Bennington (1996). Their focus was on small-scale multispecies aggregates (table 3.3). The classification is interesting because Bambach and Bennington have been careful to emphasize the differences between class-like categories (generalized community types and kinds of communities) and local manifestations (local communities) and between living and equivalent fossil units (paleocommunity types, paleocommunities, and local paleocommunities). This is a useful terminology when the chief consideration in a study is taxonomic composition of localized assemblages, and it is one of the only schemes that recognizes the subtle difference between classes of ecologic entities and their individual representations (see discussions of individuality of ecologic entities by Salthe [1985] and W. Miller [1990]).

Valentine System

In terms of classification of ecologic entities, not fossil assemblages, the system that has developed from the early work of Valentine is probably the most conceptually robust and biologically realistic (Valentine 1968, 1973; Eldredge and Salthe 1984; Eldredge 1985; W. Miller 1990, 1991, 1996). I personally favor this scheme because the levels form a true hierarchy (entities recognized at lower, less-inclusive levels form the "working parts" of those at higher, more-inclusive levels of organization) and the emphasis is on correctly scaled ecologic properties (table 3.4), not nested sets of co-occurring organisms or taxa. The fundamental properties of entities at any level include (1) identity as economic systems involved primarily in matter–energy transfers; (2) interactions with similarly scaled entities, as well as with encompassing and constituent entities; (3) scale (a matter of both spatiotemporal extent and membership and inclusiveness); and (4) the related developmental trajectories (including initial organization, intervals of relative stability and episodes of disturbance and recovery, maturation, and eventual collapse). Readers will recognize this classification as being derived from the organizational framework of Valentine's influential book, *Evolutionary Paleoecology of the Marine*

Biosphere (1973), with elaborations introduced mostly by Eldredge and Salthe (1984).

Recognition of Ecologic Entities: Some Fundamental Concepts

All of the ecologic entities specified in what I call the Valentine System are open to energy flow but relatively closed as cybernetic systems (Margalef 1968; Brooks and Wiley 1988); have developmental patterns that depend on normal pathways of energy dissipation and departures from such configurations (disturbance and recovery) (Pickett and White 1985; Pimm 1991); and consist of organisms, parts of organisms, or organism aggregates of some sort together with parts of their physicochemical contexts that have been incorporated into life processes, and the unincorporated environment (O'Neill et al. 1986). Entities are energy–material processors that interact with similarly scaled entities and simultaneously with components (providing *initiating mechanisms*) and encompassing entities (providing *boundary conditions*). Entities at different levels have different process behaviors and "predicates" (Allen and Starr 1982; Salthe 1985), meaning that they are represented by different rate constants and can accomplish different things. For example, populations within a local ecosystem may undergo seasonal cycles of expansion and contraction whereas the encompassing system appears to remain stable over decades. Such populations also may experience fluctuation during disturbance–recovery episodes, but it is the entire ecosystem that would undergo succession. The same general concepts apply to entities at lower and higher levels in the ecologic hierarchy.

Do the entities at varied scales of resolution, especially large multispecies systems, have the properties of individuals? I have discussed the criteria for recognizing individual ecologic systems at varied scales in previous essays (W. Miller 1990, 1991, 1993a, 1996), based largely on the criteria presented by Salthe (1985). Gould (1995) recently has used similar criteria to argue that evolutionary entities other than individual organisms and demes can be construed as individuals within a genealogic hierarchy. The main idea in these arguments is that it is possible to recognize individual dynamic entities by considering the interrelated criteria of boundaries, scale, integration, and continuity.

Boundaries

Ecologic systems that are larger and more inclusive than individual organisms have poorly defined boundaries except in those cases where steep environmental gradients or discontinuities produce ecoclines or ecotones. There are other ways to construe boundary. The most familiar boundaries in biology are walls of some kind: more or less permeable membranes or tissues of organ-isms and their components. For the systems that consist of organisms, a boundary can be created by the physicochemical context, disturbances, or might have more to do with internal connection or "wiring" of interactors (Allen and Starr 1982; O'Neill et al. 1986). The last view is the same one a plan-etary astronomer would adopt in delineating a planetary system. The interac-tors are stars and planets connected as a dynamic system by gravity. The expec-tation still exists that ecologic entities must have some sort of geographic border to be real, but we should anticipate that large, inclusive systems are likely to have other kinds of boundary criteria.

Paleoecologists are acquainted with the kind of biofacies that recurs with a certain sedimentary paleoenvironment. This close association of sedimentary rocks and fossils, appearing together at different localities and different strati-graphic levels within the same region or basin, may be the expression of an envi-ronmentally imposed system boundary, although the kind of ecologic entity recorded in such patterns has been difficult to interpret (W. Miller 1990, 1993b, 1996). In census assemblages (sensu Kidwell and Flessa 1995), original spatial association of organisms in local systems can be preserved (Boucot 1981, 1990c). In time-averaged deposits, however, it will be difficult to detect bound-aries where they are not controlled by abrupt change in environmental factors, although reconstruction of networks of recurrent interactors might allow recog-nition of systemic cores of ecologic entities at varied scales of resolution.

Scale

A closely related criterion is that of scale, which refers as much to hierarchical position (Eldredge and Salthe 1984; Salthe 1985, 1993) as to size and duration of a system (see recent discussions by O'Neill et al. 1986; Schneider 1994; Wu and Loucks 1995). The more inclusive entities are expected to be typically larger and more durable than the included systems. Province systems (table 3.4) should outlast the included local ecosystems and cover a larger area. In some cases spatial deployment and duration of nested systems could coincide (as with local populations within some local ecosystems), but the nestedness would still signal a difference in scale, as the term is used here.

In paleoecology, it is not always possible to specify exactly the scale of observation, but spatiotemporal dimension and relative level (apparent inclu-siveness of units resolved in data sets) can be identified and described. To claim that a widely deployed, persistent assemblage of fossils is a "community" is to ignore the fundamental property of scale (e.g., W. Miller 1997). Ecologic communities (which are usually short-lived taxonomic associations [sensu Kauffman and Scott 1976] or functional parts of local systems such as food chains or guilds) simply can not do the things many paleoecologists want them to do. They are ephemeral, localized aggregations of interactors (Bambach and Bennington 1996), not durable systems that can track changing environments over 10^5–10^6 yr (W. Miller 1990, 1993a, 1996).

Integration

This is the easiest criterion to grasp, although the nature of the things being integrated at higher levels may not be clear. Even the most ephemeral, loosely-organized multispecies assembly consists of organisms that interact in some way. In demonstrably stable assemblies, internal connections as opposed to environmental uniformity may be the source of that stability (reviewed in Roughgarden and Diamond 1986; Pimm 1991; Morris et al. 1995; W. Miller 1996). A great deal of writing in community and ecosystem ecology is devoted to exploring the complex relationships among species occurrence, connections (traditionally predator–prey and competitive population interactions are emphasized, but mutualisms and indirect interactions recently have gained prominence), and stability (meaning in most cases resilience or the ability to bounce back after a disturbance) (Pimm 1984, 1991). The favored approach in ecology involves experimental manipulation of a portion of a local ecosystem, isolated for tractability or because of interest in a particular taxonomic group. This experimental approach simply is not available in most paleoecologic studies, so evidence of integration based on static patterns and uniformitarian inferences must be used. Ecologists face the same methodologic limitations in studies of regional systems or situations in which system history is considered.

We should not, however, underestimate fossils as a record of interaction and system integration. Specimens show signs of interaction in the forms of epi- and endobiontic infestations, predation scars, skeletal inclusions and overgrowths, gut and fecal contents, and other repeated spatial associations such as tiering. Interesting new problems include the possible detection in fossil assemblages of indirect interactions (e.g., modification of a competitive interaction with introduction of a predator, parasite, or pathogen [Wootton 1994]) and the recognition of interactions between local ecosystems (e.g., Palmer, Allan, and Butman 1996; Polis and Hurd 1996) resulting in structure at the regional level (Salthe 1993; W. Miller 1996).

Continuity

For ecologic entities to qualify as individuals they must have beginnings, developmental histories, and terminations of some sort. Spatiotemporal continuity is now accepted by some evolutionary theorists as an essential property of species. Thus species-lineages are viewed as having definable beginnings (speciation), often stable histories (reflected in morphologic stasis), and eventual terminations (species-level extinction). Populations and ecosystems have beginnings owing to colonization of a local environment, histories characterized by relative stability or fluctuations in organization and function, and eventual collapse as local contexts change or as sources of recruits disappear. The histories of more inclusive entities are intimately linked to the larger patterns of climate, bathymetry, tectonics, and nutrients, and should consist of the assembly and connection of local ecosystems. This is a scale of resolution for which the traditional methods of marine paleoecology are particularly well suited.

Significance of Misidentifying the Players

Ecologic entities at different levels of organization are the players in the economy of nature. Each entity is a dynamic system that consists of smaller, faster-reacting systems and at the same time forms part of a larger, slower system. Hierarchy theorists would say that such systems exhibit the related properties of "near decomposability" and "nontransitivity," meaning that any part of a hierarchical metasystem can be extracted for study (i.e., isolation of a *focal level* of dynamic processes, together with relevant aspects of both enclosing and component systems), and that entities at different levels develop and react to disturbance in fundamentally different ways. Although paleoecologists are beginning to comprehend the significance of correct spatiotemporal scaling, there is still a tendency to anticipate process isomorphisms and to conflate levels. Nowhere is this more obvious than in the misidentification of fossil assemblages at varied scales as "communities" (figure 3.1) or in the misattribution of community (synonymous with *local ecosystem*) processes to large-scale temporal patterns in the fossil record.

Ecologic Succession

In the 1970s, it was popular to identify vertical transition patterns in fossil deposits, regardless of the scale, as examples of ecologic succession (reviewed in W. Miller 1986; Miller and DuBar 1988). The subdivisions within these sequences were recognized as serial stages or successive communities that underwent the same kind of succession described by ecologists. Even regional patterns that obviously were environmentally driven were proposed as large-scale versions of succession. It is now acknowledged that most of these patterns may be succession-like, but that the scaling is all wrong: larger, more durable entities than local communities or ecosystems are the units involved,

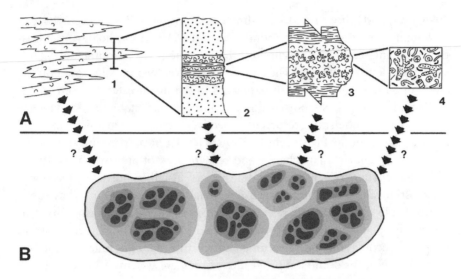

FIGURE 3.1. Almost any kind of fossil deposit or collection has been called a "community" by marine paleoecologists. (A) (1) Fossil deposits at the scales of regional facies, (2) local sequences, (3) individual bedding units, and (4) isolated samples. (B) Relationship of fossil assemblages to ecologic entities (*correspondence*) is often not evaluated in paleoecologic studies. The ecologic hierarchy is depicted here only as a "spot diagram" to emphasize inclusiveness and multiple levels. A better depiction would show the entities as dynamic systems that process energy and materials, interact with similarly scaled entities and with their components and contexts, and have developmental trajectories: an artistic feat that is beyond me.

and the within-system processes of autogenic succession [facilitation, tolerance, and inhibition (Connell and Slatyer 1977)] simply are not visible in most time-averaged assemblages.

Paleoecologists have become more careful in recognizing ecologic succession in the fossil record by considering the effects of time-averaging and mixing of ecologic units and anticipating the local conditions in which successional patterns are likely to be preserved in the first place (e.g., Wilson 1987; Taylor 1996). This new caution represents a move toward quality control and an increased ecologic sophistication.

Coordinated Stasis and Related Patterns

Brett, Baird, and co-workers have revived interest in the concept of recurrence as it applies to assemblages or biofacies that appear to track a preferred environment and maintain compositional stability for 10^5–10^6 yr (Miller, Brett, and Parson 1988; Brett, Miller, and Baird 1990; Brett and Baird 1995; Morris et al.

1995; Ivany and Schopf 1996). Coordinated stasis is recognized as recurrence of taxonomic composition, rank-abundance, and skeletal morphology of at least the dominant taxa within a recurring lithic unit, despite environmental change such as sea level fluctuation. The concept was developed initially based on the study of Silurian and Devonian sequences in the northeastern United States.

Pandolfi (1996) has reported similar compositional stability and evidence for limited membership (recurrent, selective assembly dynamics) in coral associations from the Pleistocene of Papua New Guinea, spanning an interval of high-frequency climate change and eustatic flux of approximately 10^5 yr. This kind of stability is not the kind of pattern usually reported by terrestrial ecologists for the same time interval. Terrestrial assemblages seem to have been controlled by individualistic response of taxa, compositional instability, and essentially open membership. Beyond the issues of reconciling the marine and terrestrial records and of determining relative importance of long-term stability vs. instability as dominant properties of ecologic systems, what exactly are the entities behaving in these ways?

Because of the short-lived nature of most local ecologic systems and the prevalence of time-averaging, it is unlikely that the stable assemblages described by Brett and associates and by Pandolfi equate to what an ecologist would call a community. Anticipating that larger, more durable systems are represented in the patterns begs the question of how stability works at levels of organization above that of local populations and ecosystems. Can the patterns be explained simply by matching a persistent species pool with persistent or recurring environments, or is the stability a result of autogenic processes of large ecologic systems that are comparable to the largest units of landscape ecology? The reductionist–extrapolationist view would restrict explanation to adaptive properties of organisms and provision of necessary environments, whereas the hierarchical perspective would allow the formulation of alternative models including the possibility of unfamiliar sources of stability at the level of regional ecosystems (e.g., W. Miller 1993a,b, 1996, 1997).

Phanerozoic Faunal Replacements

We have heard much about Sepkoski's concept of the three "Great Evolutionary Faunas": the Cambrian, Paleozoic, and the Modern (Sepkoski 1979, 1981, 1990, 1992, 1996; Sepkoski and Sheehan 1983; Sepkoski and Miller 1985). These grand divisions of Phanerozoic life are statistically defined mega-assemblages characterized by the prominence of certain higher taxa and are the largest units used in modern studies that trace the development of marine faunal diversity. According to Sepkoski and Miller (1985:153), ". . . all of these faunas originated early in the Phanerozoic but then diversified at different rates, with each fauna attaining a successively higher maximum diversity and appearing to displace the fauna before it. . . ." Recognized within the evolutionary faunas are the time-environment distributions of assemblages (i.e., documented collections of varied temporal scope) that appear to illustrate the actual patterns of large-scale faunal replacement. These units also are sometimes referred to as "communities." Sepkoski and Miller (1985) were careful to point out that their communities were really operational or sampling units, consisting of (p. 156) ". . . a diverse array of paleoecologic communities and assemblages as well as biostratigraphic faunules and biofacies, all of which shared the quality of being samples of the total fossil content of some restricted stratigraphic and environmental interval." Later in the same article, however, the sampling units become reified as biotic units when an attempt is made to explain the cause of the faunal replacements recorded in Paleozoic nearshore facies. When a brachiopod-dominated inner-shelf "community" is replaced by a mollusc-rich "community," what kinds of ecologic systems are actually involved? These obviously are not the same unit an ecologist would regard as a community. Are the encompassing evolutionary faunas some form of gigantic ecologic system, or simply statistical patterns?

Sepkoski's work is founded on an enormous amount of bibliographic documentation and rigorous, repeated analyses; there is no doubting the patterns. But what are the large-scale ecologic processes involved in the origination, development and elaboration, and decline of the faunas recognized in this way? I suggest that understanding the patterns involves more than evolutionary speculation and must include the specification of ecologic entities and their correctly scaled developmental dynamics. Sepkoski (1990, 1992, 1996) had begun to pursue these issues.

Conclusions

At the end of the Paleontological Society symposium on faunal stasis held in Seattle in 1994 (see Ivany and Schopf 1996), contributors and others interested in the topics covered met for a short discussion session. I had no methodologic investment to defend, and had not staked out a particular stratigraphic interval or group of favorite organisms; I was interested in the general idea of whether assemblages might remain stable for millions of years and the possibility of making a significant expansion in ecologic theory by explaining such patterns. I was perplexed by the comments made during the discussion. The few solid threads of evidence, suggestions for tests of ecologic stasis, and the conceptual advances were lost in a Babel of confused jargon and ecologic naïveté. Participants should have seen they were talking past each other, and that one of the chief reasons for this was lack of a conventionalized terminology for the things paleoecologists try to detect in the fossil record. Honors are bestowed on physical scientists for discovering the essential properties of fundamental particles; paleoecology, by comparison, has largely bypassed the issue.

The view that the structure of the living world is accurately represented as a hierarchical assembly of dynamic systems is gaining momentum in both ecology and paleoecology. At the same time it is becoming clear that pragmatic, inconsistent use of terminology and strictly reductionist–extrapolationist views have not served evolutionary paleoecologists well when they have attempted to delineate and characterize ecologic entities in the fossil record, reconstruct developmental patterns, detect processes and reactions, or recognize linkages to evolutionary dynamics. This goes beyond haggling over terminology: in these matters researchers are grappling with some of the central questions of biology. What are the essential properties of the units of study? Are the units real or convenient fictions? Do they have properties that cannot be explained by merely summing the properties of constituent parts? Do the units act as active components and causes or as backdrops of evolutionary radiations, trends, turnovers, and stasis? The Valentine System of unit classification (table 3.4) is probably the most accurate representation of the ecologic hierarchy proposed so far; it provides a tentative list of study "targets" and scaling controls and is a conceptually robust starting place for pursuing these basic questions. I conclude with a slightly altered quote from Marjorie Grene (1987:504) illustrating why terminology, properties of entities, and purpose of a discipline are so closely linked: "In general, an expanded ontology, which allows consideration of real patterns at a number of levels, could produce a search for, and discovery of, causes in quarters where classical evolutionists [and ecologists] would not have thought to seek them."

ACKNOWLEDGMENTS

I thank Warren Allmon and Dave Bottjer for their invitation to participate in this project and for their editorial efforts. Allmon, two anonymous reviewers, and Jack Sepkoski provided useful suggestions for the improvement of the text. Rosemary Hawkins typed the manuscript.

REFERENCES

Allen, T. F. H. and T. B. Starr. 1982. *Hierarchy: Perspectives for Ecological Complexity.* Chicago: University of Chicago Press.

Allmon, W. D. 1992. What, if anything, should evolutionary paleoecology be? *Palaios* 7:557–558.

Bambach, R. K. and J. B. Bennington. 1996. Do communities evolve? A major question in evolutionary paleoecology. In D. Jablonski, D. H. Erwin, and J. H. Lipps, eds., *Evolutionary Paleobiology*, pp. 123–160. Chicago: University of Chicago Press.

Boucot, A. J. 1975. *Evolution and Extinction Rate Controls.* Amsterdam: Elsevier.

Boucot, A. J. 1978. Community evolution and rates of cladogenesis. *Evolutionary Biology* 11:546–655.

Boucot, A. J. 1981. *Principles of Benthic Marine Paleoecology.* New York: Academic Press.

Boucot, A. J. 1983. Does evolution take place in an ecological vacuum? II. *Journal of Paleontology* 57:1–30.

Boucot, A. J. 1986. Ecostratigraphic criteria for evaluating the magnitude, character and duration of bioevents. In O. Wallister, ed., *Global Bioevents: A Critical Approach*, pp. 25–45. Berlin: Springer-Verlag.

Boucot, A. J. 1990a. Community evolution: its evolutionary and biostratigraphic significance. *Paleontological Society Special Publication* 5:48–70.

Boucot, A. J. 1990b. Evolution of communities. In D. E. G. Briggs and P. R. Crowther, eds., *Palaeobiology: A Synthesis*, pp. 391–394. Oxford, U.K.: Blackwell.

Boucot, A. J. 1990c. *Evolutionary Paleobiology of Behavior and Coevolution.* Amsterdam: Elsevier.

Brett, C. E. and G. C. Baird. 1995. Coordinated stasis and evolutionary ecology of Silurian to Middle Devonian faunas in the Appalachian Basin. In D. H. Erwin and R. L. Anstey, eds., *New Approaches to Speciation in the Fossil Record*, pp. 285–315. New York: Columbia University Press.

Brett, C. E. , K. B. Miller, and G. C. Baird. 1990. A temporal hierarchy of paleoecologic processes within a Middle Devonian epeiric sea. *Paleontological Society Special Publication* 5:178–209.

Brooks, D. R. and E. O. Wiley. 1988. *Evolution as Entropy: Toward a Unified Theory of Biology.* Chicago: University of Chicago Press.

Brown, J. H. 1995. *Macroecology.* Chicago: University of Chicago Press.

Connell, J. H. and R. O. Slatyer. 1977. Mechanisms of succession in natural communities and their role in community stability and organization. *American Naturalist* 111:1119–1144.

Delcourt, H. R. and P. A. Delcourt. 1991. *Quaternary Ecology: A Paleoecological Perspective.* London: Chapman and Hall.

Eldredge, N. 1985. *Unfinished Synthesis: Biological Hierarchies and Modern Evolutionary Thought.* New York: Oxford University Press.

Eldredge, N. 1989. *Macroevolutionary Dynamics: Species, Niches, and Adaptive Peaks.* New York: McGraw-Hill.

Eldredge, N. and S. N. Salthe. 1984. Hierarchy and evolution. *Oxford Surveys in Evolutionary Biology* 1:184–208.

Ereshefsky, M., ed. 1992. *The Units of Evolution: Essays on the Nature of Species*. Cambridge MA: M.I.T. Press.

Fauth, J. E., J. Bernardo, M. Camara, W. J. Resetarits Jr., J. Van Buskirk, and S. A. McCollum. 1996. Simplifying the jargon of community ecology: A conceptual approach. *American Naturalist* 147:282–286.

Gilpin, M. and I. Hanski, eds. 1991. *Metapopulation Dynamics: Empirical and Theoretical Investigations*. London: Academic Press.

Gould, S. J. 1980. The promise of paleobiology as a nomothetic, evolutionary discipline. *Paleobiology* 6:96–118.

Gould, S. J. 1995. The Darwinian body. *Neues Jahrbuch für Geologie und Paläontologie, Abhandlungen* 195:267–278.

Grene, M. 1987. Hierarchies in biology. *American Scientist* 75:504–510.

Hansson, L., L. Fahrig, and G. Merriam. 1995. *Mosaic Landscapes and Ecological Processes*. London: Chapman and Hall.

Hoffman, A. 1979. Community paleoecology as an epiphenomenal science. *Paleobiology* 5:357–379.

Hoffman, A. 1983. Paleobiology at the crossroads: A critique of some modern paleobiological research programs. In M. Grene, ed., *Dimensions of Darwinism: Themes and Counterthemes in Twentieth-Century Evolutionary Theory*, pp. 241–271. Cambridge and Paris: Cambridge University Press and Editions de la Maison des Sciences de l'Homme.

Ivany, L. C. and K. M. Schopf, eds. 1996. *New Perspectives on Faunal Stability in the Fossil Record*. Theme issue, *Palaeogeography, Palaeoclimatology, Palaeoecology* 127:1–362.

Kauffman, E. G. 1974. Cretaceous assemblages, communities, and associations: Western interior United States and Caribbean Islands. In A. M. Ziegler, K. R. Walker, E. J. Anderson, E. G. Kauffman, R. N. Ginsburg and N. P. James, *Principles of Benthic Community Analysis*, pp. 12.1–12.27. Miami: Rosenstiel School of Marine and Atmospheric Science.

Kauffman, E. G. and R. W. Scott. 1976. Basic concepts of community ecology and paleoecology. In R. W. Scott and R. R. West, eds., *Structure and Classification of Paleocommunities*, pp. 1–28. Stroudsburg PA: Dowden, Hutchinson, and Ross.

Kidwell, S. M. 1993. Patterns of time-averaging in the shallow marine fossil record. In S. M. Kidwell and A. K. Behrensmeyer, eds., *Taphonomic Approaches to Time Resolution in Fossil Assemblages*, pp. 275–300. Paleontological Society, Short Course No. 6. Knoxville TN: The Paleontological Society.

Kidwell, S. M. and D. W. J. Bosence. 1991. Taphonomy and time-averaging of marine shelly faunas. In P. A. Allison and D. E. G. Briggs, eds. *Taphonomy: Releasing the Data Locked in the Fossil Record*, pp. 115–209. New York: Plenum.

Kidwell, S. M. and K. W. Flessa. 1995. The quality of the fossil record: Populations, species, and communities. *Annual Review of Ecology and Systematics* 26:269–299.

Kitchell, J. A. 1985. Evolutionary paleoecology: Recent contributions to evolutionary theory. *Paleobiology* 11:91–104.

Margalef, R. 1968. *Perspectives in Ecological Theory.* Chicago: University of Chicago.

Mayr, E. 1970. *Populations, Species, and Evolution.* Cambridge MA: Harvard University Press.

Mayr, E. 1988. *Toward a New Philosophy of Biology: Observations of an Evolutionist.* Cambridge MA: Harvard University Press.

McIntosh, R. P. 1985. *The Background of Ecology: Concept and Theory.* Cambridge MA: Cambridge University Press.

McIntosh, R. P. 1995. H. A. Gleason's 'individualist concept' and theory of animal communities: A continuing controversy. *Biological Reviews* 70:317–357.

Miller, A. I. 1993. The science and politics of paleoecology. *Palaios* 8:409–410.

Miller, K. B., C. E. Brett, and K. M. Parson. 1988. The paleoecologic significance of storm-generated disturbance within a Middle Devonian muddy epeiric sea. *Palaios* 3:35–52.

Miller, W. III. 1986. Paleoecology of benthic community replacement. *Lethaia* 19:225–231.

Miller, W. III. 1990. Hierarchy, individuality and paleoecosystems. In W. Miller III, ed., *Paleocommunity Temporal Dynamics: The Long-Term Development of Multispecies Assemblies,* pp. 31–47. Paleontological Society Special Publication No. 5. Knoxville TN: The Paleontological Society.

Miller, W. III. 1991. Hierarchical concept of reef development. *Neues Jahrbuch für Geologie und Paläontologie, Abhandlungen* 182:21–35.

Miller, W. III. 1993a. Benthic community replacement and population response. *Neues Jahrbuch für Geologie und Paläontologie, Abhandlungen* 188:133–146.

Miller, W. III. 1993b. Models of recurrent fossil assemblages. *Lethaia* 26:182–183.

Miller, W. III. 1996. Ecology of coordinated stasis. *Palaeogeography, Palaeoclimatology, Palaeoecology* 127:177–190.

Miller, W. III. 1997. Bivalve fauna of a Plio-Pleistocene coral biostrome in eastern North Carolina: paleoecologic stability, change, and scaling. *Southeastern Geology* 37:73–90.

Miller, W. III and J. R. DuBar. 1988. Community replacement of a Pleistocene *Crepidula* biostrome. *Lethaia* 21:67–78.

Morris, P. J., L. C. Ivany, K. M. Schopf, and C. E. Brett. 1995. The challenge of paleoecological stasis: Reassessing sources of evolutionary stability. *Proceedings of the National Academy of Sciences, USA* 92:11269–11273.

O'Neill, R. V., D. L. DeAngelis, J. B. Waide, and T. F. H. Allen. 1986. *A Hierarchical Concept of Ecosystems.* Princeton NJ: Princeton University Press.

Palmer, M. A., J. D. Allan, and C. A. Butman. 1996. Dispersal as a regional process affecting the local dynamics of marine and stream benthic invertebrates. *Trends in Ecology and Evolution* 11:322–326.

Pandolfi, J. M. 1996. Limited membership in Pleistocene reef coral assemblages from the Huon Peninsula, Papua New Guinea: Constancy during global change. *Paleobiology* 22:152–176.

Pickett, S. T. A. and P. S. White. 1985. *The Ecology of Natural Disturbance and Patch Dynamics.* Orlando FL: Academic Press.

Pimm, S. L. 1984. The complexity and stability of ecosystems. *Nature* 307:321–326.

Pimm, S. L. 1991. *The Balance of Nature? Ecological Issues in the Conservation of Species and Communities.* Chicago: University of Chicago Press.

Polis, G. A. and S. D. Hurd. 1996. Linking marine and terrestrial food webs: Allochthonous input from the ocean supports high secondary productivity on small islands and coastal land communities. *American Naturalist* 147:396–423.

Roughgarden, J. and J. Diamond. 1986. Overview: The role of species interactions in community ecology. In J. Diamond and T. J. Case, eds., *Community Ecology*, pp. 333–343. New York: Harper and Row.

Salthe, S. N. 1985. *Evolving Hierarchical Systems: Their Structure and Representation.* New York: Columbia University Press.

Salthe, S. N. 1993. *Development and Evolution: Complexity and Change in Biology.* Cambridge MA: M.I.T. Press.

Schneider, D. C. 1994. *Quantitative Ecology: Spatial and Temporal Scaling.* San Diego: Academic Press.

Sepkoski, J. J. Jr. 1979. A kinetic model of Phanerozoic taxonomic diversity: II. Early Phanerozoic families and multiple equilibria. *Paleobiology* 5:222–251.

Sepkoski, J. J. Jr. 1981. A factor analytic description of the Phanerozoic marine fossil record. *Paleobiology* 7:36–53.

Sepkoski, J. J. Jr. 1990. Diversity in the Phanerozoic oceans: a partisan review. In E. C. Dudley, ed., *The Unity of Evolutionary Biology,* pp. 210–236. Portland OR: Dioscorides Press.

Sepkoski, J. J. Jr. 1992. Phylogenetic and ecologic patterns in the Phanerozoic history of marine biodiversity. In N. Eldredge, ed., *Systematics, Ecology, and the Biodiversity Crisis,* pp. 77–100. New York: Columbia University Press.

Sepkoski, J. J. Jr. 1996. Competition in macroevolution: The double wedge revisited. In D. Jablonski, D. H. Erwin, and J. H. Lipps, ed., *Evolutionary Paleobiology,* pp. 211–255. Chicago: University of Chicago Press.

Sepkoski, J. J. Jr. and P. M. Sheehan. 1983. Diversification, faunal change, and community replacement during the Ordovician radiations. In M. J. S. Tevesz and P. L. McCall, eds., *Biotic Interactions in Recent and Fossil Benthic Communities,* pp. 673–717. New York: Plenum.

Sepkoski, J. J. Jr. and A. I. Miller. 1985. Evolutionary faunas and the distribution of Paleozoic marine communities in space and time. In J. W. Valentine, ed., *Phanerozoic Diversity Patterns: Profiles in Macroevolution,* pp. 153–190. Princeton NJ and San Francisco: Princeton University Press and Pacific Division, American Association for the Advancement of Science.

Sheehan, P. M. 1991. Patterns of synecology during the Phanerozoic. In E. C. Dudley, ed., *The Unity of Evolutionary Biology,* pp. 103–118. Portland OR: Discorides Press.

Sheehan, P. M. 1996. A new look at Ecologic-Evolutionary Units. *Palaeogeography, Palaeoclimatology, Palaeoecology* 127:21–32.

Stanley, S. M. 1979. *Macroevolution: Pattern and Process.* San Francisco: W. H. Freeman and Co.

Taylor, W. L. 1996. Short-term community succession: Implications from well preserved Silurian assemblages. *Geological Society of America, Abstracts with Programs* 28(7):A-291.

Turner, M. G. 1989. Landscape ecology: the effect of pattern on process. *Annual Review of Ecology and Systematics* 20:171–197.

Valentine, J. W. 1968. The evolution of ecological units above the population level. *Journal of Paleontology* 42:253–267.

Valentine, J. W. 1973. *Evolutionary Paleoecology of the Marine Biosphere.* Englewood Cliffs NJ: Prentice Hall.

Wilson, M. A. 1987. Ecological dynamics on pebbles, cobbles, and boulders. *Palaios* 2:594–599.

Wootton, J. T. 1994. The nature and consequences of indirect effects in ecological communities. *Annual Review of Ecology and Systematics* 25:443–466.

Wu, J. and O. L. Loucks. 1995. From balance of nature to hierarchical patch dynamics: a paradigm shift in ecology. *Quarterly Review of Biology* 70: 439–466.

4

The Ecological Architecture of Major Events in the Phanerozoic History of Marine Invertebrate Life

David J. Bottjer, Mary L. Droser, Peter M. Sheehan, and George R. McGhee Jr.

THE HISTORY OF LIFE IS PUNCTUATED by mass extinctions, their recoveries, and radiations. Although the recognition and understanding of these events comes largely from taxonomic data, researchers have striven to evaluate changing ecologies associated with these events. However, evolutionary paleoecologists are still in the early stages of recognizing the particular paleoecological patterns that are associated with significant events in the history of life. Once they reach a good understanding of these patterns, they can begin to make real progress in understanding the processes that caused these paleoecological patterns. Modern ecologists find themselves in a similar position because they too are trying to recognize ecological patterns in modern communities that will allow for a better understanding and management of the ongoing modern mass extinction (e.g., Power et al. 1996; Mills, Soule, and Doak 1996).

A variety of paleoecological approaches have been used to examine or characterize large-scale temporal patterns of evolutionary paleoecology, which range in focus from Phanerozoic paleoecological patterns and trends (e.g., Valentine 1973; Ausich and Bottjer 1982; Bambach 1977, 1983; Bottjer and Ausich 1986; Bottjer and Jablonski 1989; Boucot 1983; Jablonski and Bottjer

1983; Sepkoski and Miller 1985; Sepkoski and Sheehan 1983) to the recognition of community-level patterns through time (e.g., Bretsky 1968; Boucot 1983; Sheehan 1991). Researchers have also documented temporal patterns for particular environments, such as the Phanerozoic history of reefs, which has received considerable attention (e.g., Copper 1994; Fagerstrom 1987; Wood 1995). Less common have been examinations of the distribution of biofabrics such as stromatolites (Awramik 1991; Schubert and Bottjer 1992), shell beds (Kidwell 1990; Kidwell and Brenchley 1994; Li and Droser 1997; Bottjer and Droser 1998), and ichnofabrics (Droser and Bottjer 1988, 1989, 1993).

Similarly, studies have attempted to compare the ecology of mass extinctions, recoveries, and radiations (e.g., Boucot 1990, 1996). For example, paleontologists have examined biogeographic patterns of faunas before and after mass extinctions (e.g., Sheehan 1979; Jablonski 1986, 1987; McGhee 1996; Erwin and Hua-Zhang 1996), including changes along latitudinal gradients. In addition, selectivity of extinctions for pelagic vs. benthic habitats, and whether the pelagic system crashed (as in the end-Cretaceous mass extinction), have also been studied (e.g., Paul and Mitchell 1994). Comparative paleoecology of the recoveries from mass extinctions has also begun to receive attention (e.g., Kauffman and Harries 1996; Harries, Kauffman, and Hansen 1996; Bottjer, Schubert, and Droser 1996).

There are several difficulties in evaluating ecological changes throughout the history of life. In particular, they are reflected by the variety of approaches that researchers have used. One of the biggest hurdles is the paucity of measures of paleoecological change that are comparable in a quantitative way. In large part this is because each species in an ecological context commonly has a different value in the ecosystem, thus leading to ecological measures that operate on various scales (e.g., Lamont 1995; Tanner, Hughes, and Connell 1994). Therefore, different ecological measures (like apples and oranges) are difficult to directly compare. For assessing major events such as mass extinctions or radiations, it is impossible to count ecological change as one would count taxa (even though in fact this same dilemma also exists within the taxonomic system; e.g., families of one clade may not be comparable with families of another clade).

Building on previous attempts, we have developed a comparative approach to assess major ecological changes in Phanerozoic life (Droser, Bottjer, and Sheehan 1997) (table 4.1). This method is a first attempt at integrating a number of ecological factors within a single scheme. In particular, we are interested in evaluating the level to which particular mass extinctions degraded ecological structure, as well as the level to which major radiations have completely changed preexisting ecological structure. Like other approaches, this method

TABLE 4.1. The Four Paleoecological Levels and the Characteristic Signals for Each of the Levels.

Level	Definition	Signals
First	Appearance/disappearance of an ecosystem	1. Initial colonization of environment.
Second	Structural changes within an ecosystem	1. First appearance of, or changes in ecological dominants of higher taxa. 2. Loss/appearance of metazoan reefs. 3. Appearance/disappearance of Bambachian megaguilds.
Third	Community-type level changes within an established ecological structure	1. Appearance and/or disappearance of community types. 2. Increase and/or decrease in tiering complexity. 3. "Filling-in" or "thinning" within Bambachian megaguilds.
Fourth	Community-level changes	1. Appearance and/or disappearance of paleocommunities. 2. Taxonomic changes within a clade.

is nonquantitative, but it is based on well-documented features of the fossil record.

In this chapter, we review our method of paleoecological levels and discuss some possible ecological underpinnings that have caused the criteria we use to identify these various levels to be empirically observable features of the fossil record. Similarly, we demonstrate the utility of this approach through comparative analyses of a number of the "Big 5" mass extinctions (Raup and Sepkoski 1982) and their associated recoveries, as well as the Ordovician radiation.

This method is not meant as a replacement of previous paleoecological or taxonomic approaches for the understanding of major events in the history of life, but rather as an additional means of analysis to be used in conjunction with these other approaches. In this way we can further identify a variety of paleoecological shifts and in particular the paleoecological significance of an event.

Paleoecological Levels

Major events in life's Phanerozoic history, such as mass extinctions and radiations, are typically identified by examining changes in taxonomic diversity (e.g., Sepkoski 1979, 1981). Preliminary paleoecological studies indicate, however, that the relative magnitudes of changes measured by taxonomic diversity were not the same as the relative magnitudes of associated ecological changes (e.g., Brenchley 1989). Thus, taxonomic and ecological changes may have been decoupled.

Changes in paleoecological systems are expressed so that there are scales of change, and some changes are far more important than others. This structuring provides a means to scale or rank paleoecological changes. We categorize types of paleoecological changes into four ranks that we term "paleoecological levels" (table 4.1). These paleoecological levels are not hierarchical nor additive, but they are ordered.

Changes at the first level are of the greatest magnitude and represent the advent of a new ecosystem. These types of changes include the beginning of life on planets such as Earth and Mars, and on Earth the advent of metazoan life on land, the sea floor, the deep sea, and the pelagic realm (e.g., Rigby 1997). On Earth these types of changes only happened once, and once they happened they have not been reversed (as far as we know). The best candidate for such a reversal, Seilacher's (1992) Vendobionta (Ediacaran fauna), now seems to have persisted into the Cambrian (e.g., Knoll 1996; Jensen, Gehling, and Droser 1998). In many respects, these types of ecological breakthroughs represent functional thresholds. Because they are at such a large scale and are unidirectional, first level changes will seldom play a part in an analysis of trends through time.

Changes at the second level occur within an established ecosystem and represent major structural changes at the largest ecological scale. Structural changes include the first appearance of, or changes in, ecological dominants of higher taxa within an ecosystem, such as the shift from trilobite- to brachiopod-dominated shallow soft-substrate paleocommunities in the Ordovician. Large-scale shifts in the nature of ecospace utilization are also included. Bambach (1983) introduced *adaptive strategies* as a means of evaluating paleoecological changes through time (e.g., figures 4.1 and 4.2). These include, for example, categories such as epifaunal mobile suspension feeders and pelagic carnivores. Whereas many workers have utilized the term *guild* for these categories, Bambach (1983) referred to guilds as smaller subgroups within these adaptive strategy categories. Thus, we have proposed the term *Bambachian megaguilds* for the adaptive strategies of Bambach (Droser, Bottjer, and Sheehan 1997). Addition or reduction in the number of Bambachian megaguilds serves as a useful signal of second level changes.

The development or collapse of metazoan carbonate buildups (e.g., Copper 1994; Stanley and Beauvais 1994) represents a major shift in the ecological structure of the marine ecosystem. The presence or absence of such buildups is dependent on climate and paleogeography and therefore is not as much an eco-logical signal as an environmental one. Reefs tend to disappear early in major extinction phases and thus are advance indicators of mass extinctions and global environmental crises (Copper 1994). The collapse of a reef system during a mass extinction and subsequent redevelopment of a reef system based on new higher level taxa can be considered a second level structural change within an ecosystem. The reappearance of reefs with essentially the same components is not considered a major structural shift. When metazoan reefs are lost, they are commonly replaced by a resurgence of stromatolite formation (e.g., Schubert and Bottjer 1992; Lehrmann, Wei, and Enos 1998), and so a significant increase of normal marine stromatolites also serves as a signal of structural changes. The presence of small local metazoan buildups within a largely siliciclastic or nonreefal setting would not be considered a structural shift.

Changes at the third level include community-scale shifts within an established ecological structure, in particular the appearance or disappearance of community-types. A community-type is "the aggregate of local communities and communities that have similar, but not identical, taxonomic membership and occur in similar, but not necessarily the same environments" (Bambach and Bennington 1995). Within a community-type, the filling-up of Bambachian megaguilds and an increase in tiering complexity (e.g., Bottjer and Ausich 1986) would also constitute third level changes.

Changes at the fourth level involve the appearance or disappearance of paleocommunities such as a succession of similar brachiopod communities (e.g., Boucot 1983; Harris and Sheehan 1996). These fourth level changes are common throughout the Phanerozoic and are similar in magnitude to most minor ecological changes.

Although paleoecological levels are not hierarchical or additive, there is a cascade effect, usually from the top to the bottom. If there is a second level change, it will be accompanied by third and fourth level changes. This is a "trickle-down" paleoecological effect rather than the building effect that occurs in the taxonomic system, where higher taxonomic levels are a means of grouping species according to their relatedness.

These four paleoecological levels provide a means to compare and rank ecological shifts associated with taxonomic events. How do we determine paleoecological levels for an event? Obviously the first step is the recognition of these events from taxonomic data. Some signals of paleoecological-level changes can be recognized through taxonomic shifts (e.g., the early Mesozoic transition from brachiopods to bivalves). However, in order to recognize other signals, such as an increase in tiering complexity or the addition of new paleocommunities, original paleoecological data must be collected in the field.

Similarly, in order to address the question of whether a specific ecological shift corresponds with a taxonomic shift, new paleoecological data must also be obtained.

Examples Utilizing Paleoecological Levels

As has already been noted, much discussion of the paleoecology of major events in the history of life, in particular, mass extinctions, has focused on biogeography (e.g., Sheehan 1979; Jablonski 1986, 1987; McGhee 1996; Erwin et al. 1996). Furthermore, much attention has also been paid to understanding the variations in extinctions of benthic versus pelagic organisms (e.g., Paul and Mitchell 1994; Levinton 1996). Although these measures have provided important insight into the paleoecology of mass extinctions, as discussed previously, these approaches cannot be used to recognize the extent that an extinction degraded the structure of an ecological system. In addition, the comparative paleoecological signature of radiations and recoveries has been little studied. In the following sections we use our system of paleoecological levels to analyze the Ordovician radiation and various components of the Big 5 mass extinctions and their associated recoveries.

Ordovician Radiation

The Precambrian–Cambrian radiation was the most significant event in the history of marine metazoans. Changes at all paleoecological levels occurred through this time interval as metazoans became established in Earth's seas. Clearly, a series of changes at several levels occurred as communities proceeded from Ediacaran assemblages to the Tommotian fauna ("small shellies") to typical members of the Cambrian Fauna. This radiation, regardless of its potential triggers or timing (e.g., Wray, Levinton, and Shapiro 1996), was a metazoan ecological event in which organisms were evolving into ecospace that had never before been occupied.

However, in many ways, the Ordovician radiation provides a simpler opportunity to examine paleoecological changes through the course of a radiation because there is a record of skeletal metazoans long before and after the Ordovician radiation, and there is a continuous marine record. In particular, it is not complicated by such phenomena as the advent of skeletalization or taphonomic biases associated with soft-bodied faunas. Paleoecological changes associated with the Ordovician radiation of marine invertebrates include second, third, and fourth level changes. However, evidence from both spores (Gray 1985) and trace fossils (Retallack and Feakes 1987) suggests that the initial radiation of complex life onto land occurred in the Ordovician; this constitutes a first level change.

In the marine realm, second level changes included a shift in ecological dominants from trilobite- to brachiopod-dominated shallow shelf paleocommunities (Droser and Sheehan 1995). We also observe a shift in ecological dominants in hardgrounds from echinoderm- to bryozoan-dominated paleocommunities (Wilson et al. 1992). Both of these changes resulted in the establishment of marine systems that were to last most of the Paleozoic. There were also new Bambachian megaguilds (figure 4.1), including deep mobile burrowers (Droser and Sheehan 1995). Similarly, the Ordovician witnessed the advent of stromotoporoid reefs, which dominated the reef ecosystem through the Devonian.

A major part of the Ordovician story is at the third level, where essentially Bambachian megaguilds were "filled" up to their Paleozoic levels (figure 4.1). In the Cambrian many of the Bambachian megaguilds had one or two clades, whereas by the end of the Ordovician, several megaguilds had up to eight different clades (figure 4.1; Droser and Sheehan 1995).

Additional third level changes included increases in tiering complexity from two to four levels in epifaunal suspension feeders (Bottjer and Ausich 1986), and in the shallow marine infaunal realm there were up to three tiers (as opposed to *Skolithos* piperock) (Droser, Hughes, and Jell 1994). There also was the appearance of new community types. These include a *Receptaculites*-macluritid high-energy nearshore community-type, new orthid community-types, and a bivalve-trilobite community-type in offshore muds (Droser and Sheehan 1995). The nature of the development of these new community-types still needs further study with additional field work.

Abundant fourth level changes, in the form of new paleocommunities, accompanied these second and third level changes. This demonstrates the "trickle down" effect that results from the existence of second and third level changes. The nature of these new paleocommunities is also currently under investigation.

Late Ordovician Mass Extinction and Silurian Recovery

The Late Ordovician mass extinction was the second-largest extinction in the history of metazoan life (Sepkoski 1981; Sheehan 1989). As much as 50% of all marine species became extinct (Brenchley 1989). However, ecologically, only third and fourth level changes occurred (Droser et al. 2000). Although reef communities were strongly affected by cool temperatures, the Silurian reefs that appeared soon after the extinction were mostly composed of the subfam-

A

	SUSPENSION	DEPOSIT	HERBIVORE	CARNIVORE
MOBILE		Trilobuta Ostracoda Monoplacophora	Monoplacophora Ostracoda	
ATTACHED LOW	Inarticulata Articulata			
ATTACHED ERECT	Eocrinoidea			
RECLINING	? Hyolitha			

Cambrian Fauna

	SUSPENSION	DEPOSIT	HERBIVORE	CARNIVORE
SHALLOW PASSIVE				
SHALLOW ACTIVE	Inarticulata	Trilobita "Polychaeta"		"Polychaeta"
DEEP PASSIVE			/////	/////
DEEP ACTIVE				

FIGURE 4.1. General adaptive benthic strategies that are typical of (A) the Cambrian Evolutionary Fauna and (B) the Ordovician representatives of the Cambrian and Paleozoic Evolutionary Faunas (after Bambach 1983). The shaded boxes are not biologically practical strategies. Second-level changes from A to B include the addition of new Bambachian megaguilds. Third-level changes include the "filling-in" of Bambachian megaguilds (after Droser and Sheehan 1995).

EPIFAUNA

	SUSPENSION	DEPOSIT	HERBIVORE	CARNIVORE
MOBILE	Bivalvia	Gastropoda Ostracoda Monoplacophora Trilobita	Echinodea Gastropoda Ostracoda Malacostraca **Monoplacophora**	Cephalopoda Malacostraca Stelleroidea Merostomata
ATTACHED LOW	Articulata Edrioasteroida Bivalvia Anthozoa Stenolaemata Sclerospongia **Inarticulata**			
ATTACHED ERECT	Crinoidea Anthozoa Stenolaemate Demospongia Blastoidea Cystoidea Hexactinellida **Eocrinoidea**			
RECLINING	Articulata Hyolitha Anthozoa Stelleroidea Cricooonanda			

Ordovician Representatives of the Cambrian and Paleozoic Fauna

INFAUNA

	SUSPENSION	DEPOSIT	HERBIVORE	CARNIVORE
SHALLOW PASSIVE	Bivalvia Rosroconchia			
SHALLOW ACTIVE	Bivalvia **Inarticulata**	Bivlavia Polychaeta Trilobita		Merostomata **Polychaeta**
DEEP PASSIVE			/////	/////
DEEP ACTIVE		Bivalvia		

FIGURE 4.1. *(continued).*

ilies and genera of rugose and tablulate corals and stromatoporoid sponges that were present in latest Ordovician reefs, so that the Silurian reef faunas can be regarded as Lazarus taxa (Copper 1994). Thus, the Late Ordovician extinction is not considered a major event for the reef system (Copper 1994).

In the pelagic realm, graptolites were ecologically dominant early Paleozoic filter feeders. Although during this Late Ordovician event graptolites had a huge taxonomic loss, being reduced to perhaps only a few species, they diversified quickly after the extinction and ecologically were as abundant in Silurian seas as they had been in the Ordovician (e.g., Berry and Wilde 1990; Berry 1996; Underwood 1998).

During this event, diversity of the pelagic conodonts also declined to less than 20 species but returned to near pre-extinction diversity in the Early Silurian, followed by a decline in the later Silurian (Sweet and Bergstrom 1984; Sweet 1988; Barnes and Bergstrom 1988; Barnes, Fortey, and Williams 1995; Armstrong 1996). Nonetheless, conodonts reattained a dominant ecological position following the extinction event. Similarly, Ordovician nautiloid cephalopods were a dominant species in the "pelagic carnivore megaguild," and during this extinction nautiloid diversity declined to levels not seen since their origination in the Early Ordovician. However, the mid-Silurian nautiloids had regained their ecologically prominent position in this megaguild (Crick 1990), so changes in this megaguild were only at the third and fourth level. Thus, as with the reef biota, the severe taxonomic loss during this extinction did not result in second level changes in the pelagic realm.

Brachiopods were a dominant component of benthic communities. Diversity declined among all groups of brachiopods during the Late Ordovician extinction, but none of the megaguilds were vacated (Boucot 1975; Bambach 1983; Sheehan 1991, 1996). Most changes were at the fourth level. For example, among the biconvex brachiopods, which were the most common form, species diversity declined substantially, but Silurian recovery of community dominance was rapid (Watkins 1991, 1994). An important new group of brachiopods, the wide-hinged spire-bearers, was added to the epifauna-attached suspension-feeding megaguild. The rise to prominence of spire-bearing brachiopods is an example of third level filling of Bambachian megaguilds.

Other prominent members of benthic communities such as rugose corals (Elias and Young 1998) and bryozoans (Anstey 1985; Tuckey and Anstey 1992) also declined but then recovered to be important community components in the Silurian. Similarly, the uppermost epifaunal tier levels occupied by crinoids were somewhat reduced by extinction but recovered rapidly during the Early Silurian (Ausich and Bottjer 2001). Again, only third and fourth level changes resulted from this extinction event.

Late Devonian Mass Extinction and Recovery

The Late Devonian mass extinction triggered second and associated third and fourth level ecological changes (Droser et al. 2000). Devonian reef ecosystems have been described as "the most extensive reef development this planet has ever seen" (Copper 1994), constituting almost 10 times the areal extent of reefal ecosystems present on Earth today. Reef ecosystems were virtually destroyed in the Late Devonian mass extinction, shrinking in geographic extent by a factor of 5,000 from the Frasnian Stage to the Famennian (Copper 1994; McGhee 1996). Tabulate corals and stromatoporoid sponges, major elements of the Devonian reef biota, did not recover their diversity losses or ecological dominance for the remainder of the Paleozoic (Copper 1994). The Late Devonian mass extinction thus precipitated a permanent change in the structure of global reef ecosystems in geologic time, a change at the second paleoecological level.

Cricoconarids and conodonts were two dominant elements of the Bambachian pelagic megaguilds during the Devonian. Only three species of the pelagic conodonts recovered their ecological position in the later Famennian with new species radiations (Sandberg et al. 1988). However, all of the cricoconarids were driven to extinction in the Late Devonian, representing the permanent loss of a major element of the oceanic zooplankton (McGhee 1996; Hallam and Wignall 1997). The total loss of the cricoconarids, and the major changeover in dominant conodont taxa, were permanent second and third level changes in the structure of pelagic megaguilds in geologic time.

Most elements of marine benthic ecosystems were adversely affected by the Late Devonian mass extinction (Stanley 1993; McGhee 1996). In particular, however, the Devonian was the "Golden Age" of brachiopods, which were the dominant element of benthic shellfish in Paleozoic seas, an ecological position now occupied by the molluscs. The dominant biconvex brachiopods of the Bambachian "epifauna-attached-suspension megaguild" lost more that 75% of their genera in the Late Devonian extinction (Boucot 1975; McGhee 1995) and were ecologically replaced by nonbiconvex brachiopods of the Bambachian "epifauna-reclining-suspension megaguild" in the post-Devonian Paleozoic (McGhee 1996). This shifting in ecological dominants between Bambachian megaguilds constitutes a second level change in the structure of benthic marine ecosystems precipitated by the Late Devonian mass extinction.

In the nekton, ammonoids and fish are the dominant elements of the Bambachian "pelagic-carnivore megaguild" during the Devonian. Only six genera of ammonoids survived the Frasnian Stage, though the ammonoids (like the conodonts) did recover their ecological position in the Famennian with the evolution of totally new families of clymeniids and goniatites (Becker and House 1994). Major losses also occurred in placoderm, chondrichthyan, and osteichthyan fish, and only in the Carboniferous did the fishes recover their ecological position with the evolution of entirely new fish faunas (Long 1993; Benton 1993). The Late Devonian evolution of new ammonoid familial dominants and changeover in fish faunas thus constituted a major change in the dominance structure of pelagic-carnivore megaguilds from the Devonian to the Carboniferous.

End-Permian Mass Extinction and Early Triassic Recovery

The end-Permian mass extinction is the greatest of all Phanerozoic mass extinctions (e.g., Raup 1979; Sepkoski 1992). Recent studies indicate that the aftermath of this mass extinction lasted through the Early Triassic (e.g., Hallam 1991, 1995; Schubert and Bottjer 1995), a time interval approximately 7–9 Ma long (Gradstein et al. 1995; Bowring et al. 1998). A variety of paleoecological data has been collected from Lower Triassic marine strata, which can be used to assess this event in terms of paleoecological levels. Ecospace was to a large extent emptied by the end-Permian mass extinction (e.g., Valentine 1973; Erwin, Valentine, and Sepkoski 1987; Bottjer, Schubert, and Droser 1996). Of the possible Bambachian benthic megaguilds, only four are occupied in Early Triassic paleocommunities documented from western North America, which are interpreted as typical of the Early Triassic worldwide (Schubert and Bottjer 1995; Bottjer, Schubert, and Droser 1996). Comparison with Bambach's (1983) analysis shows that there was a significant drop due to the end-Permian mass extinction, from occupation of 12 Bambachian benthic megaguilds in the Paleozoic to the 4 occupied in these Early Triassic paleocommunities (figures 4.1B and 4.2), indicating a very large change at the second level. Other significant second level changes due to this mass extinction included a major shift in ecological dominants for soft substrate shelf paleocommunities, from brachiopod-dominated in the middle and late Paleozoic to bivalve-dominated in the post-Paleozoic (e.g., Valentine 1973; Gould and Calloway 1980; Sepkoski 1981, 1996).

Furthermore, in a similar second level change, carbonate buildups were also restructured during this extinction. Although some Triassic reef taxa that occur on western North American terranes are Lazarus taxa (Stanley 1996), metazoan reefs were nevertheless restricted during the Early Triassic, and Permian-type carbonate buildups were lost (e.g., Lehrmann, Wei, and Enos

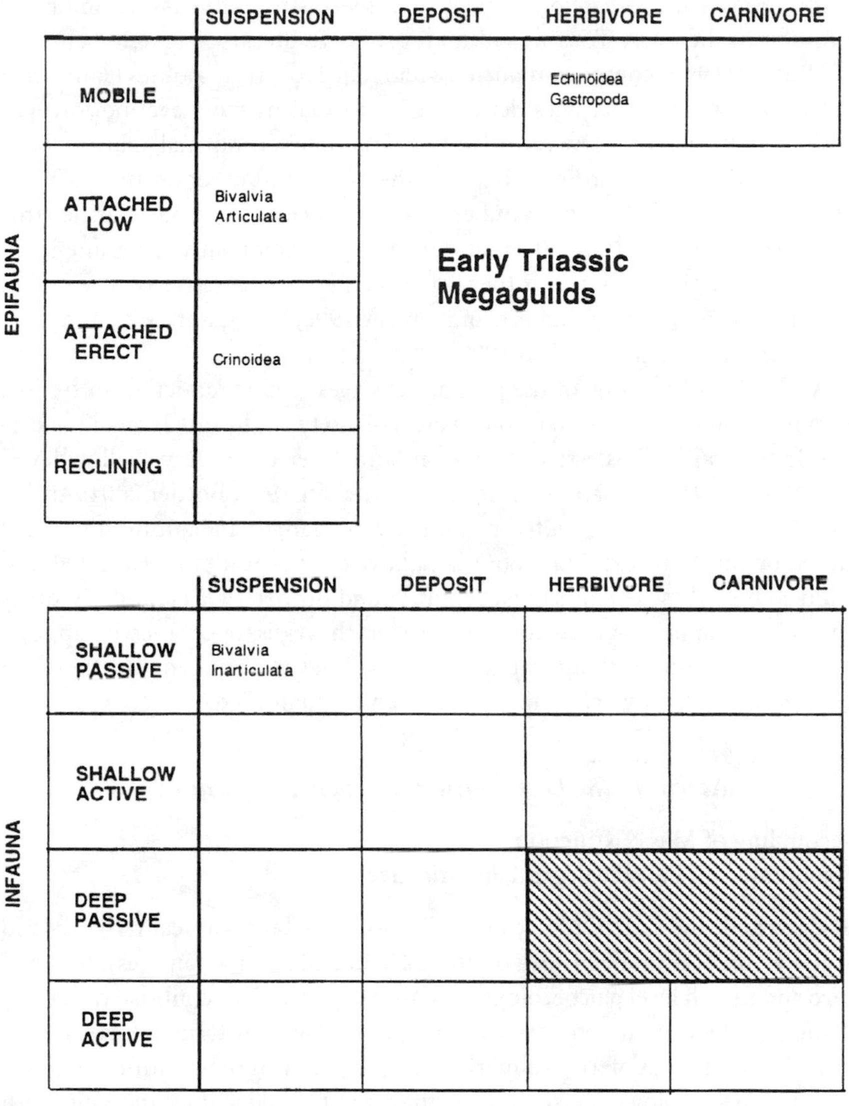

	SUSPENSION	DEPOSIT	HERBIVORE	CARNIVORE
MOBILE			Echinoidea Gastropoda	
ATTACHED LOW	Bivalvia Articulata			
ATTACHED ERECT	Crinoidea			
RECLINING				

Early Triassic Megaguilds

EPIFAUNA

	SUSPENSION	DEPOSIT	HERBIVORE	CARNIVORE
SHALLOW PASSIVE	Bivalvia Inarticulata			
SHALLOW ACTIVE				
DEEP PASSIVE			/////	
DEEP ACTIVE				

INFAUNA

FIGURE 4.2. General adaptive benthic strategies that are typical of the Early Triassic (after Bambach 1983). The shaded boxes are not biologically practical strategies.

1998). Along with restriction of metazoan reefs, stromatolites became more common in the Early Triassic. From an extensive literature search, Schubert and Bottjer (1992) compiled reported occurrences of stromatolites from strata of Silurian and younger ages deposited in normal-marine level-bottom paleoenvironments. Very few records of stromatolites in normal-marine level-bottom settings were forthcoming, with the greatest number occurring in the Early Triassic. Although the number of occurrences of normal marine stromatolites documented from the literature is small, their relative prominence in the Early Triassic, along with other microbialites, is suggestive of a real phenomenon (e.g., Bottjer, Schubert, and Droser 1996; Lehrmann, Wei, and Enos 1998; Kershaw, Zhang, and Lan 1999).

At the third level, one of the primary changes seen is reduction in tiering, from four epifaunal levels in the late Paleozoic to typically one level in the earliest Triassic, as well as a reduction of infaunal tiering to very shallow levels (e.g., Bottjer, Schubert, and Droser 1996; Twitchett 1999; Bottjer 2001; Ausich and Bottjer 2001). Undoubtedly there were changes in community-types caused by this mass extinction, but data have only begun to be collected that could address these changes (e.g., Schubert and Bottjer 1995). Similarly, other features of change for paleocommunities, such as disappearance or appearance of communities that occur on the fourth level, occurred, but they also need additional study before this feature can be quantified.

Patterns in the Distribution of Paleoecological Levels

Decoupling of Mass Extinction:
Taxonomic and Paleoecological Significance

As discussed, the Late Ordovician extinction was taxonomically the second largest extinction in the history of life (22% loss of marine families), but only third and fourth level paleoecological changes occurred. In contrast, the events of the Late Devonian extinction were nearly similar in terms of taxonomic numbers, with a loss of 21% of marine families, but this mass extinction interval was more ecologically significant than the Ordovician extinctions, with second, third, and fourth level changes (e.g., McGhee 1996; Droser et al. 2000). Thus, the Late Devonian and Late Ordovician mass extinctions represent an intriguing comparison. Taxonomically, the Late Ordovician extinction was perhaps slightly greater; however, paleoecologically, it was not as significant and it had no real lasting impact. In contrast, the Late Devonian extinction had lasting ecological impact in which ecosystem structure was changed permanently. Therefore, it appears that the ecological impact of a mass extinction can be decoupled from its taxonomic impact (Droser et al. 2000). Using a different approach, a similar effect has been demonstrated by McKinney et al. (1998) for bryozoans in the end-Cretaceous mass extinction and recovery.

The Distribution of Paleoecological Levels in Space and Time Within an "Event"

During a radiation or an extinction, we maintain that an overall pattern of paleoecological changes reflected by paleoecological levels is recorded. For example, the total summed end-Permian mass extinction had second, third, and fourth level changes. However, these level changes may have occurred in a single type or several types of environmental settings, for example, in the tropics or globally. In addition, there may be paleoecological changes occurring throughout the temporal course of an event. Thus, there are various scales in space and time at which we need to examine and determine paleoecological changes. This type of analysis is particularly instructive for our understanding of the extinction process. Much more comparative data at these scales are needed in order to truly understand the paleoecological nature of these events.

Biogeographic and Environmental Patterns

Previous paleoecological studies of mass extinctions have demonstrated that extinctions can have strong biogeographic patterns (e.g., Sheehan 1979; Jablonski 1986, 1987; McGhee 1996; Erwin et al. 1996). Indeed, as with extinctions, we would predict that changes in paleoecological levels during radiations and recoveries would proceed on a differential geographic template (e.g., figure 4.3A). For example, during the Late Ordovician extinction, extensive third and fourth level changes occurred in the temperate zone, where level-bottom communities were destroyed and restructured twice (Sheehan 1979; Brenchley 1989). However, in the tropics, only fourth level changes occurred in the very sensitive reef communities (Copper 1994). During the recovery, communities on tropical carbonate platforms reformed from open ocean settings.

Along with latitudinal variations, there are also differential effects in marine and terrestrial environments. A potential example is the end-Triassic extinction, which is another of the Big 5 mass extinctions (Raup and Sepkoski 1982), with a loss of over 20% of approximately 300 families of marine animals. A variety of researchers maintain that the vertebrate record on land shows relatively little effects from this mass extinction (e.g., Benton 1991), although this view has been contentious (e.g., Hallam and Wignall 1997). Therefore, it is likely that on land there were only third and fourth level changes. However, in the marine realm, where there was a collapse of metazoan carbonate reefs

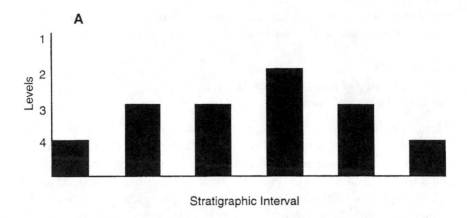

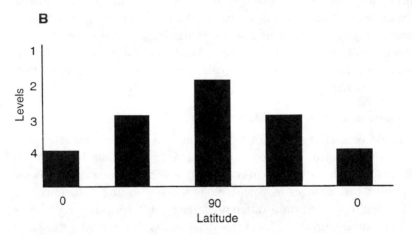

FIGURE 4.3. Hypothetical distribution of paleoecological levels through (A) a stratigraphic interval (oldest, left; youngest, right) in which a radiation is occurring; and (B) the latitudes (in degrees) during which a global radiation or mass extinction is occurring. In (A) there are paleoecological changes throughout the radiation but a second-level change occurs only once. In (B) paleoecological changes are greater at the equator than at the poles; this could be the case during either a radiation or a mass extinction.

(Hallam and Wignall 1997), second and associated third and fourth level changes appear to have occurred. Thus, the paleoecological significance of this event for the marine and terrestrial realms was likely decoupled. Although we have known that there was a different ecological history for the marine and terrestrial realms, the system of paleoecological levels provides a framework for comparing these patterns.

Temporal Patterns

Within a radiation or within a mass extinction, it may be that overall, second, third, and fourth level changes occurred. However, these changes may not occur at the same time, or there may be a temporal series of changes at different paleoecological levels (e.g., figure 4.3B). This is exemplified by the Ordovician radiation. It appears that at the base of the Middle Ordovician second, third, and fourth level changes occurred. This represents the major shift in ecology during the radiation. However, throughout the Early and Middle Ordovician, there were third and fourth level changes. By the Late Ordovician, the overall structure of the ecosystem was stabilized. The nature of this sequence of paleoecological changes is currently under investigation, with research focused on the rate at which such changes proceeded.

Researchers have also described recoveries as stepped or "fitful." In particular, both the Silurian and Triassic recoveries appear to have been stepped (e.g., Schubert and Bottjer 1995; Harris and Sheehan 1997; Bottjer, in press). This may indicate that ecological shifts as demonstrated by changes in paleoecological levels could also have been stepped or strung out over a period of time during a recovery, a phenomenon that also warrants further investigation.

Underpinnings of Mass Extinction: Taxonomic and Ecological Decoupling

Perhaps one of the most interesting aspects of this analysis of mass extinctions using our approach of paleoecological levels is the phenomenon that we term "taxonomic and ecological decoupling," where the relative level of ecological degradation is not as great as the degree of taxonomic degradation during a mass extinction event (Droser et al. 2000). This ecological decoupling appears to occur at the second paleoecological level. For example, as discussed, the Late Ordovician mass extinction, the second-largest extinction in the history of metazoan life, only has third and fourth level paleoecological changes, but the Late Devonian mass extinction has second, third, and fourth level changes. In essence, the ecological difference between these two mass extinctions is that the Late Devonian mass extinction had (1) changes in ecological dominants of higher taxa; (2) loss of metazoan reefs; and (3) loss of Bambachian megaguilds.

These differences imply that although taxonomically the two mass extinctions were relatively similar in size, the organisms lost during each mass extinction had different relative ecological importance. Such differences emphasize the "apples and oranges" relative value of taxa within an ecological context. It appears that in terms of retaining ecological structure after a mass extinction, some taxa are much more important (Droser et al. 2000).

In modern ecological studies, the importance of the differing relative ecological values of taxa is well recognized. For example, in a community, a keystone species may exist and without its presence, the whole ecological structure would collapse. This concept was introduced by Paine (1969) and has become a central principle in ecological studies. Examples include starfish in rocky intertidal communities (Paine 1969), kangaroo rats in desert shrub habitats (Brown and Heske 1990), snow geese in areas adjacent to Hudson's Bay (Kerbes, Kotanen, and Jeffries 1990) and sapsuckers in subalpine ecosystems (Daily et al. 1993).

Keystone species are typically not the most abundant species in a community (Power et al. 1996). Other types of species in a community that have relatively great ecological value are also recognized and include dominant species. Dominant species are the most abundant species in a community and play a major role in controlling the direction and rates of community processes, as well as commonly providing the major energy flow and three-dimensional structure that supports a community (Power et al. 1996).

Therefore, if a community loses 50% of its species, because each species has a different relative ecological value, it really depends on which species are lost as to whether very much damage has been done to the ecological structure. If that 50% includes the widespread loss of dominant or keystone groups, then this biotic crisis could result in major ecological changes that would be manifested at our second level. If that 50% only includes the rare taxa that do not have keystone properties, or if its effect upon dominant species is only on a regional level, then possibly very little of the ecological structure would be lost, and it may be relatively easy to rebuild the community after the crisis, with similar ecological structures as before.

Thus, second level changes can be caused primarily by the loss of ecological dominants but also possibly through the loss of keystone species in a variety of communities. Such changes in ecological dominants are exemplified by the change from dominance of brachiopods in late Paleozoic benthic level-bottom settings before the end-Permian mass extinction, to dominance by bivalves in the Mesozoic after the mass extinction. Similarly, our categorization of the loss of metazoan reefs as a second level phenomenon is similar because so much of the history of reefs is one of change from one dominant taxonomic group to another during the course of major events in life's history (e.g., Fagerstrom 1987). For example, reef change caused by the Cretaceous–Tertiary mass extinction is that of Cretaceous reefs dominated by rudist bivalves to Tertiary reefs dominated by scleractinian corals.

Ecologists and conservation biologists are struggling to develop ways to measure modern ecological degradation and to evaluate ecological changes. We have postulated herein that the most ecologically devastating mass extinctions are those where taxa of relatively high ecological value underwent selective extinction. When dealing with modern settings, identification of keystone taxa is problematical (e.g., Power et al. 1996). However, studies on modern keystones can be extended back into the relatively recent geological past, using an approach of taxonomic uniformitarianism, as done by Owen-Smith (1987) for the Pleistocene terrestrial mass extinctions. However, when viewed from 500, 350, or even 100 million years ago, we not only have a scale problem, but the actual identification of keystone taxa from the truly ancient fossil record is at best difficult. In contrast, as discussed in our analysis of various examples, the existence of dominant taxa in paleoenvironments is readily recognizable, so we can presently document patterns of large-scale second level paleoecological changes through the course of major events in the Phanerozoic.

The ultimate question to ask of this decoupling phenomenon is why one mass extinction preferentially concentrates on taxa of relatively high ecological value, such as dominant taxa, while another does not. Most likely it is just chance as to whether a particular mass extinction mechanism preferentially affects taxa of high ecological value. Thus, different mass extinction causes may lead one cause to concentrate on dominant taxa, while another mass extinction cause of equal taxonomic effect may just eliminate taxa in an ecologically nonpreferential way.

Conclusions

Traditionally, paleontologists have been interested in the disruption of ecosystems associated with mass extinctions (e.g., Valentine 1973; Jablonski 1986; McGhee 1996). Recognition of paleoecological-level changes provides a means to evaluate and compare such ecological degradation. When evaluating a mass extinction, paleoecologists can ask to what paleoecological levels did ecological degradation caused by the mass extinction proceed? In the case of the Late Ordovician event, the extinction resulted in ecological degradation only at the third and fourth levels, whereas for the Late Devonian and end-Permian extinctions, the ecological degradation proceeded to the second level. Thus, paleoecological levels can be used to determine whether the taxonomic significance of an event is decoupled from its ecological significance when compared with other events. This appears to be the case not only for some mass extinctions, but also for radiations. In addition, this system of paleoecological levels can be used to identify significant ecological innovations during recoveries.

The application of this method to the variety of events discussed in this chapter illustrates several important points. First, paleoecological changes can be ranked, and in that way the changes associated with taxonomic events can be evaluated. In addition, if there is one signal of an ecological change for any of the levels, there are commonly several, so it is a fairly robust system. For example, if we see the addition or subtraction of a Bambachian megaguild, then we also see other types of structural changes within an established eco-system. Thus, although we still cannot count paleoecological changes as we can count taxa, this system of paleoecological levels provides a means to cate-gorize and thus compare ecological changes. In this way, we can begin to bet-ter understand the role and nature of ecological changes associated with major taxonomic and evolutionary events in the Phanerozoic history of life.

In studies of the modern biodiversity crisis, biologists initially focused on reductions in diversity of taxa. Subsequently, the focus became more refined and shifted to saving critical habitats (i.e., segments of the environment with original ecological structure) in order to better conserve biodiversity. Perhaps one insight to come from this study of potential utility for modern conserva-tion biologists is that ecological structure may pass relatively unscathed through a biodiversity crisis if the taxa with relatively high ecological value are sufficiently retained after the event. Conserving a diverse group of signifi-cantly large pieces of critical natural habitat should indeed aid in preservation of the keystone and dominant taxa necessary for maintaining, in the future, ecological structures similar to those observed on Earth today.

ACKNOWLEDGMENTS

Partial funding for this work came from the Petroleum Research Fund, administered by the American Chemical Society, to Bottjer, Droser, and McGhee; National Geo-graphic Society grants to Bottjer, Droser, and McGhee; as well as National Science Foundation grants to Bottjer, Droser, Sheehan, and McGhee. Droser was supported by the White Mountain Research Station. James Valentine, Karl Flessa, Mark Wilson, Nigel Hughes, and J. John Sepkoski Jr. provided very helpful comments on earlier ver-sions of this chapter.

REFERENCES

Anstey, R. L. 1985. Bryozoan provinces and patterns of generic evolution and extinc-tion in the Late Ordovician of North America. *Lethaia* 19:33–51.

Armstrong, H. A. 1996. Biotic recovery after mass extinction: The role of climate and ocean-state in the post-glacial (Late Ordovician—Early Silurian) recovery of the conodonts In M. B. Hart, ed., *Biotic Recovery from Mass Extinction Events,* pp. 105–117. Geological Society Special Publication No. 102. London: The Geo-logical Society.

Ausich, W. I. and D. J. Bottjer. 1982. Tiering in suspension-feeding communities on soft substrata throughout the Phanerozoic. *Science* 216:173–174.

Ausich, W. I. and D. J. Bottjer. 2001. Sessile invertebrates. In D. E. G. Briggs, and P. R. Crowther, eds., *Palaeobiology II*, pp. 388–391. Oxford, U.K.: Blackwell.

Awramik, S. M. 1991. Archean and Proterozoic stromatolites. In R. Riding, ed., *Calcareous Algae and Stromatolites*, pp. 289–304. Berlin: Springer.

Bambach, R. K. 1977. Species richness in marine benthic habitats through the Phanerozoic. *Paleobiology* 3:152–167.

Bambach, R. K. 1983. Ecospace utilization and guilds in marine communities through the Phanerozoic. In M. J. S. Tevesz, and P. L. McCall, eds., *Biotic Interactions in Recent and Fossil Benthic Communities*, pp. 719–746. New York: Plenum Press.

Bambach, R. K. and J. B. Bennington. 1995. Entities in the ecological hierarchy and the comparison between neontology and paleontology. *Geological Society of America Abstracts with Programs* 27:A168.

Barnes, C. R. and S. M. Bergstrom. 1988. Conodont biostratigraphy of the uppermost Ordovician and lowermost Silurian. *Bulletin British Museum (Natural History), Geology Series* 43:325–343.

Barnes, C. R., R. A. Fortey, and S. H. Williams. 1995. The pattern of global bio-events during the Ordovician period. In O. H. Walliser, ed., *Global Events and Event Stratigraphy*, pp. 139–172. Berlin: Springer-Verlag.

Becker, R. T. and M. R. House. 1994. Kellwasser events and goniatite successions in the Devonian of the Montagne Noire with comments on possible causations. *Courier Forschungsinstitut Senckenberg* 169:45–77.

Benton, M. J. 1991. What really happened in the Late Triassic? *Historical Biology* 5:263–278.

Benton, M. J. 1993. The Fossil Record 2. London: Chapman and Hall.

Berry, W. B. N. 1996. Recovery of post-Late Ordovician extinction graptolites: A western North American perspective. In M. B. Hart, ed., *Biotic Recovery from Mass Extinction Events*, pp. 119–126. Geological Society Special Publication No. 102. London: The Geological Society.

Berry, W. B. N. and P. Wilde. 1990. Graptolite biogeography: Implications for palaeogeography and palaeoceanography. In W. S. McKerrow and C. R. Scotese, eds., *Palaeozoic Palaeogeography and Biogeography*, pp. 129–137. Geological Society Memoir No. 12. London: The Geological Society.

Bottjer, D. J. 2001. Biotic recovery from mass extinctions. In D. E. G. Briggs and P. R. Crowther, eds., *Palaeobiology II*, pp. 204–208. Oxford, U.K.: Blackwell.

Bottjer, D. J. and W. I. Ausich. 1986. Phanerozoic development of tiering in soft substrata suspension-feeding communities. *Paleobiology* 12:400–420.

Bottjer, D. J. and D. Jablonski. 1989. Paleoenvironmental patterns in the history of post-Paleozoic benthic marine invertebrates. *Palaios* 3:540–560.

Bottjer, D. J. and M. L. Droser. 1998. Tracking mass extinction recoveries with shell beds: The Lower Triassic of the western United States. *Geological Society of America Abstracts with Programs* 30:A-312.

Bottjer, D. J., J. K. Schubert, and M. L. Droser. 1996. Comparative evolutionary palaeoecology: Assessing the changing ecology of the past. In M. B. Hart, ed., *Biotic Recovery from Mass Extinction Events*, pp. 1–13. Geological Society Special Publication No. 102. London: The Geological Society.

Boucot, A. J. 1975. *Evolution and Extinction Rate Controls.* Amsterdam: Elsevier.

Boucot, A. J. 1983. Does evolution take place in an ecological vacuum? *Journal of Paleontology* 57:1–30.

Boucot, A. J. 1990. Phanerozoic extinctions: How similar are they to each other? In E. G. Kauffman and O. H. Walliser, eds., *Extinction Events in Earth History*, pp. 5–30. Lecture Notes in Earth Sciences 30. Berlin: Springer-Verlag.

Boucot, A. J. 1996. Epilogue. *Palaeogeography, Palaeoclimatology, Palaeoecology* 127:339–359.

Bowring, S. A., D. H. Erwin, Y. G. Jin, M. W. Martin, K. Kavidek, and W. Wang. 1998. U/Pb zircon geochronology and tempo of the end-Permian mass extinction. *Science* 280:1039–1045.

Brenchley, P. J. 1989. The Late Ordovician extinction. In S. K. Donovan, ed., *Mass Extinctions: Process and Evidence*, pp. 104–132. New York: Columbia University Press.

Bretsky, P. W. 1968. Evolution of Paleozoic invertebrate communities. *Science* 159:1231–1233.

Brown, J. H. and E. J. Heske. 1990. Control of a desert-grassland transition by a keystone rodent guild. *Science* 250:1705–1707.

Copper, P. 1994. Ancient reef ecosystem expansion and collapse. *Coral Reefs* 13:3–12.

Crick, R. E. 1990. Cambrian-Devonian biogeography of nautiloid cephalopods. In W. S. McKerrow and C. R. Scotese, eds., *Palaeozoic Palaeogeography and Biogeography*, pp. 147–161. Geological Society Memoir No. 12. London: The Geological Society.

Daily, G. C., P. R. Erlich, and N. M. Haddad. 1993. Double keystone bird in a keystone species complex. *Proceedings of the National Academy of Sciences, USA* 90:592–594.

Droser, M. L. and D. J. Bottjer. 1988. Trends in extent and depth of bioturbation in Cambrian carbonate marine environments. *Geology* 16:233–236.

Droser, M. L. and D. J. Bottjer. 1989. Ordovician increase in extent and depth of bioturbation: Implications for understanding early Paleozoic ecospace utilization. *Geology* 17:850–852.

Droser, M. L. and D. J. Bottjer. 1993. Trends and patterns of Phanerozoic ichnofabrics. *Annual Review of Earth and Planetary Sciences* 21:205–225.

Droser, M. L. and P. M. Sheehan. 1995. Paleoecological significance of the Ordovician radiation and end Ordovician extinction: Evidence from the Great Basin. In J. C. Cooper, ed., *Ordovician of the Great Basin*, pp. 64–106. Fullerton CA: Pacific Section of the Society of Economic Paleontologists and Mineralogists.

Droser, M. L., N. C. Hughes, and P. A. Jell. 1994. Infaunal communities and tiering in Early Paleozoic nearshore clastic environments: Trace-fossil evidence from the Cambro-Ordovician of New South Wales. *Lethaia* 27:273–283.

Droser, M. L., D. J. Bottjer, and P. M. Sheehan. 1997. Evaluating the ecological archi-tecture of major events in the Phanerozoic history of marine invertebrate life. *Geology* 25:167–170.

Droser, M. L., D. J. Bottjer, P. M. Sheehan, and G. R. McGhee Jr. 2000. Decoupling of taxonomic and ecologic severity of Phanerozoic marine mass extinctions. *Geology* 28:675–678.

Elias, R. J. and G. A. Young. 1998. Coral diversity, ecology, and provincial structure during a time of crisis: The latest Ordovician to earliest Silurian Edgewood Province in Laurentia. *Palaios* 13:98–112.

Erwin, D. H., and P. Hua-Zhang. 1996. Recoveries and radiations: Gastropods after the Permo-Triassic mass extinction. In M. B. Hart, ed., *Biotic Recoveries from Mass Extinction Events,* pp. 223–229. Geological Society Special Publication No. 102. London: The Geological Society.

Erwin, D. H., J. W. Valentine, and J. J. Sepkoski Jr. 1987. A comparative study of diver-sification events: The early Paleozoic versus the Mesozoic. *Evolution* 41:1177–1186.

Fagerstrom, J. A. 1987. *The Evolution of Reef Communities.* New York: Wiley Inter-sciences.

Gould, S. J. and C. B. Calloway. 1980. Clams and brachiopods: Ships that pass in the night. *Paleobiology* 6:383–396.

Gradstein, F. M., F. P. Agterberg, J. G. Ogg, J. Hardenbol, P. Van Veen, J. Thierry, and Z. Huang. 1995. A Triassic, Jurassic and Cretaceous time scale. In W. A. Berggren, D. V. Kent, M. P. Aubry, and J. Hardenbol, eds., *Geochronology, Time Scales and Global Stratigraphic Correlation,* pp. 95–126. SEPM Special Publication No. 54. Tulsa: Society of Economic Paleontologists and Mineralogists.

Gray, J. 1985. The microfossil record of early land plants: Advances in understanding of early terrestrialization, 1970–1984. *Philosophical Transactions of the Royal Society of London* B309:167–195.

Hallam, A. 1991. Why was there a delayed radiation after the end-Palaeozoic extinc-tions? *Historical Biology* 5:257–262.

Hallam, A. 1995. The earliest Triassic as an anoxic event, and its relationship to the end-Palaeozoic mass extinction. *Canadian Society of Petroleum Geologists Memoir* 17:797–804.

Hallam, A. and P. B. Wignall. 1997. *Mass Extinctions and Their Aftermath.* Oxford, U.K.: Oxford University Press.

Harries, P. J., E. G. Kauffman, and T. A. Hansen. 1996. Models for biotic survival fol-lowing mass extinction. In M. B. Hart, ed., *Biotic Recovery from Mass Extinction Events,* pp. 41–60. Geological Society Special Publication No. 102. London: The Geological Society.

Harris, M. T., and P. M. Sheehan. 1996. Upper Ordovician-Lower Silurian shelf sequences of the Eastern Great Basin: A preliminary report on Barn Hills and Lakeside Mountains sections. In B. J. Witzke, G. A. Ludvigson, and J. E. Day, eds., *Paleozoic Sequence Stratigraphy: Views from the North American Craton.* Geological

Society of America Special Paper 306. Boulder CO: The Geological Society of America.

Harris, M. T. and P. M. Sheehan. 1997. Carbonate sequences and fossil communities from the Upper Ordovician–Lower Silurian of the eastern Great Basin. *Brigham Young University Geology Studies* 42:105–128.

Jablonski, D. 1986. Causes and consequences of mass extinctions: A comparative approach. In D. K. Elliot, ed., *Dynamics of Extinction,* pp. 183–229. New York: Wiley.

Jablonski, D. 1987. Heritability at the species level: Analysis of geographic ranges of Cretaceous mollusks. *Science* 238:360–363.

Jablonski, D. and D. J. Bottjer. 1983. Soft-bottom epifaunal suspension-feeding assemblages in the Late Cretaceous: Implications for the evolution of benthic paleocommunities. In M. J. S. Tevesz and P. L. McCall, eds., *Biotic Interactions in Recent and Fossil Benthic Communities,* pp. 747–812. New York: Plenum Press.

Jensen, S., J. G. Gehling, and M. L. Droser. 1998. Ediacara-type fossils in Cambrian sediments. *Nature* 393:567–569.

Kauffman, E. G. and Harries, P. J. 1996. The importance of crisis progenitors in recovery from mass extinction. In M. B. Hart, ed., *Biotic Recovery from Mass Extinction Events,* pp. 15–39. Geological Society Special Publication No. 102. London: The Geological Society.

Kerbes, R. H., P. M. Kotanen, and R. L. Jefferies. 1990. Destruction of wetland habitats by lesser snow geese: A keystone species on the west coast of Hudson Bay. *Journal of Applied Ecology* 27:242–258.

Kershaw, S., T. Zhang, and G. Lan. 1999. A microbialite carbonate crust at the Permian-Triassic boundary in South China, and its palaeoenvironmental significance. *Palaeogeography, Palaeoclimatology, Palaeoecology* 146:1–18.

Kidwell, S. M. 1990. Phanerozoic evolution of macroinvertebrate shell accumulations: Preliminary data from the Jurassic of Britain. In W. Miller, III, ed., *Paleocommunity Temporal Dynamics: The Long-Term Development of Multispecies Assemblages,* pp. 305–327. Paleontological Society Special Publication 5. Pittsburgh: The Paleontological Society.

Kidwell, S. M. and P. J. Brenchley. 1994. Patterns in bioclastic accumulation through the Phanerozoic: Changes in input or in destruction. *Geology* 22:1139–1143.

Knoll, A. H. 1996. Daughter of time. *Paleobiology* 22:1–7.

Lamont, B. B. 1995. Testing the effect of ecosystem composition/structure on its functioning. *OIKOS* 74:283–295.

Lehrmann, D. J., J. Wei, and P. Enos. 1998. Controls on facies architecture of a large Triassic carbonate platform: The Great Bank of Guizhou, Nanpanjiang Basin, South China. *Journal of Sedimentary Research* 68:311–326.

Levinton, J. S. 1996. Trophic group and the end-Cretaceous extinction: Did deposit feeders have it made in the shade? *Paleobiology* 22:104–112.

Li, X., and M. L. Droser. 1997. The distribution and nature of Cambrian shell beds: Evidence from the Basin and Range Province, California, Nevada, and Utah. *Palaios* 12:111–128.

Long, J. A. 1993. Paleozoic vertebrate biostratigraphy of south-east Asia and Japan. In J. A. Long, ed., *Palaeozoic Vertebrate Biostratigraphy and Biogeography*, pp. 54–63. London: Belhaven.

McGhee, G. R. Jr. 1995. Geometry of evolution in the biconvex Brachiopods: Morphological effects of mass extinction. *Neues Jahrbuch für Geologie und Paläontologie, Abhandlungen* 197:357–382.

McGhee, G. R. Jr. 1996. *The Late Devonian Mass Extinction.* New York: Columbia University Press.

McKinney, F. K., S. Lidgard, J. J. Sepkoski Jr., and P. D. Taylor. 1998. Decoupled temporal patterns of evolution and ecology in two post-Paleozoic clades. *Science* 281:807–809.

Mills, L. S., M. E. Soule, and D. F. Doak. 1993. The keystone-species concept in ecology and conservation. *Bioscience* 43:219–224.

Owen-Smith, N. 1987. Pleistocene extinctions: The pivotal role of megaherbivores. *Paleobiology* 13:351–362.

Paine, R. T. 1969. A note on trophic complexity and community stability. *American Naturalist* 103:91–93.

Paul, C. R. C. and S. F. Mitchell. 1994. Is famine a common factor in marine mass extinctions? *Geology* 22:679–682.

Power, M. E., D. Tilman, J. A. Ester, B. A. Menge, W. J. Bond, L. S. Mills, G. Daily, J. C. Castilla, J. Lubchenco, and R. T. Paine. 1996. Challenges in the quest for keystones. *Bioscience* 46:609–620.

Raup, D. M. 1979. Size of the Permian/Triassic bottleneck and its evolutionary implications. *Science* 206:217–218.

Raup, D. M. and J. J. Sepkoski Jr. 1982. Mass extinctions in the marine fossil record. *Science* 215:1501–1503.

Retallack, G. J. and C. R. Feakes. 1987. Trace fossil evidence for Late Ordovician animals on land. *Science* 235:61–63.

Rigby, S. 1997. A comparison of the colonization of the planktic realm and the land. *Lethaia* 30:11–17.

Sandberg, C. A., W. Ziegler, R. Dreesen, and J. L. Butler. 1988. Part 3: Late Frasnian mass extinction: Conodont event stratigraphy, global changes, and possible causes. *Courier Forschungsinstitut Senckenberg* 102:263–307.

Schubert, J. K. and D. J. Bottjer. 1992. Early Triassic stromatolites as post-mass extinction disaster forms. *Geology* 20:883–886.

Schubert, J. K. and D. J. Bottjer. 1995. Aftermath of the Permian-Triassic mass extinction event: Paleoecology of Lower Triassic carbonates in the western USA. *Palaeogeography, Palaeoclimatology, Palaeoecology* 116:1–39.

Seilacher, A. 1992. Vendobiota and Psammocorallia: Lost constructions of Precambrian evolution. *Journal of the Geological Society* 149:607–613.

Sepkoski, J. J. Jr. 1979. A kinetic model of Phanerozoic taxonomic diversity: II. Early Phanerozoic families and multiple equilibria. *Paleobiology* 5:222–252.

Sepkoski, J. J. Jr. 1981. A factor analytic description of the Phanerozoic marine fossil record. *Paleobiology* 7:36–53.

Sepkoski, J. J. Jr. 1992. Phylogenetic and ecologic patterns in the Phanerozoic history of marine biodiversity. In N. Eldredge, ed., *Systematics, Ecology, and the Biodiversity Crisis,* pp. 77–100. New York: Columbia University Press.

Sepkoski, J. J. Jr. 1996. Competition in macroevolution: The double wedge revisited. In D. Jablonski, D. H. Erwin, and J. H. Lipps, eds., *Evolutionary Paleobiology,* pp. 211–255. Chicago: University of Chicago Press.

Sepkoski, J. J. Jr. and P. M. Sheehan. 1983. Diversification, faunal change, and community replacement during the Ordovician radiations. In M. J. S. Tevesz and P. L. McCall, eds., *Recent and Fossil Benthic Communities,* pp. 673–718. New York: Plenum Press.

Sepkoski, J. J. Jr. and A. I. Miller. 1985. Evolutionary faunas and the distribution of Paleozoic benthic communities. In J. W. Valentine, ed., *Phanerozoic Diversity Patterns,* pp. 181–190. Princeton NJ: Princeton University Press.

Sheehan, P. M. 1979. Swedish Late Ordovician marine benthic assemblages and their bearing on brachiopod zoogeography. In J. Gray and A. J. Boucot, eds., *Historical Biogeography, Plate Tectonics, and the Changing Environment,* pp. 61–73. Corvallis OR: Oregon State University Press.

Sheehan, P. M. 1989. Late Ordovician and Silurian paleogeography of the Great Basin. *University of Wyoming Contributions to Geology* 27:41–54.

Sheehan, P. M. 1991. Patterns of synecology during the Phanerozoic. In E. C. Dudley, ed., *The Unity of Evolutionary Biology,* pp. 103–118. Proceedings of the Fourth International Congress of Systematic and Evolutionary Biology. Portland OR: Discorides Press.

Sheehan, P. M. 1996. A new look at Ecologic Evolutionary Units (EEUs). *Palaeogeography, Palaeoclimatology, Palaeoecology* 127:21–32.

Stanley, S. M. 1993. *Exploring Earth and Life Through Time.* New York: W. H. Freeman.

Stanley, G. D. 1996. Recovery of reef communities after the Triassic mass extinction: Sixth North American Paleontological Convention Abstracts of Papers. The Paleontological Society Special Publication 8. Pittsburgh: The Paleontological Society.

Stanley, G. D. Jr. and L. Beauvais. 1994. Corals from an Early Jurassic coral reef in British Columbia: Refuge on an oceanic island reef. *Lethaia* 27:35–48.

Sweet, W. C. 1988. The Conodonta: Morphology, taxonomy, paleoecology, and evolutionary history of long-extinct animal phylum. Oxford Monographs on Geology and Geophysics No. 10. Oxford: Clarendon Press.

Sweet, W. C. and S. M. Bergstrom. 1984. Conodont provinces and biofacies of the Late Ordovician. In D. L. Clark, ed., *Conodont Biofacies and Provincialism,* pp. 69–87. Geological Society of America Special Paper No. 196. Boulder CO: The Geological Society of America.

Tanner, J. E., T. P. Hughes, and J. H. Connell. 1994. Species coexistence, keystone species, and succession: A sensitivity analysis. *Ecology* 75:2204–2219.

Tuckey, M. E. and R. L. Anstey. 1992. Late Ordovician extinctions of bryozoans. *Lethaia* 25:111–117.

Twitchett, R. J. 1999. Palaeoenvironments and faunal recovery after the end-Permian mass extinction. *Palaeogeography, Palaeoclimatology, Palaeoecology* 154:27–37.

Underwood, C. J. 1998. Population structure of graptolite assemblages. *Lethaia* 31:33–41.

Valentine, J. W. 1973. *Evolutionary Paleoecology of the Marine Biosphere.* Englewood Cliffs NJ: Prentice-Hall.

Watkins, R. 1991. Guild structure and tiering in a high-diversity Silurian community, Milwaukee County, Wisconsin. *Palaios* 6:465–478.

Watkins, R. 1994. Evolution of Silurian pentamerid communities in Wisconsin. *Palaios* 9:488–499.

Wilson, M. A., T. J. Palmer, T. E. Guensburg, C. D. Finton, and L. E. Kaufman. 1992. The development of an Early Ordovician hardground community in response to rapid sea-floor calcite precipitation. *Lethaia* 25:19–34.

Wood, R. A. 1995. The changing biology of reef-building. *Palaios* 10:517–529.

Wray, G. A., J. S. Levinton, and L. H. Shapiro. 1996. Molecular evidence for deep Precambrian divergences among metazoan phyla. *Science* 274:568–573.

5

Stability in Ecological and Paleoecological Systems: Variability at Both Short and Long Timescales

Carol M. Tang

THE FIELD OF EVOLUTIONARY PALEOECOLOGY encompasses the disciplines of evolutionary biology, ecology, geology, and paleontology. In the past, studies in evolutionary paleoecology have focused on two somewhat divergent aspects: (1) the study of the ecology of evolution and (2) the study of the evolution of ecology. The first subfield is concerned with the environmental and ecological conditions that accompany and possibly even stimulate or constrain speciation and macroevolutionary processes. On the other hand, the study of the evolution of ecology is concerned with how communities and ecological structures, in response to evolutionary and environmental parameters, have changed through time.

Although these two aspects of evolutionary paleoecology can be quite different in their approaches and philosophies, and some studies are clearly designed to address only one of these two aspects, these aspects are not mutually exclusive and can be synergistic. For either focus, to understand the environmental aspects of the fossils under investigation, evolutionary paleoecologists must be good geologists, paleoceanographers, stratigraphers, and sedimentologists. At the same time, evolutionary paleoecologists must understand the ecological functions and structures of the organisms themselves and enter the realm of neoecology so that they may evaluate models of community structure and ecosystems.

The study of long-term faunal stability in the fossil record is an example of a line of research that encompasses both aspects of evolutionary paleoecology and draws from geology, evolutionary biology, and ecology. By studying patterns of faunal stability and turnover, one can evaluate the timing of origination and extinction of lineages and communities in relationship to environmental factors (i.e., the ecology of evolution), as well as how community structure may change with regard to environmental and faunal change (i.e., the evolution of ecology). At the same time, the study of long-term faunal stasis also illustrates that the field of evolutionary paleoecology can make its own unique contributions to the more traditional aspects of paleontology, evolutionary biology, and ecology (e.g., Bretsky and Lorenz 1970; Jablonski and Sepkoski 1996).

To illustrate these ideas, this chapter will review neoecological and paleoecological concepts about faunal stability at both short and long timescales. Miller (1996) provided an early treatment of coordinated stasis within a neoecological context, and Jennions (1997) discussed the ecological implications of Pandolfi's (1996) example of coral coordinated stasis. Many paleoecologists have also discussed the evolutionary significance of long-term faunal stasis in the fossil record (e.g., DiMichele 1994; Morris et al. 1995; and articles in Ivany and Schopf 1996). In this chapter, I will incorporate paleontological data and theories with neoecological concepts in an attempt to elucidate some of the ways in which evolutionary paleoecologists can contribute to evolutionary and ecological theory.

Types of Stability

Stability is commonly examined in neoecology in terms of faunal composition and trophic structure within a community, but it can also include the analysis of nondemographic characteristics such as the rate of productivity, biomass, and nutrient flux. In the paleontological literature, stability has generally referred to the recurrence of assemblages with similar demographic characteristics: taxonomic compositions, rank abundances, and trophic structure. Most commonly, only the dominant taxa are included, and it is debated whether rare species can be expected to exhibit the same level of stability as co-occurring common taxa (e.g., McKinney, Lockwood, and Frederick 1996). Coordinated stasis itself is also characterized by nondirectional morphological change within taxa (Lieberman, Brett, and Eldredge 1995; Brett, Ivany, and Schopf 1996). The paleontological use of the term *stability* is generally similar to *persistence* referring to the ability of a system to persist as an identifiable entity (e.g., Connell and Sousa 1983; Grimm 1996).

Because the term *stability* can mean many things and its use in the ecological literature has been questioned (e.g., Grimm and Wissel 1997), it may be useful to consider some terms used to describe different types of stability that encompass many levels of analyses, timescales, and variables.

Resistance versus Resilience

Resistance is the ability of a community to withstand perturbations without significant change, whereas resilience refers to the ability of a community to rebound to initial conditions after experiencing a disturbance (Grimm and Wissel 1997, and references therein). Having high resistance does not necessarily confer high resilience upon a community and vice versa. In fact, the opposite may be true: resistance may be negatively correlated with resilience. For example, k-selected populations (equilibrium species maintaining populations close to its carrying capacity) may be resistant to perturbations but may have low resilience once disturbed (Begon, Harper, and Townsend 1996). Conversely, r-selected populations (opportunistic species that can multiply quickly) have low resistance to disturbances but can recover rapidly afterward (high resilience).

Can paleontologists distinguish between these two aspects of stability? If we can recognize single event beds and reconstruct the response of the organisms to these events, it is possible that we can recognize resistance. But as a result of time-averaging in the fossil record, paleoecologists are usually forced to focus on the resilience of individuals, populations, and communities over hundreds or thousands of years rather than on their ability for resistance. It would be difficult to evaluate whether (1) a community were truly resistant and faunal elements stayed together during sea level changes by "ecologically locking" (Morris 1995) or "tracking" their optimal environments (Brett and Baird 1995), or (2) whether the linkages between organisms were destroyed but the same community reassembled simply by employing similar assembly rules (e.g., Pandolfi 1996) and/or with the same species pool (e.g., Buzas and Culver 1994). Therefore, the patterns of stability documented by evolutionary paleoecologists could be the result of resistance or resilience or a combination of the two (for example, resistance during smaller perturbations and resilience in the face of larger disturbances).

Local versus Global Stability

Local stability refers to stability in the face of small-scale disturbances, whereas global stability refers to stability in response to major perturbations (Begon, Harper, and Townsend 1996). Clearly, these terms are relative and subjective.

They not only may have different meanings among different workers, but differences may be especially exaggerated between paleoecologists and ecologists; for example, a "100-year hurricane" may be considered a large, uncommon perturbation to a neontologist but is a small, high-frequency event to a paleontologist (one which may not even be resolvable in some paleontological systems). Thus, the local stability examined by neoecologists is almost always not relevant to paleoecologists, and the global stability examined by paleoecologists may not be as important to neoecologists.

On a theoretical level, however, the importance of differentiating between relatively local and relatively global kinds of stability is especially clear if one looks at a community experiencing a number of types of disturbances through time. A community can exhibit both high local and global stability, meaning that it is extremely stable at both short and long timescales (figure 5.1A). There are also communities with low levels of both local and global stability and thus would be unstable (figure 5.1B). More complex would be a community that exhibits high local stability but low global stability (figure 5.1C). In this case, a

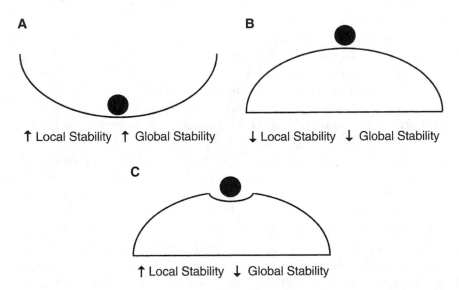

FIGURE 5.1. Schematic representation of the concepts of local versus global stability. The "ball" represents the state of a community or species. The amount of stability is related to the propensity of the community to stay in one place in the face of disturbance. (A) A system with high local and global stability. (B) A system with low local and global stability; may be representative of a stable system without coordinated turnover among taxa. (C) A system with high local stability but low global stability; may be representative of coordinated stasis systems. (After Begon, Harper, and Townsend 1996)

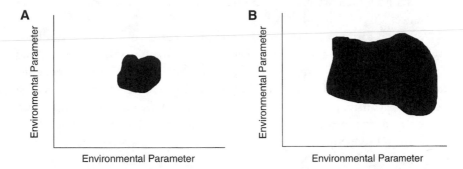

FIGURE 5.2. A two-dimensional schematic to illustrate the difference between (A) a dynamically fragile community versus (B) a robust one. A dynamically fragile community (or species) would be able to exhibit stability within a limited range of environmental conditions, whereas a robust one could withstand a wide range of conditions. (After Begon, Harper, and Townsend 1996)

community could be stable at short time intervals in the face of small disturbance events, but unstable when confronted with larger perturbations.

Dynamically Fragile versus Dynamically Robust Communities

Communities that are stable only under a limited range of environmental and ecological conditions are considered dynamically fragile, whereas communities that are stable over a wide range of conditions are dynamically robust. Figure 5.2 is a two-dimensional representation of the difference between a dynamically fragile and a dynamically robust system. Of course, a hypervolume with numerous axes would more adequately represent natural environmental conditions. A fragile community is one in which only a limited set of variables would keep the system stable (figure 5.2A), whereas a robust community can stay stable under a greater range of conditions (figure 5.2B). Again, these are subjective and scale-dependent terms, but they have significance when comparing communities and the effects of disturbances on communities under different environmental regimes.

In examining communities in this way, it has been hypothesized that predictable and stable environments will allow for the persistence of dynamically fragile communities, whereas environments with highly variable conditions would favor the presence of dynamically robust associations. It has also been hypothesized that stable environments would favor *k*-selected organisms, which are resistant to perturbations but may have low resilience. On the other hand, highly variable environmental conditions could lead to *r*-selected populations, which are predicted to have low levels of resistance and higher resilience.

Although formal definitions for generalists and specialists have not been well developed, one could apply the definition of dynamically robust and dynamically fragile not only to communities but also to species. Generalist taxa can be considered as those that are dynamically robust, whereas specialist taxa can be considered as those that are dynamically fragile. This concept may be related to the idea that generalists are more long-lived than specialists, as discussed subsequently.

Stability Over Neontological Timescales

We are currently in a period of global change in which extinctions and invasions of new habitats are proceeding quickly. Thus, it is clearly important to understand if and how ecosystems react to physical and ecological perturbations. There are documented cases in which the introduction of exotic species into new ecosystems had little effect on the native fauna (e.g., Simberloff 1981), which suggests that stability is quite predominant. In other examples, the introductions have caused local extinctions and upheaval in communities (e.g., Vitousek 1986; Atkinson 1989). With such a diversity of responses to just one type of ecological perturbation (although the deletion or addition of species is considered to be a large "global" disturbance), it seems apparent that stability in modern communities is dependent on the many environmental, biological, and ecological factors in each ecosystem.

From the 1950s to 1970s, conventional wisdom among neoecologists professed that increased ecosystem complexity, which is related to high species diversity, large number of interactions between species, and other factors, was correlated with greater levels of stability. This would appear to be intuitively correct because one would assume that the removal or addition of organisms in a highly complex food web would not greatly disrupt the entire food web, whereas a food web with few pathways would appear to be easily perturbed (MacArthur 1955). Quantitative observations were forwarded to support and explain this idea of greater stability in diverse, complex ecosystems; for example, islands with few species appear to be more vulnerable to invaders than are species-rich continental ecosystems, and crop monocultures are more vulnerable to disease and invasion than are natural mixed associations (Elton 1958).

With the advent of advanced numerical modeling in the 1970s, however, it appeared that not only was the conventional wisdom that complexity begets stability incorrect, but initial results also suggested that the complete opposite was true: increased complexity was related to greater instability in model ecosystems (May 1972). Through the succeeding years, a variety of studies have shown that community stability can be affected by a number of factors, including number of trophic levels (e.g., Pimm and Lawton 1980), energy flux (e.g., O'Neill 1976), food web dynamics (e.g., DeAngelis 1975), and life histories (e.g., Pimm and Rice 1987). In addition, follow-up models have shown mixed results when considering larger perturbations (global rather than local stability) (e.g., Pimm 1979). The bottom line appears to be that numerical modeling has shown that no definitive relationship between complexity and stability can be clearly demonstrated.

Unfortunately, field-based studies also have not been able to provide definitive clarification on the relationship between stability, complexity, and other ecological factors. Tilman and Downing (1994) documented that increased species richness enhanced community stability (both resistance and resilience) in response to drought conditions in a terrestrial plant community, thus providing support for the original conventional wisdom (complexity begets stability). Similarly, Death (1996) found that more complex stream invertebrate communities were more resilient than less complex communities, and many recent studies have challenged the idea that complexity begets instability based on analysis of connectance (Fonseca and John 1996), food chain models (e.g., Sterner, Bajpai, and Adams 1997), and density-dependent populations (Dodds and Henebry 1996). In another example, however, high-diversity terrestrial plant communities exhibited lower resilience in response to changes in nutrient flux and grazing than did low-diversity associations (McNaughton 1977).

Therefore, results from both field studies and theoretical modeling are equivocal and seem to indicate that, as one ecology textbook states, "[N]o single relationship will be appropriate in all communities. . .[t]he relationship between the complexity of a community and its inherent stability is not clear-cut. It appears to vary with the precise nature of the community, with the way in which the community is perturbed and with the way in which stability is assessed" (Begon, Harper, and Townsend 1996). Despite problems with differences among field studies in the level of analysis, scope of study, and choice of analyzed variables (see Connell and Sousa 1983), as well as problems with taxonomic resolution (Hall and Raffaelli 1993), it generally appears that modern communities exhibit a range of stability.

Stability Over Paleontological Timescales

Long-term faunal stability has been recognized for some time in the fossil record and is in fact the fundamental basis of the development of the geological timescale. The initial recognition of the geological periods, epochs, and ages was really an acknowledgment that there was an underlying similarity of faunas through some time periods separated by short intervals of faunal turnover (e.g., d'Orbigny 1840–1846). For many years, Boucot has proposed a pattern of community evolution in which the structure of communities is stable for millions of years during Ecologic–Evolutionary Units (EEUs) (Boucot 1983, 1990; Sheehan 1996). Although his initial work on this pattern was directed toward biostratigraphic questions, Boucot began to focus his attention on the ecological and evolutionary implications of this pattern in the late 1970s (Boucot 1996). At this time, other paleoecologists were also examining the environmental and evolutionary factors controlling long-term faunal stability and relating paleontological patterns to ecological theory and models (e.g., Bretsky 1968, 1969; Bretsky and Lorenz 1970).

The documentation of "coordinated stasis" in the Silurian–Devonian Appalachian Basin, which began as an empirical field-based observation (Brett and Baird 1992), has become a focal point for evolutionary paleoecology and has brought paleontological ideas further into the realm of modern ecology and evolutionary biology. Although the original example of coordinated stasis has spawned many attempts to look at long-term faunal stability (reviewed in Brett, Ivany, and Schopf 1996), at this time, the only other published study that uses comparable temporal, spatial, and taxonomic coverage is by Tang and Bottjer (1996); the two studies have some differences in analytical procedures (Schopf and Ivany 1997; Tang and Bottjer 1997). Patzkowsky and Holland (1997) employed a similar temporal and spatial scale with an Ordovician brachiopod fauna and concluded that there was no long-term stability in species and paleocommunities and no pulses of species turnover.

To explore the possible factors that control stability in the fossil record, the following discussion will focus on comparing and contrasting Brett and Baird (1995) with Tang and Bottjer (1996), with some allusions to other faunal stability studies and reviews. The differences, as well as the similarities, between the taxonomic composition, environmental conditions, and stasis patterns (table 5.1) provide much insight into the dynamics of community stability and the factors that may control it. In the following discussion, the term paleocommunity is used similarly to the way the authors used it in their original articles to refer to recurrent taxonomic associations. Because these are fossil assemblages analyzed under the usual taphonomic limitations, these paleocommunities most likely represent time-averaging of the dominant faunal components.

Both Brett and Baird (1995) and Tang and Bottjer (1996) examined faunal persistence through several marine sequences that spanned tens of millions of years and found that some benthic species and paleocommunities exhibited

TABLE 5.1. Comparison Between Long-Term Stability Patterns Documented in Coordinated Stasis (Brett and Baird 1995) and Uncoordinated Stasis (Tang and Bottjer 1996; Tang 1996)

	Brett and Baird (1995)	Tang (1996), Tang and Bottjer (1996)
Temporal scale of units	3–7 million years	2–6 million years
Time interval	Silurian–Devonian	Jurassic
EEU interval (Sheehan 1996)	P3 (stable familial diversity)	M2 (increasing familial diversity)
Spatial scale of study	Appalachian Basin	Western Interior seaway
Paleoenvironments	Shallow shelf to deep water	<100 m in depth
Sediments	Mixed carbonate-clastic system	Mixed carbonate-clastic system
Taxonomic composition	Brachiopods common; corals sometimes common; bryozoans, bivalves, crinoids, trilobites present	Bivalves most common; corals, crinoids, bryozoans present; gastropods and brachiopods rare
General stability pattern	High local stability Low global stability	Variable local stability Variable global stability (can be very high)
Holdover % between units	9.6%–55%	34%–100%
Carryover % between units	4.7%–34.5%	17%–93%

stability over millions of years. The patterns of stability, however, differed; in the Silurian–Devonian Appalachian Basin system, long intervals of stability were interrupted by short intervals of highly synchronized turnover (Brett and Baird 1995). In contrast, the Jurassic–U.S. western interior system was characterized by some long-lived taxa and paleocommunities (some of which extended for the entire 20 m.y. span of the epicontinental seaway), but these taxa and associations did not exhibit synchronized turnover or origination (Tang and Bottjer 1996).

Both studies documented "outages" when specific taxa and paleocommunities disappeared for short intervals before reappearing in the rock record. This would indicate that resilience, not resistance, is the factor being analyzed in these evolutionary paleoecological studies.

For Brett and Baird's example, one could classify the pattern as one of very high local stability within each evolutionary ecologic (EE) subunit. The paleocommunities appear to be resilient after major storms and invasions by epibole taxa (defined by Trueman 1923, refined by Brett and Baird 1997). What keeps this paleocommunity locally stable has been the subject of much debate (see review of proposed causes for *coordinated stasis* in Ivany 1996). But global stability between EE subunits is low, as evidenced by the fact that there are major near synchronous turnovers at times associated with worldwide oceanographic events or basin-wide sea level changes and anoxia. Of interest here is what ecological or environmental trigger caused these taxa or communities to become globally unstable near synchronously across lineages (see Ivany 1996 for a review of proposed causes for *coordinated turnover*). When, however, the turnover event is associated with such an extreme disturbance as anoxia, it is difficult to predict what the global stability and resilience levels would have been in the face of less extreme environmental perturbations.

Although the characteristics of high local stability and low global stability appear to be valid for the ecosystem as a whole, there may be different levels of local and global stability at different timescales for different environments and taxa within the Silurian–Devonian Appalachian Basin (Brett and Baird 1995). For example, although the majority of species go extinct at EE subunit boundaries, a fraction of the fauna does persist through these boundaries. In addition, the nearshore siliciclastic biofacies can exhibit higher levels of global stability between EE subunits than other biofacies (Brett and Baird 1995). These cases also illustrate that the dynamic fragility and robustness of the paleocommunities and species seem to vary across the ecosystem.

In Tang and Bottjer's (1996) Jurassic example, they found it is not possible to assess the local or global stability or robustness of the entire ecosystem because a large range of patterns is exhibited by both taxa and associations. On the extreme end are some taxa and paleocommunities that have both high local and global stability. These long-lived entities not only persist through and are resilient in the face of small perturbations, such as the shifting of barrier bars, storms, or invasions of other taxa, but they also do so with respect to major sea level and environmental shifts.

Factors That May Control Faunal Stasis
Over Paleoecological Timescales

The fundamental differences seen in these two patterns may be related to a number of factors. For example, the Silurian–Devonian P3 EEU examined by Brett and Baird (1995) is characterized by a relatively stable level of familial diversity, whereas faunal diversity increased rapidly during the Jurassic–Cretaceous M2 EEU (Sheehan 1996). In addition, the Jurassic is a time of rapid escalation, infaunalization, and ecological change (Vermeij 1977). This background of great global diversification and global change may have affected patterns of stability, although it is unclear what this relationship may have been.

Another difference of the geological periods is that there is a significant statistical difference between faunal constituents of the Paleozoic Fauna and the Modern Fauna (Sepkoski 1981). The Silurian–Devonian faunal system is dominated by articulate brachiopods, whereas the Jurassic fauna is heavily dominated by bivalves. The differences in faunal components may play a role in affecting patterns and strength of stability. For example, bivalves have been shown to possess relatively long species durations (Stanley 1979). If one takes this into account, it may be expected that some of the intervals of stability exhibited by Jurassic western interior bivalve-dominated assemblages are longer than those seen in the Silurian–Devonian Appalachian Basin. In modern neoecological studies, it has been acknowledged that long-lived individuals must be accounted for in evaluating stability (Connell and Sousa 1983). At paleoecological scales, Tang and Bottjer (1996) proposed that it is just as important to account for long-lived taxa in explaining patterns of stability in the fossil record.

Related to this hypothesis regarding the importance of intrinsic species durations in determining stability, Westrop (1996) has suggested that Upper Cambrian trilobites do not exhibit coordinated stasis because they had inherently higher turnover (speciation and extinction) rates than the brachiopod-dominated assemblages described by Brett and Baird (1995). If inherent turnover rates control stasis within lineages, long-term stability would be more likely to occur in post-Paleozoic systems as the level of constraint increased through time. Tang and Bottjer's (1996) results do indicate that the overall levels of stasis in some Jurassic species and paleocommunities may be higher than in the Silurian–Devonian (i.e., longer intervals of stasis). However, the higher level of stability did not lead to punctuated turnover events associated with coordinated stasis. Thus, lower turnover rates in individual lineages does not necessarily lead to a pattern of coordinated stasis (sensu stricto).

Another factor that controls species durations and thus could potentially control patterns of faunal stasis is the ecological nature of the species themselves. For example, as discussed earlier, one would intuitively suspect that dynamically robust taxa could have longer species durations simply because they can withstand larger environmental fluctuations. Eldredge and Cracraft (1980:304) have hypothesized that eurytopic species (which have "relative breadth of tolerance to specifiable parameters of the physical and biotic environment") have low extinction rates, react to interspecific competition by mutual exclusion, and occur over wide geographic ranges. In studies of Cenozoic mammals (Vrba 1987) and Paleozoic crinoids (Kammer, Baumiller, and Ausich 1997), eurytopic taxa have been shown to have longer durations than stenotopes.

In the case of the Jurassic faunas, species and paleocommunities can often be found in a number of different paleoenvironments, which indicates that they are eurytopic (Tang 1996). In addition, paleocommunities have low levels of alpha and beta diversity, which suggests low levels of both habitat and niche and resource specialization (Tang 1996). The generalist nature of the Jurassic fauna supports the idea that this system may have been predisposed to exhibiting high levels of stasis simply by containing long-lived taxa. It would be worthwhile to examine whether generalist taxa also dominate the most stable Silurian–Devonian Appalachian biofacies.

Last, if one looks at the specific environmental context for stable paleocommunities in both systems, one can see that there in fact may not be major discrepancies between the Paleozoic and Mesozoic examples of long-term faunal stability. Brett and Baird (1995) have shown that there appear to be varying levels of stability exhibited in different environments within the Appalachian Basin; for example, nearshore siliciclastic systems with lower species diversities than carbonate shelf environments exhibited higher levels of stability. Bretsky (1968, 1969) also documented a pattern of long-term stability in low-diversity, nearshore Upper Ordovician communities of the central Appalachians, whereas more rapid compositional changes occurred in offshore environments. Thus, it is possible that the pattern of long-term stasis without synchronous turnover seen in the Jurassic shallow epicontinental seaway are comparable to those seen in corresponding shallow-water environments of the Paleozoic and that the lack of deeper, offshore environments in the Jurassic North American seaway is responsible for the different patterns. Although more study is needed to confirm this observation, an environment-specific analysis could do much to explain the dynamics of faunal stasis and answer some of the fundamental ecological questions about community stability.

Contributions of Evolutionary Paleoecology to Neoecological Theories

Although it has been suggested that the main reason coordinated stasis has not been reported from the Quaternary or from neoecological studies is due to differences in scale (Schopf and Ivany 1998), there is evidence to suggest that scaling alone is not a viable explanation. There are studies at Pleistocene timescales that do show coordinated stasis (e.g., Pandolfi 1996) and similarly, there are examples of a lack of coordinated stasis in pre-Pleistocene systems studied with the same timescale as given by Brett and Baird (1995) (e.g., Tang and Bottjer 1996; Patzkowsky and Holland 1997).

Thus, it appears that the patterns of faunal stasis can be variable at a variety of scales and that coordinated stasis, although present under some circumstances, is not characteristic of all systems examined at the scale of 10^6–10^7 years. Given that a range of stability patterns exists over neoecological timescales, depending on individual environmental, taxonomic, and ecological conditions, it is illustrative to look at the context for paleoecological stability patterns as well. It is especially interesting that both coordinated stasis (Pandolfi 1996) and continuous ecological shifts (e.g., Valentine and Jablonski 1993; and Bennett 1997) can be found in the Quaternary, a time of extreme environmental shifts.

It appears that in both the Jurassic and Silurian–Devonian examples, low-diversity, shallow-water communities exhibit higher levels of stability. The strong stasis could be a result either of the low-diversity nature of the fauna or of the variable environmental conditions under which the communities existed. Although the relationship between diversity and stability is still unclear and probably quite complex, the results from both Brett and Baird (1995) and Tang and Bottjer (1996) support previous paleontological studies that suggest that low diversity and low complexity are correlated with increased stability (e.g., Bretsky 1968, 1969; Bretsky and Lorenz 1970).

On the other hand, these low-diversity assemblages are found within shallow-water environments, and in these settings, it will be difficult to tease apart the effects of low diversity and environmental context on patterns of stability. The possibility that environments with high levels of disturbance could lead to greater stability has also been widely considered by neoecologists and paleoecologists (e.g., Bretsky 1969; Bretsky and Lorenz 1970; Sepkoski 1987; Chesson and Huntley 1989; Sheldon 1996). For example, it has been proposed that the reason biological invasions of marine organisms into the San Francisco Bay has not had devastating effects on the native fauna is that the Bay is a highly variable estuarine system that creates a harsh, fluctuating environment, and thus organisms are already able to withstand perturbations (Valentine and Jablonski 1993). Paleoecologists have proposed that high-frequency disturbance may play a role in the development of stability in ecosystems (e.g., Sheldon 1996; Plotnick and McKinney 1993; Morris et al. 1995). Numerical modeling, however, has shown that short-term instability is not consistently correlated with long-term stability; in fact, as discussed earlier, short-term instabilities can either promote long-term stability, long-term instability, or have no long-term effects, depending on other biological factors including life histories. Initial comparisons between the studies by Brett and Baird (1995) and Tang and Bottjer (1996) support the idea that low-diversity, high-frequency disturbances, or a combination of the two factors, is correlated with increased levels of community stability.

Because resilience (and not resistance) is being analyzed at the timescale of tens of millions of years, explanations forwarded to explain these patterns of faunal stasis, whether coordinated or uncoordinated, should focus on how paleocommunities become established and reassembled rather than how they stay together. In this sense, studies such as Pandolfi (1996) and the current neoecological debate about community assembly rules (see Jennions 1997) may hold a clue to understanding the mechanisms behind long-term faunal stasis as seen by evolutionary paleoecologists.

Conclusions

At this time of high disturbance and extinctions in natural systems, ecologists, paleoecologists, and conservation biologists must understand the underlying factors involved in species, community, and ecosystem stability (Grimm and Wissel 1997; Palmer, Ambrose, and Poff 1997). The examination of long-term faunal stability patterns in the fossil record can provide insights into some of the fundamental questions in ecology regarding community structure and the relationship between stability and diversity, environmental disturbance, and generalist versus specialist life histories.

Mathematical models and neoecological field studies appear to show that faunal stability depends on many different factors, such as connectivity, species diversity, number of trophic levels, life histories,—and that these relationships are not clear-cut. The level of stability appears to be highly dependent on the biological and environmental conditions of each system. In this light, should paleontologists expect to see only one pattern of faunal stability in the fossil record and search for only one mechanism for producing these different patterns? I would propose that instead of debating the merits of one all-encompassing theory, the study of coordinated and uncoordinated stasis should focus on elucidating the different patterns and underlying environmental, ecological, and phylogenetic conditions so that evolutionary paleoecologists may begin to examine the range of possible controls of faunal stability.

The topic of faunal and community stability in the fossil record is a perfect example of how evolutionary paleoecologists can bridge the fields of paleontology, ecology, and evolutionary biology in developing and testing models of how species and species interactions change through time. The time perspective that paleontologists work with allows them to test models that neoecologists may not be able to address. Interestingly enough, however, in order to contribute to neontology, evolutionary paleoecologists are required to become good geologists because without the environmental framework and time constraints of evolutionary and ecological change that field-based paleoecological, paleoenvironmental, and stratigraphic work provide, our models will continue to be regarded as "mind exercises" with little impact on neontological theory.

ACKNOWLEDGMENTS

I'd like to thank W. Allmon and D. Bottjer for inviting me to participate in the Evolutionary Paleoecology symposium and to contribute to this volume. I developed the ideas and research presented here as a Chancellor's Postdoctoral Fellow in the Department of Integrative Biology at the University of California-Berkeley, and as a doctoral candidate in the Department of Earth Sciences at the University of Southern California. Helpful reviews were provided by W. Allmon, W. Miller III, J. Sepkoski, and P. Sheehan. Stimulating discussions were provided by D. Bottjer, C. Brett, the ecology group at the University of California-Berkeley and Arizona State University, and participants at the Geological Society of America Penrose Conference on Spatial and Temporal Patterns in Ecology and Paleoecology (May 1998).

REFERENCES

Atkinson, I. 1989. Introduced animals and extinctions. In D. Western and M. C. Pearl, eds., *Conservation for the Twenty-First Century*, pp. 54–75. Oxford, U.K.: Oxford University Press.

Begon, M., J. L. Harper, and C. R. Townsend. 1996. *Ecology*, 1068 pp. Oxford, U.K.: Blackwell Science.

Bennett, K. D. 1997. *Evolution and Ecology: The Pace of Life*, 241 pp. Cambridge MA: Cambridge University Press.

Boucot, A. J. 1983. Does evolution take place in an ecological vacuum? *Journal of Paleontology*. 57:1–30.

Boucot, A. J. 1990. *Evolutionary Paleobiology of Behavior and Coevolution*, 725 pp. Amsterdam: Elsevier.

Boucot, A. J. 1996. Community evolution. *Palaeontology Journal* 30:634–636.

Bretsky, P. W. Jr. 1968. Evolution of Paleozoic marine invertebrate communities. *Science* 159:1231–1233.

Bretsky, P. W. Jr. 1969. Evolution of Paleozoic benthic marine invertebrate communities. *Palaeogeography, Palaeoclimatology, Palaeoecology* 6:45–59.

Bretsky, P. W. Jr. and D. M. Lorenz. 1970. An essay on genetic-adaptive strategies and mass extinctions. *Geological Society of America Bulletin* 81:2449–2456.

Brett, C. E. and G. C. Baird. 1992. Coordinated stasis and evolutionary ecology of Silurian-Devonian marine biotas in the Appalachian Basin. *Geological Society of America Abstracts with Programs* 24:139.

Brett, C. E. and G. C. Baird. 1995. Coordinated stasis and evolutionary ecology of Silurian to Middle Devonian faunas in the Appalachian Basin. In D. H. Erwin and R. K. Anstey, eds., *New Approaches to Speciation in the Fossil Record*, pp. 285–315. New York: Columbia University Press.

Brett, C. E. and G. C. Baird. 1997. Epiboles, outages, and ecological evolutionary bioevents: Taphonomic, ecological, and biogeographic factors. In C. E. Brett and G. C. Baird, eds., *Paleontological Events*, pp. 249–285. New York: Columbia University Press.

Brett, C. E., L. C. Ivany, and K. M. Schopf. 1996. Coordinated stasis: An overview. *Palaeogeography, Palaeoclimatology, Palaeoecology* 127:1–20.

Buzas, M. A. and S. J. Culver. 1994. Species pool and dynamics of marine paleocommunities. *Science* 264:1439–1441.

Chesson, P. and N. Huntley. 1989. Short-term instabilities and long-term community dynamics. *Trends in Ecology and Evolution* 4:293–298.

Connell, J. H. and W. P. Sousa. 1983. On the evidence needed to judge ecological stability or persistence. *American Naturalist* 121:789–824.

DeAngelis, D. L. 1975. Stability and connectance in food web models. *Ecology* 61:764–771.

Death, R. G. 1996. The effect of patch disturbance on stream invertebrate community structure: The influence of disturbance history. *Oecologia* 108:567–576.

DiMichele, W.A. 1994. Ecological patterns in time and space. *Paleobiology* 20:89–92.

Dodds, W. K. and G. M. Henebry. 1996. The effect of density dependence on community structure. *Ecological Modelling* 93:33–42.

d'Orbigny, A. D. 1840–1846. *Paléontologie Française, Terraines Jurassiques*. Paris: Masson.

Eldredge, N. and J. Cracraft. 1980. *Phylogenetic Patterns and the Evolutionary Process*, 349 pp. New York: Columbia University Press.

Elton, C. S. 1958. *The Ecology of Invasion by Animals and Plants*. London: Methuen.

Foncesca, C. R. and J. L. John. 1996. Connectance: A role for community allometry. *Oikos* 77:353–358.

Grimm, V. 1996. A down-to-earth assessment of stability concepts in ecology: Dreams, demands, and the real problems. *Senckenbergiana Maritima* 27:215–226.

Grimm, V. and C. Wissel. 1997. Babel, or the ecological stability discussions: An inventory and analysis of terminology and a guide for avoiding confusion. *Oecologia* 109:323–334.

Hall, S. J. and D. G. Raffaelli. 1993. Food webs: Theory and reality. *Advanced Ecology Research* 24:187–239.

Holling, C. S. 1973. Resilience and stability of ecological systems. *Annual Review Ecological Systematics* 4:1–24.

Ivany, L. C. 1996. Coordinated stasis or coordinated turnover? Exploring intrinsic vs. extrinsic controls on pattern. *Palaeogeography, Palaeoclimatology, Palaeoecology* 127:239–256.

Ivany, L. C. and K. M. Schopf. 1996. New perspectives on faunal stability in the fossil record. *Palaeogeography, Palaeoclimatology, Palaeoecology* 127:1–359.

Jablonski, D. and J. J. Sepkoski Jr. 1996. Paleobiology, community ecology and scales of ecological pattern. *Ecology* 77:1367–1378.

Jennions, M. D. 1997. Stability in corals: A natural experiment. *Trends in Ecology and Evolution* 12:3–4.

Kammer, T. W., T. K. Baumiller, W. I. Ausich. 1997. Species longevity as a function of niche breadth: Evidence from fossil crinoids. *Geology* 25:219–222.

Lieberman, B. S., C. E. Brett, and N. Eldredge. 1995. A study of stasis and change in two species lineages from the Middle Devonian of New York state. *Paleobiology* 21:15–27.

MacArthur, R. H. 1955. Fluctuations of animal populations and a measure of community stability. *Ecology* 36:533–536.

May, R. M. 1972. Will a large complex system be stable? *Nature* 238:413–414.

McKinney, M. L., J. L. Lockwood, and D. R. Frederick. 1996. Does ecosystem and evolutionary stability include rare species? *Palaeogeography, Palaeoclimatology, Palaeoecology* 127:191–208.

McNaughton, S. J. 1977. Diversity and stability of ecological communities: A comment on the role of empiricism in ecology. *American Naturalist* 111:515–525.

Miller, W. III. 1996. Ecology of coordinated stasis. *Palaeogeography, Palaeoclimatology, Palaeoecology* 127:177–190.

Morris, P. J., L. C. Ivany, K. M. Schopf, and C. E. Brett. 1995. The challenge of ecological stasis: Reassessing sources of evolutionary stability. *Proceedings of the National Academy of Sciences* 92:11269–11273.

O'Neill, R. V. 1976. Ecosystem persistence and heterotrophic regulation. *Ecology* 57:1244–1253.

Palmer, M. A., R. F. Ambrose, N. L. Poff. 1997. Ecological theory and community restoration ecology. *Restoration Ecology* 5:291–300.

Pandolfi, J. M. 1996. Limited membership in Pleistocene reef coral assemblages from the Huon Peninsula, Papua New Guinea: Constancy during global change. *Paleobiology* 22:152–176.

Patzkowsky, M. E. and S. M. Holland. 1997. Patterns of turnover in Middle and Upper Ordovician brachiopods of the Eastern United States: A test of coordinated stasis. *Paleobiology* 23:420–443.

Pimm, S. L. 1979. Complexity and stability: Another look at MacArthur's original hypothesis. *Oikos* 33:351–357.

Pimm, S. L. and J. H. Lawton. 1980. Are food webs divided into compartments? *Journal of Animal Ecology* 49:879–898.

Pimm, S. L. and J. C. Rice. 1987. The dynamics of multi-species, multi-life-stage models of aquatic food webs. *Theoretical Population Biology* 32:303–325.

Plotnick, R. E. and M. L. McKinney. 1993. Ecosystem organization and extinction dynamics. *Palaios* 8:202–212.

Schopf, K. M. and L. C. Ivany. 1997. Long-term faunal stasis without evolutionary coordination: Jurassic benthic marine paleocommunities, western interior, U.S.A.—Comment. *Geology* 25:470.

Schopf, K. M. and L. C. Ivany. 1998. Scaling the ecosystem: A hierarchical view of stasis and change. In M. L. McKinney and J. A. Drake, eds., *Biodiversity Dynamics,* pp. 187–211. New York: Columbia University Press.

Sepkoski, J. J. Jr. 1981. A factor analytic description of the Phanerozoic marine fossil record. *Paleobiology* 7:36–53.

Sepkoski, J. J. Jr. 1987. Environmental trends in extinction during the Paleozoic. *Science* 235:64–66.

Sheehan, P. M. 1996. A new look at ecologic evolutionary units (EEUs). *Palaeogeography, Palaeoclimatology, Palaeoecology* 127:21–32.

Sheldon, P. R. 1996. Plus ça change: A model for stasis and change in different environments. *Palaeogeography, Palaeoclimatology, Palaeoecology* 127:209–228.

Simberloff, D. S. 1981. Community effects of introduced species. In M. H. Nitecki, ed., *Biotic Crises in Ecological and Evolutionary Time,* pp. 53–81. New York: Academic Press.

Stanley, S. M. 1979. *Macroevolution: Pattern and Process,* 332 pp. San Francisco: W. H. Freeman and Company.

Sterner, R. W., A. Bajpai, and T. Adams. 1997. The enigma of food chain length: Absence of theoretical evidence for dynamic constraints. *Ecology* 78:2258–2262.

Tang, C. M. 1996. Evolutionary paleoecology of Jurassic marine benthic invertebrate assemblages of the western interior, U.S.A. Ph.D. thesis, University of Southern California, Los Angeles, CA.

Tang, C. M. and D. J. Bottjer. 1996. Long-term faunal stasis without evolutionary coordination: Jurassic benthic marine paleocommunities, western interior, U.S.A. *Geology* 24:815–818.

Tang, C. M. and D. J. Bottjer. 1997. Long-term faunal stasis without evolutionary coordination: Jurassic benthic marine paleocommunities, western interior, U.S.A.—Reply. *Geology* 25:471–472.

Tilman, D. and J. A. Downing. 1994. Biodiversity and stability in grasslands. *Nature* 367:363–365.

Trueman, A. E. 1923. Some theoretical aspects of correlation. *Proceedings of the Geological Association* 36:11–25.

Valentine, J. W. and D. Jablonski. 1993. Fossil communities: Compositional variation at many time scales. In R. E. Ricklefs and D. Schluter, eds., *Species Diversity in Ecological Communities,* p. 341–349. Chicago: University of Chicago Press.

Vermeij, G. J. 1977. The Mesozoic marine revolution: Evidence from snails, predators, and grazers. *Paleobiology* 3:245–258.

Vitousek, P. 1986. Biological invasions and ecosystem properties: Can species make a difference? In H. A. Mooney and J. A. Drake, eds., *Ecology of Biological Invasions in North America and Hawaii,* pp. 163–176. Berlin: Springer-Verlag.

Vrba, E. S. 1987. Ecology in relation to speciation rates: Some case histories of Miocene-Recent mammal clades. *Evolutionary Ecology* 1:283–300.

Westrop, S. R. 1996. Temporal persistence and stability of Cambrian biofacies: Sunwaptan (Upper Cambrian) trilobite faunas of North America. *Palaeogeography, Palaeoclimatology, Palaeoecology* 127:33–46.

6

Applying Molecular Phylogeography to Test Paleoecological Hypotheses: A Case Study Involving Amblema plicata *(Mollusca: Unionidae)*

Bruce S. Lieberman

A SUBJECT THAT HAS ATTRACTED considerable debate in paleontology is the issue of what happens to communities over long periods of time. Are they obdurately stable entities, as some have argued (e.g., Jackson 1992; Morris et al. 1995; Jackson, Budd, and Pandolfi 1996), or are they ephemeral entities, transitory over long time intervals, and representative of a set of species whose broad environmental preferences happen to overlap in a given area (e.g., Davis 1986; Huntley and Webb 1989; Bambach and Bennington 1996)? Because the debate on this topic involves data from a variety of fields, including ecology, evolutionary biology, and paleontology, there are a variety of ways to approach this problem. Here I will discuss some of the conceptual issues related to testing ecological hypotheses with data from evolutionary biology using a case study involving molecular phylogeographic analysis. In so doing I will pay particular attention to the contributions that evolutionary paleobiology and hierarchy theory can make in this area.

Avise (1992), Zink (1996), and references therein outline how phylogenetic analysis of molecular data can be used to ascertain whether groups of species that occur in a particular geographic region evolve in a similar manner over time. They term this line of research "comparative phylogeography," the study of how individual species are divided up evolutionarily across their geographic range. In principle, this involves determining how populations of a single species are related to one another using cladistic analysis of molecular data. Cladograms for several species that occur in similar regions can be compared for patterns of congruence. For example, concerning aquatic organisms, the following question, which relates to congruence, might be of interest: are populations from the Hudson River always the closest relatives of populations from the Housatonic River in species A, B, and C, and so forth. It has been argued that this type of pattern would provide evidence of a long association in the same region, with common evolutionary responses to geological or climatic processes (Brooks and McLennan 1991). Furthermore, it has been argued that a pattern of congruence, when augmented with detailed ecological studies of community interactions, would provide strong support for the hypothesis that communities made up of populations of different species are stable over long periods of time (Avise 1992; Zink 1996). By contrast, it has been argued that if different species shared different patterns of population subdivision, such that in different species the populations from the Hudson River were closely related to populations from a variety of different river systems, we would conclude that the species in that area did not respond as a unit to geological or climatic changes, and did not have a long, close association (Avise 1992; Zink 1996). Such phylogeographic studies are relatively new and primarily consist of analyses of terrestrial and marine organisms. They demonstrate a range of patterns, with close coupling and community stability indicated in some instances, but not in others (see Avise 1992; Zink 1996; and references therein).

Using Phylogenetic Studies to Test Paleoecological Hypotheses

One area in which phylogenetic studies have made an important contribution to the understanding of the evolution of ecological interactions is in the study of the coevolution of hosts and their parasites (Brooks and McLennan 1991). In these studies, the search for congruence between host and parasite phylogenies does not imply unwavering verisimilitude of ecological interactions but rather patterns of constant association and isolation with concomitant diversification in host and parasite. These patterns of association, isolation, and diversification ensure some ongoing ecological interaction between host and parasite organism, but do not specify its nature.

Studies of molecular phylogeography share an obvious kinship with coevolutionary studies. Separate analyses are conducted on different species with overlapping geographic ranges to look at how different populations are related to one another. Two intergrading results of such studies are possible. If populations in different species always show the same pattern of biogeographic differentiation, then these populations were continually associated, became isolated at roughly the same time, and underwent concomitant intraspecific differentiation (if a non–ad hoc approach to the analysis of biogeographic patterns is accepted). I will term this phylogeographic association. This pattern indicates the important role that earth history factors play in structuring evolution. By contrast, if different phylogenies show different patterns of intraspecific differentiation, then these populations were not continually associated and did not become isolated and undergo differentiation at the same time. Population 1 of species A might be associated with population 2 of species B at time *x* and with population 3 at time *y*. I will term this phylogeographic nonassociation. This pattern indicates that earth history factors do not play an important role in structuring evolution. Let us assume further that we knew that in each of these cases, populations and organisms of these species were interacting ecologically. What would either of these patterns tell us about the nature of ecological interactions through time?

The conclusions depend on the way that scientists believe nature is structured. Some work in hierarchy theory as applied to paleobiology has divided life up into two hierarchies, the genealogical and the ecological (see Eldredge 1985, 1989, and references cited therein). These contain largely separate entities. For example, species and clades belong to the genealogical hierarchy, and ecosystems and the biosphere belong to the ecological hierarchy. However, in certain instances, entities can appear in both hierarchies. For example, organisms and populations both interact ecologically, as members of the ecological hierarchy, and replicate, as members of the genealogical hierarchy.

Considering this, a pattern of phylogeographic association implies coincidence between genealogical descent, and potentially, ecological interactions at the population level through time. A pattern of phylogeographic nonassociation implies disjunction between genealogical descent and ecological interactions at the population level through time. However, if different populations of a single species tend to be ecologically commensurate, then even without genealogical coincidence between populations across geographic space, similar ecological interactions may be preserved through time. This would go against the notion that species are typically broken up into different populations that have their own distinctive adaptations and ecological preferences (Eldredge 1985, and references cited therein). Therefore, only if we view the species rather than the population as the entity that provides the significant context for ecological interactions can we say that incommensurate phylogeographic patterns among populations of different associated species imply maintenance of ecological interactions through time. Of course, when comparing a phylogeny at one level of the genealogical hierarchy (for example, the population level) with phylogenies at a different hierarchical level (for example, the species level), difficulties can arise. If there is phylogeographic nonassociation, there would be no evidence for consistency of ecological interactions through time, regardless of the hierarchical level at which one views significant ecological interactions initiating. By contrast, phylogeographic association, even between clades of populations on the one hand and species on the other, would provide evidence for coincidence of ecological interactions through time (though not necessarily their similarity, of course). However, the groups would show differences in their propensity to speciate.

Thus, it is clear that phylogeographic studies of populations have the potential to reveal something about the constancy of ecological interactions through time. Without a pattern of phylogeographic association, ecological interactions cannot have been maintained through time. How can these phylogenetic studies be extended to the analysis of paleoecological hypotheses such as coordinated stasis? Previously, Lieberman (1994) and Lieberman and Kloc (1997) conducted phylogenetic studies involving genealogical entities at the species level to test aspects of the hypothesis of coordinated stasis. In particular, the hypothesis of coordinated stasis as set out in Brett and Baird (1995) and Morris et al. (1995) predicted that the establishment of the different faunas defined in Brett and Baird (1995) should be a roughly singular event associated with a particular episode of biogeographic emigration following extinction. For one of the paradigm examples of coordinated stasis, the Middle Devonian Hamilton Group fauna, phylogenetic evidence indicated that the initiation of at least part of the Hamilton Group fauna could not be confined to a single event. Rather, different taxa that comprised the Hamilton Group fauna actually arrived from different regions at different times.

The hypothesis of coordinated stasis as discussed in Morris et al. (1995) also invoked the mechanism of ecological locking, the close coupling of ecological interactions through time, as a process that might preserve the stability of faunas recognized to prevail over long periods of time by Brett and Baird (1995) and Morris et al. (1995). It is clear that phylogeographic analyses of populations offer a partial test of this aspect of coordinated stasis, but the nature of the fossil record makes the phylogenetic analysis of populations of species extremely difficult or perhaps impossible. Thus, a study of extant taxa is required, with molecular methods offering a potential means of looking at evolutionary relationships at the population level. In the study presented herein, the search was for patterns of phylogeographic association, based on the view that the population level, rather than the species level, is the hierarchical level from which significant ecological interactions are initiated. Failure to recover patterns of phylogeographic association, although not a complete refutation of the coordinated stasis hypothesis, would be counter to the predictions of that model in the sense that ecological interactions were not maintained over long periods of time. Results from phylogenetic studies also have additional bearing on the hypotheses that are discussed further in the following sections.

The Case Study

This study searches for phylogeographic association using molecular techniques and extant organisms. Phylogeographic patterns are elucidated in an aquatic species that occurs in a region powerfully affected by climatic changes since the Neogene. The patterns in this species are compared with other codistributed aquatic species to determine if congruence in evolutionary response prevails across different species, therefore implying community stability. Because this study focuses on responses during times of major environmental change, it provides a potentially stringent test of the hypothesis of community stability. However, the time frame considered in this study is shorter than that discussed by Morris et al. (1995). Further, this study focuses on population level taxa, whereas Morris et al. (1995) considered taxa at higher levels of the genealogical hierarchy in their discussion of coordinated stasis. Therefore, the nature of the test of coordinated stasis provided by the use of molecular methods and extant taxa needs to be qualified. First, it allows finer resolution than what is available in most paleontological studies. If results can be extrapolated between the shorter and longer time scales and between the lower and higher levels of the genealogical hierarchy considered herein and in Brett and Baird (1995) and Morris et al. (1995), respectively, then the test performed of that hypothesis herein may be adequately constructed. By contrast, inability to extrapolate between these scales and levels would qualify the test presented herein.

The species analyzed is the unionid bivalve mollusk *Amblema plicata* (Say 1817). The unionids are a group of freshwater mussels that attain their greatest diversity in North America, and recently they have been the subject of a comprehensive molecular phylogenetic analysis by Lydeard, Mulvey, and Davis (1996). Subsets of unionid taxa have also been analyzed in a phylogenetic framework (e.g., Hoeh et al. 1995). In the past few decades the unionid fauna of North America has come under intense stress from environmental degradation mediated by humans, as well as from exotic pest species such as the zebra mussel *Dreissena polymorpha* (Pallas); as a result many species have gone extinct or are seriously endangered, though some species remain abundant (Lydeard and Mayden 1995).

Amblema plicata, like other unionids, has a parasitic relationship with freshwater fish taxa, which they depend on for reproduction. *Amblema plicata* probably has several host taxa, though these have not as yet been identified. The species is distributed throughout the Ohio–Mississippi River drainage system (figure 6.1), a set of freshwater habitats that were powerfully affected by geologically recent environmental changes (Calkin and Feenstra 1985; Clarke and Stansbery 1988). In particular, the structure of river drainages in the Ohio–Mississippi River drainage system has changed significantly since the Pliocene, mainly due to the major environmental changes in the Pleistocene and Recent (Mayden 1988). Because patterns of phylogeographic differentia-

FIGURE 6.1. Map of the central and eastern United States showing distributions of major river courses, approximate sites from which specimens were obtained (large circles) with numbers from table 6.1 presented beside them to provide geographic context, and historical distribution of species (small dots) (potential distributions in North Dakota not shown and not sampled).

tion in aquatic species are potentially controlled by the courses of rivers, and because some of these have changed significantly in North America during the Neogene, biogeographic studies of unionids and other freshwater organisms are of intrinsic interest (e.g., Johnson 1980; Smith 1982; Wiley and Mayden 1985; Mayden 1988; Strange and Burr 1997). When Wiley and Mayden (1985) and Mayden (1988) analyzed biogeographic patterns in the freshwater fish fauna of the Ohio–Mississippi River drainage, they found that it had a biogeographic signature that is more congruent with the distribution of pre-Pleistocene river drainages than present river drainages. Strange and Burr (1997) found patterns that agreed with certain aspects of these studies, though not with others. This provides evidence for the stability of this fauna, even in the face of major environmental changes. If biogeographic patterns in *A. plicata* are congruent with those of the freshwater fish, it would provide further evidence for a shared common evolutionary history by the fauna in this region, and thus support the hypothesis of community stability. By contrast, a very different biogeographic pattern would indicate that different types of organisms in this region did not share a common evolutionary history and thus would refute the hypothesis of community stability. In addition, this study makes it possible to consider the relative contributions of Pleistocene versus post-Pleistocene environmental and geographic changes to patterns of intraspecific differentiation.

Phylogeographic patterns across a large portion of the geographic range of *A. plicata* were assessed using the randomly amplified polymorphic DNA (RAPD) technique (Williams et al. 1990), alternatively known as arbitrarily primed polymerase chain reaction (AP-PCR) (Welsh and McClelland 1990). This is a technique that uses randomly generated, short primer sequences in conjunction with the polymerase chain reaction (PCR), to search the genome for complementary regions of DNA. These primers allow amplification of the region intercalated between these sequences, providing potentially homologous markers. The strength of this technique and the reason it was chosen is that these genetic markers have proven quite useful in revealing patterns of intraspecific differentiation in many taxa (Chalmers et al. 1992; Hadrys, Balick, and Schierwater 1992; Jones, Okamura, and Noble 1994; Allegrucci et al. 1995; Stewart and Porter 1995; Stiller and Denton 1995; Stewart and Excoffier 1996) including freshwater mollusks (Kuhn and Schierwater 1993; Langand et al. 1993).

Materials and Methods

Specimens were collected or obtained from a large number of localities throughout much of the geographic range of *A. plicata* (table 6.1; figure 6.1),

TABLE 6.1. Localities from Which Specimens Were Obtained

Body of Water	State	County	City/Site
(1) Dunkard Creek	PA	Greene	Mount Morris
(2) French Creek	PA	Venango	Utica
(3) Lake Erie	PA	Erie	Presque Isle
(4) Muskingum River	OH	Washington	Devola
(5) Ohio River	WV	Cabell	Huntington
(6) Duck River	TN	Marshall	Lillard Mill Dam
(7) Tippecanoe River	IN	White/Carrol	near Monticello
(8) Ohio River	IL	Pulaski	Lock and Dam 53
(9) Mississippi River	WI	Vernon	Turtle Island
(10) Mississippi River	IA	Muscatine	Muscatine
(11) St. Croix River	MN	Chisago	Lindstrom Falls
(12) Big Piney River	MO	Pulaski	near Big Piney
(13) Ouachita River	AR	Ouachita	Camden
(14) White River	AR	Jackson/ Independence line	upstream of confluence with Black River, near Newport
(15) Illinois River	AR	Washington/ Benton line	near Elm Springs
(16) Neosho River	KS	Neosho	near St. Paul
(17) Pottawatomie Creek	KS	Franklin	Lane

in an attempt to sample a broad amount of the genetic differentiation in the species. Specimens were stored in liquid N_2 upon collection and were later deposited in a $-80°C$ freezer. Tissues from specimens, as well as valves, are housed at the Yale University Peabody Museum of Natural History, Division of Invertebrate Zoology. DNA was isolated using the CTAB protocol of Saghai-Maroof et al. (1984), modified for mollusks as described in Lieberman, All-mon, and Eldredge (1993). The conditions used to create RAPD markers by PCR are given in Allegrucci et al. (1995) but were slightly modified for use with unionid DNA. The concentrations of $MgCl_2$, DNA, dNTP, and Taq polymerase were varied and then standardized to see how this affected the size and number of amplified products, following the protocols and suggestions of Hadrys, Balick, and Schierwater (1992), Langand et al. (1993), Schierwater and Ender (1993), Jones, Okamura, and Noble (1994), Allegrucci et al. (1995), and Stiller and Denton (1995), to ensure reproducibility of RAPD products. Amplifications used a Hybaid thermal cycler with a 25 μl solution containing 11.1 μl dH_2O, 2.5 μl reaction buffer (Boehringer Mannheim), 1–10 ng of template DNA, 0.075 μl each dNTP at 2.5 mM, 1 μl of a single 10-mer RAPD primer at half manufacturer's (Operon Technologies, Alameda, CA) concentration, and 0.1 μl of Taq polymerase (Boehringer Mannheim). Reactions were overlain with mineral oil. Amplification conditions included: preheating block to 80°C; one cycle of 25 seconds at 94°C; 45 cycles of 15 seconds at 94°C, 45 seconds at 34°C, 45 seconds at 72°C; and one cycle of 5 minutes at 72°C, all with the fastest possible transitions between each temperature. Amplifications were followed by a 4°C soak.

RAPD primers were purchased as kits containing 20 random 10-mer primers from Operon Technologies, Alameda, CA. Kits A and E were used, as these proved most efficacious in amplifying unionid DNA, and all primers within these kits that successfully amplified unionid DNA were utilized. PCR samples were subjected to electrophoresis in 1.4% agarose gels containing 0.7 μl/ml of EtBr. Gels were run at approximately 100 volts for 6 hours. Bands were visualized using UV fluorescence and photographed, and the pictures were used in subsequent analysis. Band size was determined by comparisons with the DNA size standard 1 kb ladder (BRL). To confirm markers, amplifications were performed twice for each locality. Only reproducible bands, regardless of intensity, were utilized as characters for phylogenetic analysis, following the procedure of Allegrucci et al. (1995). Although only a limited number of individuals could be considered from each locality in this analysis, molecular data from two other species of unionids from the Ohio–Mississippi River drainage in White, McPheron, and Stauffer (1996) suggest that there may be low levels of within-site variability for other unionid taxa. Bands were scored as present or absent and treated as independent phenotypic markers. Only polymorphic bands were included in the analysis, and a large number of bands were identified. Because the interpretation of bands as homologous traits has been questioned in some instances by Smith et al. (1994), Rieseberg (1996), and Stothard and Rollinson (1996), a large number of bands from many different primers were considered, following Allegrucci et al. (1995), who stated that increasing the scope of the data analyzed may tend to reduce artifactual patterns.

A total of 169 potential synapomorphies were recovered. Character data were coded into a matrix (table 6.2), and phylogenetic patterns were evaluated using a cladistic parsimony analysis that employed the algorithm PAUP 4.0 (Swofford 1998). A heuristic search was used with a stepwise addition sequence that employed 100 random replications. RAPD data have frequently been subjected to cladistic analysis (e.g., Ralph et al. 1993; Yang and Quiros 1993; Stewart and Porter 1995) or have been cited as appropriate types of cladistic character data (e.g., Hadrys, Balick, and Schierwater 1992). Smith et al. (1994), Rieseberg (1996), and Stothard and Rollinson (1996), however, have questioned the use of RAPDs to assess phylogenetic patterns among different

TABLE 6.2. Data Matrix of RAPD Fragments Used in Phylogenetic Analysis

Dunkard Creek, PA
??????????100000000001111101010001111010011000111100101010000001011001100110
0000101000011011101000110010111111000010111111100000000011100011111010000101
110???????110000

French Creek, PA
1000000000000010000000???????110001????01000001100000100010000000?????????1100000
11011000101001011011000000000??????1010110010000000010100000000111101011100001
100001111000

Lake Erie, PA
??????????0001010100001111111??????110010011000110101101100000000?????????11110011
11100011011011001?????????111111101011111000100101?????????0111110101???1000000?
?????

Muskingum River, OH
1010011011011110??????????00111010100011100100000001011001?????????1000111000110000
0110101001????????01100110100000000?????????1011010011000000001000011010001000000010000
00000?????????

Ohio River, WV
????????????????????????1111110000000011????0110100111111100007000????1000011000111001001
010001??????10111110010100111011111?????????10001010000100001001001010111100000000
0100000

Duck River, TN
1000000000000001001100101101000000001101101010000??????11?0000000??????????????0
001000011010101000010000000011110110100000100000100101010000??????????????????????
?111000

Tippecanoe River, IN
??????????0000010000001111110??????0100??????????????1000000000110001000000?????????????????
?110010?????100000000010011010100000???????????????????0010000100000??????111010

Ohio River, IL
10110000000110100110101000000001100111110000100077777777171111111?????????111010
0?????????1100000100011001000001110101011011171???????010110111001110110101111100011
0100111101

Misssissippi River, WI
1010001011010001100010000110101000011007777777711000000101000000101011000077777
????????????0010000100000100000001111011000010010000000000000000001110100000???0
000000110111

Misssissippi River, IA
10010010000010100000000011111017777770101000011001101000010000????1000111000?????
??0101100110100010100010000000011111010110000100010000100000000011000101117??
?0010000111000

(*Continued on next page*)

TABLE 6.2. (*continued*)

St. Croix, River, MN
1011000001110011110000100010001111111010100110011010000 1?000000011000100001
010100010110101000101100001011000001111101000000010000001?????????????????????100
0000000111101

Big Piney River, MO
??????????0011011001001011010111111111000110001000000001?000000010010000001111
0100101100111100101000010100000011111110110100100100000000100000101110000101
???1011000110101

Ouachita River, AR
??????????0001001110001001010100011100000000000000000001 10000000100001000 0?????
??0001100011000111100?????????1111101100000010000000100100110001010000000???00
00000??????

White River, AR
00100110100010001100 00??????01000101110010100011110010001111111110011010111
100101011001100000111011110001101110111101001001100110110010000000111111011
01001000100111000

Illinois River, AR
10110011001110100000101010100010001110110101100111100000?000????000000101010
01000???????10000010100100000000001000000110000100000000000000000001010000100
100?????????????

Neosho River, KS
100000000000000111010010101100000011111101010001111000 0?????????0100001101000
000001011000??????11100110111100011011111101000100000000000100001001010110100?
??0000000111000

Pottawatomie Creek, KS
1011001000000000001000001001010111 10????000111001100000011000000000 0110100110
100010000001011001101010100000001111111110010001000100011110010011111101101
00000001000111101

Note: 0 = band absent, 1 = band present, and ? means primer failed to amplify samples.

species or distantly related taxa, arguing that bands of similar size are less likely to be homologous. Thus, phylogenetic analysis was confined to members of *A. plicata,* and the tree for the species was rooted using the midpoint rooting option of PAUP 4.0 (Swofford 1998) rather than using a different species as the outgroup.

Results

Analysis of character data recovered five most parsimonious trees, 426 steps in length. These had a retention index of 0.33 and a consistency index of 0.37 when uninformative characters were excluded. A strict consensus of these trees is shown in figure 6.2. To assess the support of the various nodes of this tree a bootstrap analysis was conducted following the recommendations of Rieseberg (1996) specific to RAPD data. Bootstrap analysis used a heuristic search with 100 bootstrap replications, and for each replication a stepwise addition sequence was employed using five random replications. Groups were retained that were compatible with a 50% majority-rule consensus tree. Bootstrap values are shown in figure 6.2. To further assess the quality of and overall phylogenetic signal within the character data of table 6.2, permutation tail probability (PTP) tests (Faith 1991; Faith and Trueman 1996) were performed using PAUP. The PTP test compares the length of trees generated using randomized data (character states are assigned randomly to taxa) with the length of the most parsimonious tree(s). The proportion of the randomized data trees having a most parsimonious cladogram length equal to or less than that of the original most parsimonious cladogram is tabulated. This is referred to as the cladistic PTP and treated as equivalent to a p-value at which the character data differ from random data. This method is described in detail in Faith (1991), Swofford et al. (1996), and Faith and Trueman (1996). In the PTP test, the character data for all taxa were randomized 100 times, and in each of these replications a heuristic stepwise search with a random addition sequence and five replications was used to find the most parsimonious cladogram based on the random data. For each replication, the difference between the tree length of the random data set and the original set was calculated. In this test the PTP value was 0.02, a highly significant value, implying good cladistic structure and phylogenetic signal in the database.

The most thorough studies of biogeographic patterns on aquatic organisms in the region of the Ohio–Mississippi River drainage include the analyses of Wiley and Mayden (1985) and Mayden (1988) on fish taxa. Mayden (1988) identified two major biogeographic regions within that drainage system, as well as several other smaller biogeographic regions. His terminology is followed in this analysis. For instance, the species considered in this study is distributed throughout what has been termed the Central Highlands of eastern North America (sensu Mayden 1988). This region can be further divided into two major biogeographic areas: the Eastern Highlands, east of the Mississippi River; and the Interior Highlands, comprising the Ouachita and Ozark Highlands west of the Mississippi River (Mayden 1988). [The Upper Mississippi River is treated as belonging to the Interior Highlands following Wiley and

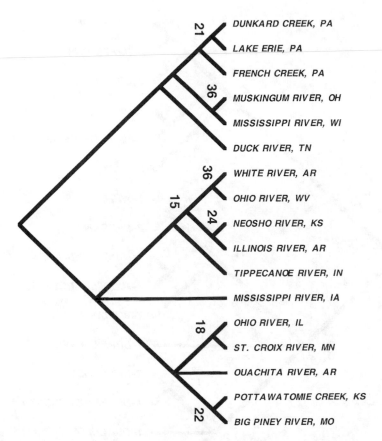

FIGURE 6.2. A strict consensus cladogram of the five most parsimonious trees of length 426 steps produced from analysis of character data in table 6.2 with PAUP 3.1.1 (Swofford 1993). Cladogram constructed using a stepwise addition sequence with 100 random replications. The retention index is 0.33, and the consistency index when uninformative characters are excluded is 0.37. Bootstrap confidence values (see text for bootstrapping procedure utilized) shown on cladogram.

Mayden (1985)]. These regions were largely continuous before the Pleistocene, and during the early Pleistocene, environmental conditions would have facilitated dispersal between these two regions.

Using these biogeographic precepts, the phylogeny in figure 6.2 is converted to an area cladogram by placing the geographic occurrence of the taxon, in this case a river drainage, at the appropriate terminal (figure 6.3). The position of this river drainage in either the Interior or Eastern Highlands biogeographic regions is also noted at the terminal. These biogeographic states are then optimized to ancestral nodes using the Fitch (1971) parsimony algorithm following

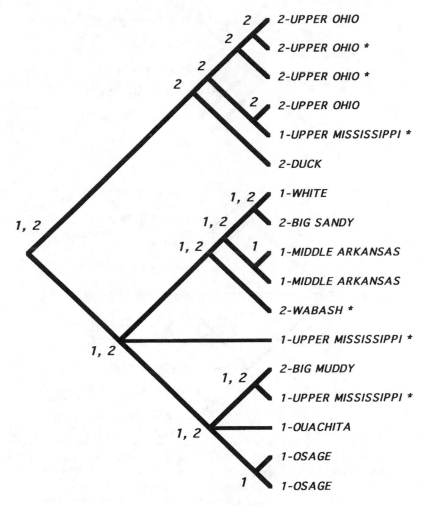

FIGURE 6.3. An area cladogram produced by substituting river drainage system (from Mayden 1988) for collection site. River drainages were assigned to the two major biogeographic regions in the Central Highlands; the Interior Highlands = 1 and the Eastern Highlands = 2. These are placed at the appropriate terminal and then optimized to nodes using the Fitch (1971) algorithm. An asterisk is placed next to those taxa that occur in regions that are north of the line of maximal glacial advance and thus reflect recent dispersal events.

the discussion in Lieberman and Eldredge (1996). The ancestral state of the entire species was assumed to be present in both regions. Some of these regions were covered by large ice sheets less than 20,000 yr B.P., and unionid populations must have invaded these regions since then (though they may have also been there during previous interglacials).

Two major clades are revealed, one chiefly comprising Eastern Highlands populations. Within this clade, there is evidence of postglacial dispersal into the Upper Mississippi River Basin from the Eastern Highlands. Also, the clade must have expanded its range into the Upper Ohio River drainage at or around this time. The second clade is primitively present throughout the Central Highlands region. Populations within this clade appear to have differentiated by vicariance, as ranges of populations (terminal taxa) have contracted relative to their ancestral states (figure 6.3). In this clade there are three Interior Highlands populations in the Upper Mississippi River drainage that occur in regions that were inhospitable during times of maximal glaciation. These populations could have dispersed into these areas either from the Eastern Highlands or from the Interior Highlands.

In some instances there is good concordance between geographic proximity and phylogenetic position. Most of the populations from the Upper Ohio River drainage are sister taxa, as are populations from the Middle Arkansas Drainage and the Osage Drainage. There are also several cases, however, in which geographically disparate taxa share evolutionary propinquity. For instance, the population from the White River drainage is more closely related to a population from the Big Sandy drainage, in the Upper Ohio River region, rather than to some other drainage system, for example, the Middle Arkansas.

Discussion

Although patterns of relationship shared among populations of a single species should not be used to draw general biogeographic conclusions (Brooks and McLennan 1991), area relationships in *A. plicata* can be instructive if compared with results of other codistributed species to consider more general biogeographic patterns, patterns of phylogeographic association, and thus also patterns of ecological interaction over long periods of time. To do this, it is necessary to assume that the populations sampled adequately reflect underlying diversity within the species, that the hypothesis of phylogenetic relationship can be accepted, and that there is some relationship between present and ancestral biogeographic states. In addition, where the root of the tree is placed at least partly affects the biogeographic patterns.

Phylogeographic and biogeographic patterns from several other codistributed groups recovered by other studies allow meaningful comparison with *A. plicata*. When this is done, two major patterns are evident. First, studies of intraspecific differentiation in salamanders (Routman, Wu, and Templeton 1994) found little congruence between geographic proximity and phylogenetic propinquity. Zink and Dittman (1993) and Routman, Wu, and Templeton (1994) hypothesized that this general pattern could be due to recent colonization events postdating the Pleistocene. This would point to the ephemeral nature of community assemblages in the Central Highlands.

In *A. plicata* there was some congruence between geographic proximity and phylogenetic propinquity, particularly in the cases of more narrowly circumscribed biogeographic regions such as the Upper Ohio and Middle Arkansas drainages. Such congruence is quite incomplete, however, with several instances of sister group relationships between populations from the two major biogeographic regions in the Central Highlands to the exclusion of sister group relationships between populations within a single one of these regions. This could be taken as further evidence for the transitory nature of community assemblages in the Central Highlands. Based on the second chief pattern that emerges from comparative phylogeography and biogeography, however, this conclusion is not supported.

Instead, phylogeographic patterns in *A. plicata* are resonant with interspecific biogeographic patterns from numerous clades of aquatic fish of the Central Highlands given in Wiley and Mayden (1985) and Mayden (1988). One salient feature found by this study and that of Wiley and Mayden (1985) and Mayden (1988) is a sister group relationship between taxa in the Interior and Eastern Highlands, assuming parsimonious optimizations of biogeographic states. These regions have been split since the Illinoian glaciation (Wiley and Mayden 1985). To explain this evolutionary pattern as the result of late or post-Pleistocene environmental change, one must posit a large number of dispersal events across biogeographic barriers. A more parsimonious view of the phylogeographic patterns would have *A. plicata* occurring across the Central Highlands prior to the Illinoian glaciation, with one largely Eastern Highlands clade and one clade homogeneously distributed throughout the Central Highlands. This indicates that *A. plicata*, like the freshwater fishes considered in Wiley and Mayden (1985) and Mayden (1988), persisted for a long period of time in the Central Highlands without experiencing fundamental evolutionary and biogeographic alteration during the major environmental changes at the end of the Pleistocene. A limited amount of post-Illinoian dispersal did occur in *A. plicata*, but the overall pattern is of a long association with other taxa in the region, such as fish. That is, evidence exists for phylogeographic association, although it would be desirable to have a greater range of intraspecific phylogeographic patterns from other taxa in the same region. As mentioned previously, unionid taxa share a parasitic relationship with freshwater fish, but the precise host of *A. plicata* is not known; indeed, the species probably has several hosts. Therefore, the phylogeographic association between population level geographic divergence within *A. plicata* and the interspecific geographic divergence replicated in several fish clades may provide evidence for strict coevolution or phylogeographic association. The similar patterns of geographic differentiation hold across different levels of the genealogical hierarchy, perhaps due to differences between unionids and fish in their propensity to speciate.

The patterns of coevolution and phylogeographic association and the evidence for biogeographic differentiation with an early Pleistocene signature in modern unionids and fish suggests a long period of maintained ecological interactions associated with evolutionary differentiation. These results provide at least some support for the hypotheses of Jackson (1992), Morris et al. (1995), and Jackson, Budd, and Pandolfi (1996) in the sense that they posited that communities made up of populations of different species can be stable over long periods of time, and in this study populations showed phylogeographic association even when environments changed significantly. Additional data are of course necessary on the nature of these communities, and the types of interactions that prevailed. Association need not be equated with stability, but without long-term association as evidenced by phylogeographic studies, there could not have been community stability. Thus, studies that concentrate on members of the genealogical hierarchy can still make contributions to our understanding of paleoecological hypotheses.

However, the hypotheses of Jackson (1992), Morris et al. (1995), and Jackson, Budd, and Pandolfi (1996) do not receive unambiguous support from this study. Over the Quaternary, the unionid taxon showed evidence of intraspecific differentiation, whereas the fish clades discussed in Wiley and Mayden (1985) and Mayden (1988) speciated. Thus, members of the genealogical hierarchy, the entities that provided the participants in the ecological hierarchy, were not obdurately stable. If a crucial component of the hypothesis of coordinated stasis is complete stability of the members of the genealogical hierarchy that make up a fauna, then these results would have to be seen as a challenge to the hypothesis. Similarly, Lieberman (1994) and Lieberman and Kloc (1997) challenged the predictions that coordinated stasis made about members of the genealogical hierarchy. By contrast, the possibility of ecological association, and thus potentially community stability, even in the face of evolutionary change is indicated by the results of this study, although community stability has not yet been demonstrated, nor can it be demonstrated by this type of study alone. Further, Bambach and Bennington (1996) have suggested that it is unlikely that ecological communities can be stable over long periods of time.

The notion that life is divided up into two largely nonequivalent hierarchies implies that aspects of the coordinated stasis hypothesis can be valid for one but not both of these hierarchies. In addition, this hierarchical structure of nature implies that studies that rely on information from a single hierarchy, such as this one, can provide a means, though not the sole means, of testing a hypothesis that attempts to describe patterns within each of the two hierarchies. Thus, the value of a research program in evolutionary paleoecology, though complicated by the hierarchical structure of nature, seems clear.

ACKNOWLEDGMENTS

I thank J. Powell for providing space and materials in his lab during the course of this study, E. Vrba for discussions and scientific advice, and W. Allmon and D. Bottjer for allowing me to participate in this symposium volume. The following provided help with fieldwork and/or in obtaining specimens: W. Sage, R. Anderson, C. Barnhart, D. Berg, J. Garner, J. Harris, M. Hove, E. Masteller, A. Miller, M. Mulvey, B. Obermeyer, B. Sietman, C. Thompson, D. Waller, T. Watters, and L. White. A. Caccone, and J. Gibbs, and J. Gleason provided assistance with lab work. T. Collins, H. Hadrys, P. Jarne, B. Schierwater, and L. White gave advice with molecular work. W. Allmon, H. Hadrys, C. Lydeard, J. J. Sepkoski Jr., L. White, and two anonymous reviewers provided comments on an earlier version of this chapter. D. Hodgins and I. Calderon gave logistical assistance. The Conchologists of America, the Long Island Shell Club, the American Museum of Natural History Theodore Roosevelt Fund, and the Ecosave Fund of Yale University provided financial support for this project. This research was also supported by NSF Grant EAR 9505216.

REFERENCES

Allegrucci, G., A. Caccone, S. Cataudella, J. R. Powell, and V. Sbordoni. 1995. Acclimation of the European sea bass to freshwater: Monitoring genetic changes by RAPD polymerase chain reaction to detect DNA polymorphisms. *Marine Biology* 121:591–599.

Avise, J. C. 1992. Molecular population structure and the biogeographic history of a regional fauna: A case history with lessons for conservation biology. *Oikos* 63:62–76.

Bambach, R. K. and J. B. Bennington. 1996. Do communities evolve? A major question in evolutionary paleoecology. In D. Jablonski, D. H. Erwin, and J. H. Lipps, eds., *Evolutionary Paleobiology,* pp. 123–160. Chicago: University of Chicago Press.

Brooks, D. R. and D. A. McLennan. 1991. *Phylogeny, Ecology, and Behavior: A Research Program in Comparative Biology.* Chicago: University of Chicago Press.

Calkin, P. E. and B. H. Feenstra. 1985. Evolution of the Erie-Basin Great Lakes. In P. F. Karrow and P. E. Calkin, eds., *Quaternary Evolution of the Great Lakes,* pp. 149–170. Geological Association of Canada Special Paper 30. Ottawa: Geological Association of Canada.

Chalmers, K. J., R. Waugh, J. I. Sprent, A. J. Simons, and W. Powell. 1992. Detection of genetic variation between and within populations of *Gliricidia sepium* and *G. maculata* using RAPD markers. *Heredity* 69:465–472.

Clarke, A. H. Jr. and D. H. Stansbery. 1988. Are some Lake Erie mollusks products of post-Pleistocene evolution? In J. F. Downhower, ed., *The Biogeography of the Island Region of Western Lake Erie*, pp. 85–91. Columbus: Ohio State University Press.

Davis, M. 1986. Climatic instability, time lags, and community disequilibrium. In J. Diamond and T. Case, eds., *Community Ecology*, pp. 269–284. New York: Harper and Row Publishers.

Eldredge, N. 1985. *Unfinished Synthesis.* New York: Oxford University Press.

Eldredge, N. 1989. *Macroevolutionary Dynamics.* New York: McGraw-Hill.

Faith, D. P. 1991. Cladistic permutation tests for monophyly and non-monophyly. *Systematic Zoology* 40:366–375.

Faith, D. P. and J. W. H. Trueman. 1996. When the Topology-Dependent Permutation Test (T-PTP) for monophyly returns significant support for monophyly should that be equated with (a) rejecting a null hypothesis of nonmonophyly, (b) rejecting a null hypothesis of "no structure," (c) failing to falsify a hypothesis of monophyly, or (d) none of the above? *Systematic Biology* 45:580–586.

Fitch, W. M. 1971. Toward defining the course of evolution: minimum change for a specific tree topology. *Systematic Zoology* 20:406–416.

Hadrys, H., M. Balick, and B. Schierwater. 1992. Applications of random amplified polymorphic DNA (RAPD) in molecular ecology. *Molecular Ecology* 1:55–63.

Hoeh, W. R., K. S. Frazer, E. Natanjo-Garcia, and R. J. Trdan. 1995. A phylogenetic perspective on the evolution of simultaneous hermaphroditism in a freshwater mussel clade (Bivalvia: Unionidae: *Utterbackia*). *Malacological Review* 28:25–42.

Huntley, B. and T. Webb III. 1989. Migration: Species' response to climatic variations caused by changes in the earth's orbit. *Journal of Biogeography* 16:5–19.

Jackson, J. B. C. 1992. Pleistocene perspectives on coral reef community structure. *American Zoologist* 32: 719–730.

Jackson, J. B. C., A. F. Budd, and J. M. Pandolfi. 1996. The shifting balance of natural communities? In D. Jablonski, D. H. Erwin, and J. H. Lipps, eds., *Evolutionary Paleobiology*, pp. 89–122. Chicago: University of Chicago Press.

Johnson, R. I. 1980. Zoogeography of North American Unionacea (Mollusca: Bivalvia) north of the maximum Pleistocene glaciation. *Bulletin of the Museum of Comparative Zoology* 149:77–189.

Jones, C. S., B. Okamura, and L. R. Noble. 1994. Parent and larval RAPD fingerprints reveal outcrossing in freshwater bryozoans. *Molecular Ecology* 3:193–199.

Kuhn, K. and B. Schierwater. 1993. Strain diagnostic RAPD markers in *Ancylus fluviatilis. Verhandlungen der Deutschen Zoologischen Gesellschaft* 86:54.

Langand, J., V. Barral, B. Delay, and J. Jourdane. 1993. Detection of genetic diversity within snail intermediate hosts of the genus *Bulinus* by using Random Amplified Polymorphic DNA markers (RAPDs). *Acta Tropica* 55:205–215.

Lieberman, B. S. 1994. Evolution of the trilobite subfamily Proetinae and the origin, evolutionary affinity, and extinction of the Middle Devonian proetid fauna of Eastern North America. *Bulletin of the American Museum of Natural History* 223:1–176.

Lieberman, B. S. and N. Eldredge. 1996. Trilobite biogeography in the Middle Devonian: Geological processes and analytical methods. *Paleobiology* 22:66–79.

Lieberman, B. S. and G. Kloc. 1997. Evolutionary and biogeographic patterns in the Asteropyginae (Trilobita, Devonian). *Bulletin of the American Museum of Natural History* 232:1–127.

Lieberman, B. S., W. D. Allmon, N. Eldredge. 1993. Levels of selection and macroevolutionary patterns in the turritellid gastropods. *Paleobiology* 19:205–215.

Lydeard, C. and R. L. Mayden. 1995. A diverse and endangered aquatic ecosystem of the southeast United States. *Conservation Biology* 9:800–805.

Lydeard, C., M. Mulvey, and G. M. Davis. 1996. Molecular systematics and evolution of reproductive traits of North American freshwater Unionacean mussels (Mollusca: Bivalvia) as inferred from 16sr RNA gene sequences. *Philosophical Transactions of the Royal Society of London Series B* 351:1593–1603.

Mayden, R. L. 1988. Vicariance biogeography, parsimony, and evolution in North American freshwater fishes. *Systematic Zoology* 37:329–355.

Morris, P. J., L. C. Ivany, K. M. Schopf, and C. E. Brett. 1995. The challenge of paleoecological stasis: Reassessing sources of evolutionary stability. *Proceedings of the National Academy of Sciences, USA* 92:11269–11273.

Ralph, D., M. McClelland, J. Welsh, G. Baranton, and P. Perolat. 1993. *Leptospira* species categorized by arbitrarily primed polymerase chain reaction (PCR) and by mapped restriction polymorphisms in PCR-amplified rRNA genes. *Journal of Bacteriology* 175:973–981.

Rieseberg, L. H. 1996. Homology among RAPD fragments in interspecific comparisons. *Molecular Ecology* 5:99–105.

Routman, E., R. Wu, and A. R. Templeton. 1994. Parsimony, molecular evolution, and biogeography: the case of the North American giant salamander. *Evolution* 48:1799–1809.

Saghai-Maroof, M. A., K. M. Soliman, R. A. Jorgensen, and R. W. Allard. 1984. Ribosomal DNA spacer length in barley: Mendelian inheritance, chromosomal location, and population dynamics. *Proceedings of the National Academy of Sciences, USA* 81:8018–8021.

Schierwater, B. and A. Ender. 1993. Different thermostable DNA polymerases may amplify different RAPD products. 1993. *Nucleic Acids Research* 21:4647–4648.

Smith, D. G. 1982. The zoogeography of the freshwater mussels of the Taconic and southern Green Mountain region of northeastern North America. *Canadian Journal of Zoology* 60:261–267.

Smith, J. J., J. S. Scott-Craig, J. R. Leadbetter, G. L Bush., D. L. Roberts, and D. W. Fulbright. 1994. Characterization of Random Amplified Polymorphic DNA (RAPD) products from *Xanthomonas campestris* and some comments on the use of RAPD products in phylogenetic analysis. *Molecular Phylogenetics and Evolution* 3:133–145.

Stewart, C. N. Jr. and L. Excoffier. 1996. Assessing population genetic structure and variability with RAPD data: application to *Vaccinium macrocarpon* (American Cranberry). *Journal of Evolutionary Biology* 9:153–171.

Stewart, C. N. Jr. and D. M. Porter. 1995. RAPD profiling in biological conservation: an application to estimating clonal variation in rare and endangered *Iliamna* in Virginia. *Biological Conservation* 74:135–142.

Stiller, J. W. and A. L. Denton. 1995. One hundred years of *Spartina alterniflora* (Poaceae) in Willapa Bay, Washington: random amplified polymorphic DNA analysis of an invasive population. *Molecular Ecology* 4:355–363.

Stothard, J. R. and D. Rollinson. 1996. An evaluation of randomly amplified polymorphic DNA (RAPD) for the identification and phylogeny of freshwater snails of the genus *Bulinus* (Gastropoda: Planorbidae). *Journal of Molluscan Studies* 62:165–176.

Strange, R. M. and B. M. Burr. 1997. Intraspecific phylogeography of North American highland fishes: a test of the Pleistocene vicariance hypothesis. *Evolution* 51:885–897.

Swofford, D. L. 1993. PAUP (Phylogenetic analysis using parsimony) version 3.1.1. Smithsonian Institution, Washington, D.C.

Swofford, D. L. 1998. PAUP (Phylogenetic Analysis Using Parsimony), version 4.0. Sunderland MA: Sinauer Associates.

Swofford, D. L., J. L. Thorne, J. Felsenstein, and B. M. Wiegmann. 1996. The Topology-Dependent Permutation Test for monophyly does not test for monophyly. *Systematic Biology* 45:575–579.

Welsh, J. and M. McClelland. 1990. Fingerprinting genomes using PCR with arbitrary primers. *Nucleic Acids Research* 18:7213–7218.

White, L. R., B. A. McPheron, and J. R. Stauffer Jr. 1996. Molecular genetic identification tools for the unionids of French Creek, Pennsylvania. *Malacologia* 38:181–202.

Wiley, E. O. and R. L. Mayden. 1985. Species and speciation in phylogenetic systematics, with examples from the North American fish fauna. *Annals of the Missouri Botanical Garden* 72:596–635.

Williams, J. G. K., A. R. Kubelik, K. J. Livak, J. A. Rafalski, and S. V. Tingey. 1990. DNA polymorphisms amplified by arbitrary primers are useful as genetic markers. *Nucleic Acids Research* 18:6531–6535.

Yang, X. and C. Quiros. 1993. Identification and classification of celery cultivars with RAPD markers. *Theoretical and Applied Genetics* 86:205–212.

Zink, R. M. 1996. Comparative phylogeography in North American birds. *Evolution* 50:308–317.

Zink, R. M. and D. L. Dittmann. 1993. Gene flow, refugia, and evolution of geographic variation in the song sparrow (*Melospiza melodia*). *Evolution* 47:717–729.

7

Nutrients and Evolution in the Marine Realm

Warren D. Allmon and Robert M. Ross

IT IS A CENTRAL THEME OF ECOLOGY that energy flow is one of the most important aspects of any biological community or ecosystem (e.g., Ricklefs 1990; Begon, Harper, and Townsend 1996). Under various terms (e.g., *trophic structure, nutrient cycling,* etc.), the causes and effects of energy transfer and how it is accomplished among organisms and taxa are virtually universally viewed as among the basic organizing factors of the biosphere. This is an *ecological* view. When this view is expanded to longer or *evolutionary* timescales, it frequently has been assumed that because energy has such an important role in organizing ecological relationships, it must have an equally important role in affecting evolutionary processes. It has also been generally supposed that energy flow works in evolutionary time in more or less the same way as it does in ecological time, that is, by allowing for higher biological productivity in a clade or environment, which somehow leads to evolutionary activity. As discussed subsequently, there is substantial evidence that this is often the case. Exactly how it works, however, has received less attention.

One major obstacle to an adequate understanding of the role of nutrients and energy flow in evolution may be the way we have viewed evolution (Allmon 1994). As long as evolution is seen primarily as the transformation of lineages by natural selection, it is seen largely as change driven by adaptation. Evolutionary processes, in this view, consist literally of the scaling up of processes examined in ecological time. This has been called the "transformational view" by Eldredge (1982). If, on the other hand, evolution is seen primarily as comprised of the appearance and disappearance of species, then the ecological processes that "matter" to evolution are principally those that affect speciation and extinction rather than those that affect adaptive transformation. Eldredge (1982) has called this the "taxic view" of evolution. Nutrients and energy flow thus may be important to evolution because they affect adaptation in the transformational or anagenetic mode, or because they affect the origin and extinction of species, or both.

Vermeij (1987b, 1995) has discussed an "ecological" or "economic" view of the history of life that has come closest to integrating the effects of nutrients on ecological and evolutionary scales. In his 1987 book, Vermeij (1987b) suggests that "scope of adaptation" and "opportunity for selection" are greatest in environments of high primary productivity. In his 1995 article, Vermeij focuses on "revolutions" in life's history and attributes them principally to nutrient, temperature, and sea level changes associated with submarine volcanism. Although he briefly discusses applying this perspective to "more normal times" and to a consideration of differing modes of speciation, Vermeij does not explicitly frame the discussion around how nutrients affect the specific evolutionary processes responsible for particular events, revolutionary or otherwise. Does productivity affect adaptation or speciation? When and how? Vermeij has made important observations about the general coincidence between evolutionary innovation and productivity conditions and has speculated in general about potential mechanisms, but without a clear linkage between particular evolutionary mechanisms and extrinsic influences, it will be difficult to test specific instances of cause and effect.

In this chapter, we present an ecological view of the history of life in the seas that takes much from earlier discussions (e.g., Vermeij, 1978, 1987b, 1995; Hallock 1987; Allmon 1992a, 1994; Bambach 1993). We build on and expand these previous studies, however, by focusing on the process of speciation and attempting to synthesize understanding of ecological processes with explicit models of evolutionary process over geologically significant time spans in the marine biosphere. Our objective is to propose an explicit explanatory framework that can be applied to a wider spectrum of examples than those we consider here. We suggest that this perspective may offer an approach to the longstanding problem of scale in evolutionary ecology (e.g., Aronson 1994; Martin 1998b) and contribute to an understanding of major extrinsic influences on macroevolutionary patterns.

We first summarize data that appear to support a close connection between nutrients and evolution. We then present a general theoretical framework for examining such patterns and determining what processes might have been responsible for them. We next specifically explore plausible processes that might link nutrient supply to processes of diversification and extinction, in both "revolutions" and "normal" times. We conclude with two case studies of the application of this approach.

Definitions

Nutrients are materials "essential to the structure and/or function of organisms" (DeAngelis 1992:9). Of particular importance in considering the effect of nutrients and their variation on ecological and evolutionary processes is whether one or more nutrients limit the growth of primary producers in an ecosystem. In the marine environment, "macronutrients" include N, P, and (to a lesser degree) Si; less important "micronutrients" include Fe, Zn, Cu, and Mn. Carbon dioxide may even be said to be limiting if it is present in low enough amounts (Riebesell, Wolf-Gladrow, and Smetacek 1993). In theory, all bioelements could be limiting; in practice, in any particular case, usually more than one is limiting.

Not all these elements are equally limiting, and each environment tends to be characterized by its own particular set of limitations (e.g., DeAngelis 1992:40; Valiela 1984:56). Open marine and coastal environments, for example, are generally N-limited (Ryther and Dunstan 1971; Howarth 1988, 1993; Smith and Atkinson 1984; Vitousek and Howarth 1991); upwelling systems are sometimes Si-limited; and some southern ocean environments are limited in micronutrients such as Fe (Martin 1995, 1996). Furthermore, different organisms are limited by different nutrient requirements: phytoplankton with siliceous tests (e.g., diatoms) are frequently Si-limited. Thus, in saying that a particular marine environment is silica-limited, we are actually saying that the dominant phytoplankton in that environment are limited by silica (Smayda 1989).

Nutrients are not independent of the biota that depend on them. The amount of available nutrients has much to do with the rates of uptake and recycling and thus is related to the nature of the food web. Food webs vary significantly according to environment and sometimes by season. Rates of recycling tend to be much higher in ecosystems with low ambient nutrient concentrations such as open ocean and coral reef environments; inefficient recycling in the marine environment often leads to leakage of nutrients out of the system in the form of sinking organic matter, which then provides a food source for consumer communities outside the photic zone (Valiela 1984).

Production is usually measured as the amount of newly created biomass or individuals and is expressed as the biomass or population increase per unit area or volume. Primary production is the rate of production of autotrophic (generally photosynthetic) organisms, and secondary production is the rate of all higher trophic level consumers (Valiela 1984; Ricklefs 1990).

Consumers are generally said to be limited in food or trophic resources rather than in nutrients per se. There is, however, a strong though complex relationship between secondary production and nutrients: this relationship depends on growth of the nearest photosynthetic community (which may be several kilometers of water column away) and on the dispersion, degradation, and recycling of materials from this primary production (e.g., Hargrave 1980). Consumers may, in fact, obtain their trophic resources from several sources, for example, from detritus from the overlying surface, from detritus washed in from terrestrial ecosystems, from occasional food falls from large vertebrates, and from material washed in from other productive marine ecosystems. In spite of these many complications, there remains a fairly close relationship between nutrient concentration in the overlying water column and food resources of the underlying consumer communities, and thus we refer liberally to "nutrient-limitation" when discussing heterotrophic organisms (Valiela 1984; Ricklefs 1990).

The degree to which marine communities and ecosystems are actually nutrient or food-limited is a subject of considerable debate. It may be the case that most marine communities are in fact nutrient-limited in one way or the other, that is, most ecosystems will eventually change qualitatively if the volume of one or more nutrients is increased. Yet a particular ecosystem might nevertheless not be using all of the nutrients available to it and may be more structured by other environmental factors. That is, community change in the presence of changing nutrients may be step-like: As nutrients are added, there is little additional uptake or community change until a threshold is reached at which a fast-growing (invading or previously subdominant) competitor can use the nutrients and overgrow the old community (Valiela 1984; Ricklefs 1990).

The Nutrient Paradox

In ecological time, total taxonomic diversity tends to be inversely correlated with nutrient concentration, a phenomenon known as the "nutrient paradox" because it seems to contradict our intuition that more resources should permit greater, not less, diversity (Rosenzweig and Abramsky 1993). For example, diversity is relatively low in both eutrophic habitats, such as ponds and estuaries, and in many highly oligotrophic settings, whereas highest taxonomic diversities are observed in communities with low-to-moderate nutrient status, such as coral reefs. Modern marine plankton diversity is lowest in nutrient-rich regimes and highest in oligotrophic waters (Martin 1996, and references therein). Deep-sea diversity is higher than previously believed and is in fact higher than most shallow-water temperate habitats (Grassle 1989, 1991; Grassle and Maciolek 1992); deep-sea biotas beneath high-fertility waters, however, are less diverse than those beneath low-fertility waters (Rex et al. 1993; Rex, Etter, and Stuart 1997). The explanation for this paradox appears straightforward (Rosenzweig and Abramsky 1993): At very high nutrient levels, one or a few species usually come to dominate resources and outcompete other species, thereby reducing diversity. The paradox in fact breaks down in the most oligotrophic settings, which are also low diversity because energy supplies are too low to maintain stable population sizes of many species. A more general caveat is that although this pattern holds for total diversity and some taxa, it does not hold for all taxa; each taxon has its own optimum diversity with respect to nutrient concentration. Different taxa will therefore show different patterns of diversity in relation to productivity (e.g., Rosenzweig and Abramsky 1993; Vermeij 1995; Taylor 1997).

In evolutionary time, on the other hand, numerous examples have been cited of positive correlation between productivity and diversity (although Valentine [1971, 1973] argued that the reverse might just as easily be the case). These are briefly summarized in the following sections.

The "Diversity Pyramid"

Within many clades, the species richness of various trophic levels parallels productivity at those levels (figure 7.1). As noted long ago by Elton (1927), the ecological efficiency of a community dictates that only a fraction of the productivity at one trophic level will be transferred to the next highest level, producing a pyramid; herbivore production may, for example, be only 20% of plant production and first-level carnivore production may be only 15% of herbivore production, and therefore only amounts to 3% of plant production. Similarly, within at least some clades of terrestrial vertebrates (reptiles, birds, dinosaurs), species diversity varies inversely with trophic level (Allmon 1992a), implying some positive correlation between trophic resources available and resulting species richness.

Similar patterns are indicated by higher rates of origination among some taxa with particular trophic modes. Roy, McMenamin, and Alderman (1990), for example, found that suspension-feeding benthos (crinoids, bivalves, bryozoans) showed higher familial origination rates than nonsuspension feeders

second carnivore

first carnivore

herbivore

plant

Energy flux

FIGURE 7.1. An "ecological pyramid" representing the net productivity of each trophic level in an ecosystem. This particular pyramid represents transfers of 20, 15, and 10%, respectively, among trophic levels. From Ricklefs (1990). By analogy, a "diversity pyramid" may also exist, with greater taxonomic diversity among lower trophic levels.

(cephalopods and arthropods) in the late Cretaceous. Allmon et al. (1992) found the reverse for Paleozoic gastropods; suspension-feeding genera show lower origination rates than nonsuspension feeders.

Shorter-Term Temporal Patterns (10 millions of years)

The later part of the Cenozoic Era (especially the interval between the mid-Miocene and mid-Pliocene, ca. 15–3 Ma) witnessed dramatic diversification in many clades of marine organisms. These include taxa as different as sea grasses (Domning 1981, 1982), bivalve and gastropod mollusks (Allmon et al. 1991), and cetacean and pinniped mammals (Lipps and Mitchell 1976). It is noteworthy that this same interval also saw the deposition of massive phosphorite deposits, especially in the Western Atlantic region, as well as high availability of particulate P in the Pacific (Delaney 1990; Delaney and Filippelli 1994) and an influx of terrestrial runoff from the northern Andes (Domning 1981, 1982). These observations have been used by several authors to argue for regionally, and perhaps globally, high oceanic productivity during this time (Lipps and Mitchell 1976; Carter and Kelley 1989; Allmon et al. 1996a,b). An apparent collapse of this high productivity pattern in the Middle to Late Pliocene (perhaps due to oceanic circulation changes associated with the formation of the Central American Isthmus) has been cited as contributing to a period of extinction in marine birds and mammals and high taxonomic turnover (increased extinction and origination) in mollusks and corals in the Western Atlantic (Allmon et al. 1993, 1996a,b; Johnson, Budd, and Stemann 1995; Budd, Johnson, and Stemann 1996).

At the other end of the Phanerozoic, the Cambrian explosion has also been connected to a change in the nutrient status of the world's oceans (Cook and Shergold 1984; Brasier 1991, 1992a,b; Brasier et al. 1994; Cook 1992; Tucker 1992; Butterfield 1997). Vermeij (1995) has similarly argued that enhanced marine nutrient supply and higher global temperatures produced by increased submarine volcanism were primarily responsible for biotic "revolutions" in the Cambrian–Ordovician and the middle to late Mesozoic. Brasier (1995) gives a table of at least 26 "turning points in evolutionary biology relevant to nutrient and carbon cycles of the oceans," although exactly how nutrients caused these turning points is unspecified. Vermeij (1987a, 1995) also speculates that the often cited onshore–offshore gradient in evolutionary innovation (e.g., Jablonski and Bottjer 1990) may have been driven by patterns of nutrient supply because nearshore waters usually have higher primary productivity.

Longer-Term Temporal Patterns (100 millions of years)

Several authors have noted that patterns of productivity parallel patterns of taxonomic diversity across the entire Phanerozoic and have drawn causal connections between the two. Vermeij (1978, 1987b, 1995) has suggested that primary productivity, and therefore the supply of food, has increased across the whole spectrum of marine habitats during the Phanerozoic. Bambach (1993) has elaborated on this idea and has suggested that this increase in food supply led to the well-documented increase in marine diversity through the Phanerozoic. The rise in food supply has also led, suggested Bambach, to many fundamental features of Phanerozoic marine ecological history, such as the increase in bioturbation and epifaunal tiering, the expansion in modes of life, the increase in predation intensity, and the shift from typical Paleozoic benthic macrofauna. This shift is characterized by taxa with relatively low individual biomass and metabolic demands to typical modern (Mesozoic and Cenozoic) faunas that are characterized by taxa with greater "fleshiness" and more active life habits such as deep burrowing, swimming, and predation. Martin (1995, 1996, 1998a,b) has also argued for a secular increase in nutrient levels in the world's oceans throughout the Phanerozoic and suggested that this increase explains many aspects of the history of marine plankton, such as the late Paleozoic disappearance of acritarchs, the Mesozoic expansion of dinoflagellates and planktonic foraminifera, and the Cenozoic expansion of diatoms.

A Theoretical Framework

The resolution of the nutrient paradox lies in the problem of scale and in distinguishing between the *maintenance* and the *origin* of taxonomic diversity (Allmon 1992a, 1994). On ecological timescales, factors such as nutrient supply, predation, habitat, and competition act as mechanisms of diversity maintenance, affecting the number of species that can "fit" into a given community at any given time (Allmon, Morris, and McKinney 1998, and references therein). On evolutionary timescales, however, these same factors may also have effects on the origin of new species. An understanding of the way in which this can occur requires a sufficiently detailed and explicit theoretical framework for the way in which speciation happens (Allmon 1992a, 1994; McKinney and Allmon 1995; Allmon, Morris, and McKinney 1998).

Every occurrence of allopatric speciation requires at least three stages: an isolated population must form, that is, become separated from the parental population; it must persist long enough to differentiate into an new species; and it must actually undergo that differentiation (Allmon 1992a). This framework for viewing the speciation process is very useful for examining the effects of nutrient conditions on the origins of diversity, for it focuses attention on individual evolutionary hypotheses at various stages in the process. Specifically, this framework suggests that nutrients (and the resulting patterns of productivity) may potentially affect speciation in at least three ways (figure 7.2).

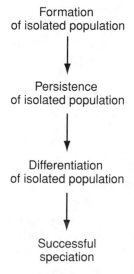

FIGURE 7.2. A three-stage explanatory framework for analysis of allopatric speciation (from Allmon 1992a).

Effects on Isolate Formation

Geographically isolated populations can form in one of two principal ways (e.g., Valentine and Jablonski 1983; Allmon 1992a): by dispersal or by vicariance. Dispersal probability is governed mainly by intrinsic factors (Stanley 1990; Allmon 1992a); vicariance is mainly a function of environmental disturbance (Nelson and Platnick 1981; Allmon, Morris, and McKinney 1998). Disturbance may be defined as an environmental perturbation that, at the temporal and spatial scale under consideration, removes all of the organisms under consideration from a particular area (Allmon, Morris, and McKinney 1998). The temporal and spatial magnitude of environmental change that qualifies as a disturbance will vary with the taxon [particularly with the dispersal ability of the taxon (Stanley 1990; Allmon 1992a)] and with the hierarchical level in the ecosystem; disturbance is both scale- and taxon-dependent (McKinney and Allmon 1995; Allmon, Morris, and McKinney 1998). Disturbance creates vi-cariance by eliminating the individuals of a species from a portion of its range, geographically separating formerly continuous populations (for discussion of possible examples from the fossil record, see Cronin 1985; Cronin and Ikeya 1990; Cronin and Schneider 1990; Stanley 1986b; Johnson, Budd, and Stemann 1995).

In the marine realm, disturbances can include discrete phenomena such as storms, fresh water incursions, sedimentation events, invasions of a new predator, or local loss of a food source or habitat space. Disturbances may also include overall environmental deterioration so that tolerable conditions for a given species may gradually disappear from its geographical range. Large scale environmental variables that can change in this way include temperature, nutrient supply, sea level, and turbidity.

Low environmental disturbance will (all other factors being equal) result in low production of isolated populations. Yet high disturbance may reduce the probability of the persistence of such isolated populations (see following sections). Thus, speciation should be expected to be maximal at intermediate levels of disturbance (Allmon, Morris, and McKinney 1998).

The stability of nutrient supplies may affect the probability that isolated populations will form. If nutrient and productivity conditions change relatively rapidly, they may cause local extinction within a formerly continuous geographic range of a species and so potentially increase the probability of isolate formation by vicariance. [This is the same process designated by Stanley (1986b) as the "fission effect," although he discussed fission occurring only due to predation (see Allmon 1992a, 1994).]

Rate of isolate formation may also be affected by the distribution of nutrient resources within the dispersal range of an organism. Multiple islands, whether actual physical islands or simply areas of locally modified resources, with appropriate nutrient or food resources may promote isolate formation. Examples include areas of localized upwelling in otherwise oligotrophic areas that provide oases for nutriphilic organisms (e.g., Petuch 1981; Vermeij and Petuch 1986), and oceanic islands that provide not only substrate but sufficient nutrient retention for nutriphilic benthos in oligotrophic tropical seas.

Effects on Isolate Persistence

The survival or persistence of an isolated population can depend on many environmental factors (Allmon 1992a). These factors are almost all reducible to the continued survival and reproduction of individual organisms within that population. These environmental factors can include: temperature (Clarke 1993), food supply (Vermeij 1995), predation pressure (Vermeij 1978, 1987a; Stanley 1986b), habitat space (Allmon 1992a, and references therein), and abiotic disturbances such as storms. As is the case with isolate formation by disturbance, a particular environmental change can affect isolate persistence in different taxa in different ways. A temperature decrease, for example, may benefit species at the equatorward limit of their range but may be deleterious to species at their poleward limit (Valentine 1984; Stanley 1986a). An increase in available nutrients, and so in primary productivity, may benefit suspension feeders (e.g., Allmon 1988), but may have a negative effect on forms with photosymbionts (e.g., Hallock and Schlager 1986; Hallock 1988; Edinger and Risk 1994).

Environmental effects on the persistence of isolated populations are part of a continuum, ending with environmental effects on the persistence of established species (Stanley 1986b, 1990; Allmon 1992a; Johnson, Budd, and Stemann 1995; Allmon, Morris, and McKinney 1998). Environmental changes that at one level of magnitude or intensity can lead to the extirpation of a local population or metapopulation can at a higher level lead to the extinction of the entire species; it is therefore a prediction of this "intermediate disturbance" model that maximal diversity is produced at low to moderate levels of environmental disturbance, and that an increase in origination may precede an increase in extinction (McKinney and Allmon 1995; Allmon, Morris, and McKinney 1998).

Nutrient supply may affect the survivorship of isolated populations. If nutrient levels, and therefore primary productivity, increase within an ecosystem, the probability that an isolated population would become extinct decreases, and the probability that isolates would survive by providing them with more food increases. Clearly this applies only to nutriphilic taxa, such as suspension feeders, and only in situations in which food is limiting to population size. Although there is a lower threshold of trophic resources below which populations cannot survive, increased resources may bolster population sizes sufficiently to enhance the probability of survival during ecological disturbances. Enhanced food supply may also indirectly benefit some marine taxa; e.g., those that depend on vegetation with high nutrient uptake, such as sea grasses and some macroalgae, for their habitat.

Taxa that suffer in the presence of high nutrient levels, such as zooxanthellate corals (and other taxa with photosymbionts, such as fusulinid foraminifera and rudistid bivalves), may respond in the opposite way (Brasier 1995); a decrease in nutrients might be expected to enhance the probability of isolate persistence. Indeed, this appears to have been the case at the end of the Eocene (Brasier 1995) and in the Plio-Pleistocene of the Western Atlantic (Johnson, Budd, and Stemann 1995; Budd, Johnson, and Stemann 1996; see following discussion). If food is not a limiting factor for populations, this explanation would not apply.

Instability of trophic supplies on an ecologic timescale may also affect isolate persistence. This may be the case in some deep-sea environments, in which the delivery of organic matter from the shallow shelf and surface may be sporadic and annually variable (Grassle and Maciolek 1992; Aronson 1994, and references therein). Too much or too rapid change (whether caused by alteration of nutrient conditions or some other factor), however, will reduce the probability that isolates can survive. Therefore, net speciation is highest at low to intermediate levels of disturbance (McKinney and Allmon 1995; Allmon, Morris, and McKinney 1998). Continued or increased intensity of disturbance may eventually lead to extinction of established species (see further discussion of extinction, below).

On a global scale, the total availability of nutrients in the world's oceans may dramatically affect the amount of habitable ecospace and thereby the probability that isolated populations would persist to become successful new species. The overall linkage between global productivity and diversity discussed by Bambach (1993) may be explicable in this way. Hallock (1987) has proposed that global diversity may respond not so much to changes in nutrient status within single habitats, but to the overall array of nutrient conditions in the sea. This Trophic Resource Continuum described by Hallock expands and contracts, depending largely on global tectonics, and creates or destroys habitable space not only at the high-nutrient end but also at the low-nutrient end.

Effects on Isolate Differentiation

The environment can affect population differentiation to the degree that such differentiation is controlled by selection (Allmon 1992a), Thus, virtually any environmental variable might be potentially important at one time or another for one taxon or another. Food supply may thus be a source of selection pressure. Some isolated populations may diverge genetically through adaptation to local trophic conditions. More generally, total nutrient availability may also affect probability and magnitude of differentiation by altering ecological constraints (Vermeij 1995). When more trophic resources are available, adaptive trade-offs may be lessened or altered, "enabling traits or combinations of traits to become established that otherwise would have been purged from the population because of unacceptable functional conflicts" (Vermeij 1995:132). Thus, Vermeij suggests, high-nutrient conditions should favor rapid or large differentiation, or both.

Nutrients and Extinction

Nutrients and productivity may affect diversity on evolutionary timescales not only through their effects on the origin of species, but also through their effects on the extinction of species. A number of studies have noted a correlation between trophic pattern and susceptibility to extinction in benthic marine invertebrates, specifically the apparently higher extinction rate of suspension feeders compared with nonsuspension feeders, although interpretations for the cause of these correlations vary. Levinton (1974) noted that suspension-feeding bivalves showed lower generic survivorship (and therefore higher extinction probability) over the entire Phanerozoic than deposit-feeding bivalves and attributed this difference to the greater "instability" of the suspension feeders' food source. (He reversed himself in 1996, arguing that differences in extinction rates among bivalve trophic patterns may relate to other, nontrophic factors.) Similarly, Paleozoic gastropod genera inferred to have been suspension feeders show higher extinction rates (during both mass and background extinction) than do nonsuspension feeders (Allmon et al. 1992). For Cretaceous mollusks, Kauffman (1972, 1977) found that suspension feeding bivalves and gastropods show much shorter species durations than deposit-feeding bivalves. Sheehan and Hansen (1986) note that suspension feeders show higher extinction rates than deposit feeders across the Cretaceous–Tertiary boundary and suggest that this was due to decimation of plankton in the water column. This interpretation is disputed by Levinton

(1996). This line of reasoning has also been applied at higher trophic levels. Jeppson (1990) suggests that diversity of Silurian conodonts (presumed to be nektonic carnivores) was negatively impacted by decline in nutrient supplies around Baltica.

The strongest evidence for a linkage between nutrients and extinction comes not from studies of individual clades, but from more general studies attempting to understand paleoenvironmental conditions around mass extinction boundaries. Many such studies have concluded that at least some mass extinction events were associated with massive collapses of marine productivity (e.g., Vermeij 1995). A number of authors have concluded that the Cretaceous–Tertiary boundary is coincident with a dramatic reduction in primary productivity in the oceans (Bramlette 1965; Tappan 1968, 1986; Sheehan and Hansen 1986; Arthur, Zachos, and Jones 1987; Corfield and Shackleton 1988; Zachos, Arthur, and Dean 1989; Paul and Mitchell 1994; Martin 1998a,b; Smith and Jeffery 1998), a condition that persisted for at least several hundred thousand years (Caldeira and Rampino 1993; Hollander, McKenzie, and Hsü 1993) and perhaps as long as three million years (D'Hondt et al. 1998). Similar nutrient decreases of various magnitude have also been claimed for the Permo-Triassic extinction event (Wang et al. 1994; Martin 1996, 1998a,b), the Cenomanian–Turonian event (Paul and Mitchell 1994), the Paleocene–Eocene event (Rea et al. 1990), and for the Pliocene event in the Western Atlantic (Allmon 1992b; Allmon et al. 1996a,b). Increased instability in seasonal productivity has been implicated in the Late Eocene event (Purton and Brasier 1997). As mentioned previously, for taxa favoring oligotrophic conditions, a rapid increase in nutrient levels might lead to increased extinction; this has been suggested for photosymbiont-bearing foraminifera in the late Paleozoic (Martin 1995, 1998a,b) and Late Eocene (Brasier 1995).

Two Examples

We give two examples of evolutionary change in biotically diverse marine communities that we believe have been profoundly affected by nutrient concentrations. The first example shows the complexity of determining the origin and importance of nutrients for a modern ecosystem (reefs), followed by the possible effect of nutrient variation on isolate formation and persistence on island reef systems, with special attention paid to ostracodes. The second example is late Neogene faunal change in the Western Atlantic, with particular attention paid to the relationship among diversification, extinction, and nutrient changes associated with closure of the Central American Seaway.

Reefs and Reef Systems

Modern coral reefs represent the low-nutrient half of the nutrient paradox. A large proportion of the taxonomic diversity of the modern ocean occurs within or immediately adjacent to tropical coral reefs in some of the most oligotrophic waters in the oceans; reefs may in fact be decimated by an increase in nutrients (Hallock and Schlager 1986). There is, however, ambiguity about the relationship between diversity and nutrients in these systems, because among reef communities themselves the highest diversities (both for corals and the entire community) are near the relatively nutrient-rich waters off continental and high islands (oceanic islands with a large emergent landmass) rather than the low islands of atolls (e.g., Birkeland 1987; Risk, Sammarco, and Schwarcz 1994; Taylor 1997). Since reefs are so significant in terms of both diversity and diversification and fossil record, we review here more closely some of the current controversy surrounding the actual nutrient needs of this ecosystem. As in other areas of evolutionary paleoecology, determining the evolutionary implications of paleoecologic change in nutrient conditions is hampered by our lack of full understanding of how reefs and reef diversity react to nutrients today. We will use the term *reef system* to include not only the reef but also backreef lagoons and associated mangrove and sea grass communities. The interactions among these reef system communities may have implications for both the diversity and nutrient status of reefs and reef systems.

Reefs as Environments Without Nutrient Limitations

Presently there exist strong differences in opinion regarding the nutrient recycling efficiency of reefs, and thus the need for reefs to obtain additional nutrients beyond those obtained from ambient ocean water. Resolution of this disagreement has obvious implications for our understanding of the effect of nutrient changes through geological time. One widely accepted view is that cycling within reefs is high enough and net production is low enough that there is little flux into or out of the reef or reef-lagoon system (e.g., Lewis 1989; Crosslands, Hatcher, and Smith 1991). At its most extreme, this hypothesis suggests that reefs are almost never nutrient limited, no matter how low the ambient concentrations (Wiebe 1987). Anecdotal reports support this view. Falkowski et al. (1993), for example, found that high concentrations of dissolved inorganic N lead to unbalanced growth of zooxanthellae and poorer C transfer to the host.

Do reefs obtain resources from neighboring communities in the reef system? Associated sea grass and mangrove communities may have very high production rates, but the production that is not recycled in these communities is not known to be removed to the reef community. Available information suggests that mangroves do not provide nutritious detritus to bordering communities, but act to trap nutrients in their own community near the shore (Alongi 1990; Boucher and Clavier 1990). Sea grasses trap additional nutrients between the coast and the reef, thus inhibiting rather than aiding nutrification of the reef (Wiebe 1987). In sum, in this view reefs are not nutrient limited, and if marginal communities such as sea grasses and mangroves play any role at all with respect to nutrients, it may be to maintain oligotrophy over the reef complex.

Additional evidence that reefs themselves do not fare better with increased nutrients is the observation that nutrient levels above that of oligotrophic waters tend to stimulate a change in community structure rather than result in obviously increased coral growth (Smith et al. 1981). Conflicting views exist on whether nutrients positively or negatively affect the growth of coral and calcareous algae. A number of studies have suggested that nutrients, particularly P, may inhibit calcification (Risk and Sammarco 1991; Falkowski et al. 1993; Delgado and Lapointe 1994), but other evidence suggests that up to a point corals under isolated conditions grow faster in greater nutrient concentrations through increased growth of zooxanthellae (Muscatine et al. 1989).

Existing evidence suggests that nutrients most heavily influence reefs not through reef builders such as coral, but through increasing growth rate of soft algae competing for the same space. That is, changes in community structure occur through the extrinisic affects of overgrowth and bioerosion or reefs. Reefs are overgrown by algae, first macrophytic benthic algae, and then at very high nutrient levels by phytoplankton. Littler and Littler (1984) proposed a model in which community dominance changes with increased nutrients from corals to coralline algae and, if herbivory declines, to frondose macroalgae. Simultaneously, the fauna changes from one dominated by animals in symbiosis with zooxanthellae to dominance of herbivorous grazers and finally to dominance of heterotrophic suspension feeders (Birkeland 1987; D'Elia and Wiebe 1990), and the biomass of the grazers and suspension feeders becomes larger with higher production rates of sea grass and phytoplankton (Birkeland 1987; Littler et al. 1984).

In the past decade a number of authors have pointed out that reef destruction through bioerosion is as significant in considering reef system and community structure as reef construction (e.g., Hallock and Schlager 1986; Sammarco and Risk 1990; Edinger and Risk 1994). Bioerosion increases with nutrient concentration because the organisms causing bioerosion are frequently planktivorous heterotrophic filter feeders such as endolithic sponges and bivalves, polychaete and sipunculid worms, and boring barnacles (Sammarco and Risk 1990). Bioerosion may make reefs more susceptible to catastrophic high-energy events such as storms, through weakening of the framework (Sammarco and Risk 1990), or more susceptible to drowning by overwhelming the rate of carbonate production (Hallock and Schlager 1986).

Reefs and Reef Systems as Nutrient-Limited Environments

In contrast, other workers believe that reefs are limited by nutrients in at least some contexts, and that as in other marine environments, growth and diversity are strongly controlled by pulses of nutrient input. Some evidence for positive nutrient enrichment from external sources is provided by correlations between local diversity and local nutrient enrichment. For example, several authors have noted correlations between reef system biomass and upwelling (Hiatt and Strasburg 1960; Kimmerer and Walsh 1981; Andrews and Gentien 1982; but see also Glynn and Stewart 1973; Pujos and Javelaud 1992) or input from marginal settings such as mangroves or sea grass beds (Birkeland 1987; Risk, Sammarco, and Schwarcz 1994). In addition, though many reef-lagoon complexes seem to show little or no net organic matter export, quantities of organic detritus have been found leaving some reefs, suggesting that at these locations there must be nutrients imported by means other than the oligotrophic surface water. Nutrient limitation is also implied by hypothetical considerations that suggest that because N fixation (D'Elia and Wiebe 1990) makes N more readily available than P and because P is easily bound in carbonate sediments, P must be limiting for aquatic vegetation (Short, Dennison, and Capone 1990; Littler, Littler, and Titlyanov 1991; Fourqurean, Zieman, and Powell 1992). Such nutrient limitations can be especially important for adjacent sea grass and mangrove communities.

Nutrient limitations are important for many organisms in environments in close association with reefs, such as lagoons (Birkeland 1987; Taylor 1997). For example, suspension-feeding bivalve mollusks, nassariid gastropods, and barnacles are far less abundant and diverse on nutrient-poor atolls than along continental margins and islands with nutrient-rich settings (Vermeij 1990; Taylor 1997). Some organisms commonly found in reefs, such as endolithic borers, also benefit by increased nutrient concentrations.

The most general consensus between these two views might be that reef organisms are not significantly nutrient limited, but vegetation and animals in nearby associated reef system environments may be. These adjacent communities may affect the maintenance of reef diversity not only via altered nutrient and food conditions directly, but also through increased habitat diversity or community interactions, such as providing a haven for fish juveniles and other reef taxa (e.g., Hallock 1987). Thus diversity maintenance may be highest in areas in which a spectrum of reef system environments develop. Such a spectrum requires: (1) the proper geomorphology, and (2) nutrient concentrations sufficient for macrophytic vegetation, but not sufficient for phytoplankton or fleshy macroalgae that might overgrow the reef. Maintenance of total reef system diversity is increased through habitat heterogeneity associated with a spectrum of nutrient concentrations (Hallock 1987; Wilkinson 1994).

These discussions of ecological influences on diversity implicitly assume that species diversity is a function of local availability of niches and do not consider the origin of the reef system species or the temporo-spatial variability in regional probability of successful speciation in reef systems. Considering again the theoretical framework for speciation (figure 7.2), isolate formation will be enhanced when this spectrum of reef system environments is modified in time and space by shifts in nutrient distribution. Because the processes affecting the frequency of isolate persistence are effectively those affecting maintenance of diversity, we might expect isolate persistence to be maximized in complex reef systems with a range in nutrient availability.

Oceanic Islands as a Special Case

Oceanic islands may be appropriate places to more closely examine the relationship among nutrients, diversity, and evolution, not only methodologically because the system is relatively closed compared to continental areas, but also substantively because island reef systems have been suggested as ideal localities for isolate formation and speciation (e.g., Ladd 1960). One might imagine then that some oceanic islands may provide the right conditions both for isolate formation (particularly through chance long-distance dispersal) and persistence, and that they may have played a role in diversification of some or many western equatorial Pacific marine organisms (Vermeij 1987a).

Generally, the ambient nutrient concentrations surrounding tropical oceanic islands are extremely low, approximately that of oligotrophic open-ocean surface water. The center of the large Pacific oceanic gyres are known as "oceanic deserts" for their lack of resources (e.g., Ryther 1969). A variety of factors, however, do create variation in local nutrient concentrations around and between oceanic islands. For example, the shape of the island lagoon and the nature of the passages that connect it to the open ocean affect the flushing rate of water and associated nutrients in the lagoon (Charpy and Charpy-Roubaud 1990); similarly, wind speeds and their seasonality control the rate of flux into and out of the lagoon (Furnas et al. 1990) and control recirculation within the lagoon (Arx 1954; Atkinson, Smith, and Stroup 1981). The water surrounding some islands apparently increases in nutrient concentration as it approaches and passes around the island, a phenomenon known as the "island mass effect" (Doty and Oguri 1956). The reason for this effect has not been well demonstrated, but some have suggested that internal waves breaking against the slope of an island shelf inject subeuphotic zone nutrients into the photic zone (Sanders 1981; Andrews 1983). Hamner and Hauri (1981) suggest that minicurrents caused by an island mass effect contain enriched plankton and associated planktivorous fish that stimulate growth of reef colonies. Submarine discharge of groundwater has also been reported to increase nutrients in various carbonate settings (Kohout and Kolipinski 1967; Marsh 1977; D'Elia, Webb, and Porter 1981; Lewis 1987). Storms may create nutrient injection by mixing of deep water. Although coastal upwelling at oceanic islands is normally insignificant, upwelling may occur at oceanic divergences, for example at the equator. Particularly important is the size of the island and the local climate, which affect the supply of terrestrially derived nutrients by erosion.

The nutrient concentrations near substantial landmasses ("high" islands) are different from those near "low" islands (in which the islands are mostly <10 m above sea level and <0.5 km wide), especially in lagoonal areas immediately surrounding larger landmasses. High islands are generally associated with a greater spectrum of values, ranging from that of ambient open ocean sea water to fluvially enriched values (Lewis 1989; D'Elia and Wiebe 1990). Near high islands, nutrients enter the lagoon via fluvial input. These nutrients are sufficient to support aquatic vegetation, which tends to be relatively abundant off larger landmasses, for example, mangroves and sea grasses. These lagoonal areas with aquatic vegetation may be relevant to reefs because they (1) provide areas where feeding larvae of reef organisms and developing juvenile fish have high rates of survival (Ogden 1986; Parrish 1986); (2) provide a place for fauna living on the reef to graze (especially fish and echinoids); (3) trap nutrients and sediment that might interfere with clear oligotrophic water over the reef; (4) create relatively quiet sediment environments; and (5) provide trace nutrients that are not generally present in substantial quantities in oceanic waters (Matson 1989) or perhaps small enough quantities of nutrients to stimulate reef production without changing the ecosystem ecology. They also add substantial habitat diversity for organisms that require, for example, aquatic vegetation as a substrate or food resource, low-energy environments and bottom water, and pockets of low-oxygen or brackish bottom water.

A relatively recent idea on the origin of new nutrients to reef systems could add a twist to current views on the importance of recycling. French coral researchers have put forth the idea that reefs obtain a high flux of nutrients through pore water via a phenomenon termed "endo-upwelling" (Rougerie and Wauthy 1986; Rougerie, Wauthy, and Andrie 1990). Endo-upwelling involves the ascension through the atoll flanks of intermediate water to the surface, by thermoconvective advection. One implication is that reefs and their associated environments store or export large quantities of organic matter. Rougerie, Wauthy, and Andrie (1990) suggest that, based on the flux of upwelled nutrients, only 30% of primary production comes from recycled nutrients. Andrie et al. (1992) attempt to account for the missing organic matter by suggesting that it is trapped in biomass at the top of the outer shelf, an area that is often neglected because of the difficulties of working in a zone of such intense wave energy. An ongoing aspect of this research is to determine how endo-upwelled nutrients change with variations in the thickness and mineralogy of the carbonates and volcanics through which aqueous solutions pass. Endo-upwelling, if significant, may be at equilibrium at ecological time-scales, but show important variations in geologic time.

In summary, endo-upwelling aside, island reef systems with the greatest nutrient concentrations and number of communities are found off high islands that possess a significant landmass above the surface. These provide the closest analog to the most diverse reef system faunas off larger continental islands. Following is a specific example, using marine ostracodes, speculating upon opportunities for isolate formation and persistence on oceanic islands, considering variations in island size and nutrient conditions.

Micronesian Ostracodes: A Closer Look at Isolates and Nutrients in Island Reef Systems

Marine speciation with co-existing parent species seems to occur rarely, if ever, *within* a particular island; thus each island potentially represents one isolated population of a given species. To what extent a given island or island group is actually genetically isolated from populations from other landmasses depends greatly on the dispersal capabilities of the species in question (Palumbi 1992). Among those organisms that can arrive at an island occasionally by chance dispersal and that are effectively genetically isolated from other populations, the likelihood of speciation will depend on the probability of successful colonization (isolate formation), the persistence of the population, and the likelihood of permanent genetic divergence (figure 7.2). In the context of this example, if trophic resources are a key to the distribution of some species, we may imagine that colonization will be influenced by the probability of landing in a trophically appropriate habitat upon arrival, and that persistence will be influenced by existence of and fluctuations in nutrient-related environmental conditions through time. At a given island the number of colonizers (isolate formation) and degree of persistence (isolate persistence), which together account for maintenance of local species richness, will control the number of opportunities for long-term isolate divergence and speciation.

It is well known that species richness drops on islands from west to east across the Pacific in both terrestrial and marine organisms of a wide variety of taxonomic groups (e.g., Kay 1980). Determining the primary cause of this diversity drop is complicated by correlations among the causal variables: as one moves east, the distance from high-diversity mainland faunas increases, and in general the size, ecological heterogeneity, and nutrient re-sources of the islands decrease. The effect of these variables will, of course, vary by type of organism. Distance controls the likelihood of isolate formation, whereas available resources, including nutrients, control isolate persistence. Our contention is that the confluence of evidence suggests that nutrients are ultimately the primary determinant of long-term persistence of ostracode populations on islands and that as this has changed through time, so have rates of speciation and local extinction.

Ostracodes, small bivalved crustaceans, are among the better known organisms from the late Cenozoic of the western equatorial Pacific (e.g., Holden 1976; Ross 1990; Cronin et al. 1991) because their valves are relatively abundant and well preserved in cores taken from atolls and in samples of sediment taken from modern environments. The fossil and modern records (for all organisms) of oceanic islands are, however, understudied and incomplete, and the actual degree of genetic isolation between living island populations has been poorly documented. We present this example to suggest the possible significance of nutrients in the steps of speciation, with recognition of our current limitations in data.

Isolate Formation and Persistence: Maintenance of Diversity

Benthic ostracodes have no planktonic dispersal stage and are assumed to colonize islands via chance dispersal on floating benthic vegetation (Teeter 1973); substantial populations have been found in shallow water on algae uprooted after storms (Suzuki 1997). To the extent that most or all ostracode species on a given oceanic island can be considered isolates, species richness of ostracodes is a measure of isolate formation and persistence. Modern ostracodes, like many other marine organisms, are most diverse on reef complexes off the Indo-West Pacific continental landmasses, less diverse on high islands in western Micronesia, and least diverse in atoll lagoons in the Marshall Islands (Weissleader et al. 1989). Most benthic ostracodes are deposit feeders or feed on microalgae; ostracodes are most abundant and generally most diverse in fine-grained sediments rich in organic matter and on macroalgae and sea grasses. Although taxa have a tendency to live on a particular part of the reef, the boundary lines of distribution are indistinct; distribution correlates best with the presence of appropriate substrate and trophic resources. Moreover, the west to east drop-off in species diversity is not easily explained by decreases in habitat heterogeneity because none of the ostracodes are known to be restricted to environments more frequently associated with larger islands, such as sea grass beds and mangroves. Some are restricted to areas with macroalgae, which occurs in variable abundance on all oceanic islands in Micronesia. In fact, few of the ostracodes present off high islands are actually restricted to these islands; rather, the higher diversity off high islands seems to represent simply a larger "random" draw from the pool of species present throughout the western Pacific islands. Ostracodes also appear to be much more abundant in sediments off the high islands than low islands (Ross 1990).

The best fossil record among the Pacific islands comes from Anewetak (formerly Eniwetok, Enewetak) Atoll, where cores taken in the mid-1980s permit investigation of shallow marine lagoonal faunas through about the past 10 million years. Today Anewetak faunas are typical of the low diversity faunas elsewhere in the Marshall Islands. The most significant historical change in the faunas in these cores from Anewetak occurs at approximately the start of high amplitude sea level fluctuations associated with repeated glacial cycles. Prior to these fluctuations multiple lines of evidence suggest that Anewetak had taxonomically richer marine and terrestrial faunas (Ross 1990) associated with a significant emergent island, higher habitat heterogeneity, and more sedimentary organic matter. These richer faunas include ostracode faunas of a diversity and abundance similar to that of high islands today such as Guam (Ross 1990). Among the taxa present are bivalves presently found only in relatively high nutrient environments (Vermeij 1990). The dramatic drop in diversity of ostracodes at a major event in the environmental history of the island suggests that differences between high and low island environments, rather than distance from continental islands, has played the strongest role in determining the nature of marine ostracode assemblages at this particular island.

A primary difference between high and low islands that might explain these diversity differences, other than habitat heterogeneity, is the ambient nutrient concentration in the water surrounding the island and the amount of organic-rich substrate, some of which is contributed from erosion of the island landmass. Significant increases in sedimentation and nutrient runoff would kill the reefs; smaller amounts facilitate the growth of sea grass, macroalgae, and mangroves in backreef lagoons. The implications for island reef system ostracodes is that for at least some species, the probability of colonization and of weathering stochastic variations in population size on the island may be bolstered by available trophic resources, potential population size (cf., Vermeij 1995), and increased vegetation substrates. The prerequisite ecological conditions for speciation seem to have been better satisfied at Anewetak prior to frequent high amplitude sea level fluctuations.

In contrast, atoll lagoons have been disturbed numerous times over the past three million years, as the lagoon completely or nearly disappeared during numerous sea level drops associated with glacial advances. The disturbances caused by sea level drop are likely to have temporarily extirpated the populations of many lagoonal organisms (Paulay 1990). For example, many ostracodes at Anewetak disappear temporarily at unconformities representing sea level drops, sometimes reappearing later in the section. Insufficient time for recolonization does not appear to be the reason for low diversity Quaternary faunas, as diversity does not increase through time between the sea level drops. Glacial-interglacial cycles promote repeated erosion of islands and shallow marine sediments (and associated nutrients) followed by rapid upward growth of the reef, hampering the development of broad shallow sedimentary deposits where macroalgae, sea grasses, and mangroves can accumulate and bind fine sediment and nutrients. The result is relatively steep-walled nutrient-poor lagoons; this is the state of most atoll lagoons today (Ross 1990).

Isolate Differentiation: Speciation in the Indo-Pacific

Because of the large number of oceanic islands and the rapid dispersion of many marine groups, locating the island or island group of origin of a particular species can be problematic. The next best evidence of differentiation may be to look for morphologic divergence as a guide to potential incipient speciation in fossil and modern species. If persistence time increases the probability or degree of divergence, then we should expect greater spatial and temporal morphologic variation in conditions facilitating isolate persistence. The number of studies of morphologic divergence in marine species on tropical oceanic islands is, however, quite low. This may seem surprising in light of the interest in morphologic diversity in these regions, but most of the research has been on diversity at the taxonomic levels of species or genera.

Ross (1990) has studied morphologic divergence in species of *Loxoconcha* from the same fossil and modern samples used for studies of species richness. The ostracode genus *Loxoconcha* is abundant and widespread on oceanic islands in both Recent and fossil deposits (Ross 1990). Its five most abundant species with sufficient data for analysis show remarkably little intraspecific variation, either within one lagoon or even across Micronesia, and the few variations that occur are at high islands. This notable lack of variation is also characteristic of most fossil populations throughout the interval of high amplitude sea level fluctuations at Anewetak.

In contrast, intralagoonal variation is generally greater within individual samples from the Late Miocene and Early Pliocene, and some, but not all, of this variation seems to be temporally directional. An interesting contradiction to these data is that in Late Pliocene sediments, amidst high amplitude sea level fluctuations, there is apparent directional morphologic change within one lineage in the core that resembles a speciation event. Most of the available data, however, are highly consistent with greater divergence, or at least greater morphologic variation, in preglacial environments at Anewetak. The pattern of temporal changes in geographic variation among lagoons is generally unknown due to paucity of material from other atolls.

If conditions promoting isolate formation, persistence, and differentiation at Anewetak were common to dozens or hundreds of other atoll localities prior to high amplitude sea level fluctuations, then one might also expect to see higher speciation rates in the Indo-Pacific region at this time. Circumstantial evidence, consisting of estimates of a slowdown in late Cenozoic speciation during high amplitude sea level fluctuations in several groups of Indo-Pacific marine invertebrates, suggests that this may indeed be the case. Both the evidence and its link to nutrients remain open to interpretation, but we can provide a preliminary model for testing.

If oceanic islands such as Anewetak were a greater source for isolate formation and persistence prior to high amplitude sea level fluctuations, we should expect a decline in speciation from the Late Pliocene in ostracodes and other groups that might be similarly ecologically affected. Over 60% of ostracode species found at Anewetak first appeared before the Late Pliocene (Ross 1990); this is a minimum estimate since those that appeared first after the Late Pliocene may also have originated before the Late Pliocene but had some delay in colonizing Anewetak (or merely were not sampled in the cores) (Ross 1990). Ostracodes are not known to be competitively excluded from regions, thus it seems likely that new species would have appeared at some point between deposition of the Late Pliocene and modern sediments if ecologically suitable taxa had arisen within the region. The implication is that most species first occur prior to major glaciation, consistent with our hypothesis.

Data from some other groups also seem to suggest a slowdown in speciation since the Late Pliocene and enhanced speciation in the Miocene or Early Pliocene. Corals show species-level slowdown of both speciation and extinction (Potts 1984, 1985, Indo-Pacific corals; Veron and Kelly 1988, Papua New Guinea; Fagerstrom 1987, reef communities in general [with the exception of the dendroid corals]; Chevalier 1977). Larger foraminifera (Adams 1990) exhibit similar evolutionary patterns to corals. In contrast, in one of the most detailed studies of a single taxon from the region, Kohn (1985) found increased late Neogene speciation rates in the reef gastropod *Conus,* and New-man (1986), using Ladd's (1982) compiled data, also found an increase in evo-lutionary rates in mollusks in the Quaternary of the western Pacific islands.

The apparent discrepancy in rates of speciation might be an artefact of the fossil record, since pre-Pleistocene fossil records of macrofossils are patchy. If the discrepancy is real, it might be explained by consideration of which groups would be most affected by the rapid drops in sea level and associated changes in nutrient conditions, substratum geomorphology, and other ecological factors. Paulay (1990), for example, showed that marine bivalves that prefer lagoonal soft sediments fare poorly in terms of species diversity at islands where this habitat is absent. He hypothesized that infaunal bivalves and other organisms with similar ecological needs may go extinct locally at oceanic islands with each major sea level drop and recolonize again when lagoonal conditions reappear. Conversely, it may seem curious that corals and larger foraminifera would show a speciation pattern similar to that of ostracodes, since these groups are ecolog-ically so distinct. Sessile organisms in photosymbiosis may, however, have reacted to the same disturbance events, but through different processes, per-haps through temporary nutrification during sea level drops and erosion or sea level rise and leaching of soils (Hallock 1988). Organisms that were mobile, and less affected by either loss of lagoon or changes in nutrient concentrations, would be most likely to survive through high amplitude sea level fluctuations. Data are currently insufficient for us to understand how corals and foraminifera reacted in diversity to historical changes at the atoll.

Summary

Many populations of species on oceanic islands can be considered isolates that are candidates for differentiation. Probability of isolate differentiation is related (among other factors) to survival time of the population (isolate per-sistence). The probability of both colonization (isolate formation) and sur-vival (persistence) on oceanic islands can be explained, for at least some taxo-nomic groups such as benthic ostracodes, most simply as a function of trophic resources and of substrates that are themselves related to trophic resources. Drastic environmental disturbances such as repeated glacial sea level drops cause essential island substrates to disappear and seem to deplete available nutrient resources, thus significantly decreasing rates of isolate formation and lengths of intervals of isolate persistence of certain organisms. Since these fun-damental changes to atoll ecology would occur simultaneously throughout the Indo-Pacific at the commencement of high amplitude sea level fluctuations, we might expect to see associated historical trends in speciation rates accompanying this event. A variety of evidence from records of ostracodes and other groups seem to be consistent with this hypothesis, but more explicit data on trophic needs, Late Neogene Pacific island ecological changes, and rates and biogeography of speciation are necessary to confirm it. The effect of glacial-interglacial fluctuations of sea level on marine speciation, through destruction of evolving isolates, is likely to be a global phenomenon (Cronin 1985, 1987; Bennett 1990).

Reefs in particular, and reef systems in general, maintain an enormous diversity of organisms. Though explanations of this diversity are frequently in terms of available niche space and resources, the origin of the diversity must occur from opportunities for genetic differentiation. We gave an example in which variations in nutrients on oceanic island reefs may have affected formation and persistence of isolates. Nutrient dynamics, and thus reef systems, off continents are also very dynamic at a variety of timescales and may control isolate formation via either vicariance or founder populations. Thus, although reefs themselves thrive in oligotrophy, settings with greater nutrient availability provide a greater spectrum of environments and enhance isolate formation and persistence at evolutionary timescales.

Late Neogene Faunal Change in the Western Atlantic

The late Cenozoic (i.e., Pliocene–Recent, the last five million years) marine fossil record of the Western Atlantic region, especially in the Caribbean, Central America, and the Coastal Plain of the southeastern United States is a particularly appropriate record in which to study the connection between environmental and faunal change because of the dramatic paleoceanographic changes associated with the rise of the Central American Isthmus (CAI). Of special interest in much work on the Western Atlantic Plio-Pleistocene has been the relationship between different environmental variables (particularly the relative roles of temperature and nutrient levels) in the marine realm and patterns of origination and extinction (Vermeij 1978, 1987b, 1989, 1990; Stanley 1986a; Allmon 1992b, in press; Allmon et al. 1993, 1996a,b; Jackson et al. 1993, 1997; Jackson, Jung, and Fortunato 1996; Johnson, Budd, and Stemann 1995; Budd, Johnson, and Stemann 1996; Roopnarine 1996), but these studies have been without much resolution.

In this section, we examine patterns of origination and extinction of three of the better known marine groups in this region [mollusks, corals, and vertebrates (birds and mammals)], followed by an analysis of changes in nutrient supply and their effects on stages in the speciation process.

Late Cenozoic Change in the Western Atlantic: Faunas

MOLLUSKS. It was long believed that mollusks suffered a late Cenozoic extinction that was more severe in the Western Atlantic than in the Eastern Pacific, especially among so-called paciphile taxa (formerly Atlantic taxa now surviving only in the Pacific; Woodring 1966), producing a modern Western Atlantic molluscan fauna depauperate relative to that of the Eastern Pacific (Dall 1892; Olsson 1961; Woodring 1966; Keen 1971; Vermeij 1978, 1991a; Stanley and Campbell 1981; Stanley 1986a; Vermeij and Petuch 1986; Stanley and Ruddiman 1995). More recent studies, however, have revised this view. It now appears that, although molluscan extinction was indeed higher in the Western Atlantic than elsewhere in the tropics (Vermeij 1991b), molluscan diversity was constant or even increased in the Western Atlantic during the last five million years, with extinctions more or less balanced by originations (Allmon et al. 1993, 1996b; Jackson et al. 1993; Jackson, Jung, and Fortunato 1996), and that the modern Western Atlantic mollusk fauna is approximately as diverse as that of the Eastern Pacific (Allmon et al. 1993, 1996b).

Total molluscan diversity on the Caribbean coast of Costa Rica and Panama shows no reduction since the Late Miocene; on the contrary, diversity increased from the Early to the Late Pliocene (Jackson et al. 1993). Paciphile taxa make up only 9% of the large sample analyzed by Jackson et al. (1993), but do not show a decrease in diversity until just before the Plio-Pleistocene boundary (ca. 2.0 Ma), during a major turnover event characterized by increased rates of both origination and extinction. Extinction rates for paciphiles are, however, more than two times greater during this event than for the fauna as a whole (Jackson et al. 1993).

Gastropods as a whole in the southeastern U.S. Coastal Plain show no significant change in diversity since the Pliocene (Allmon et al. 1993, 1996b). Gastropod diversity is essentially the same today in the region as it was in the mid-Pliocene, despite approximately 70% extinction since that time. These high rates of extinction are evidently balanced by high rates of origination, although the relative timing of these two processes is not known. South of Cape Hatteras (and best represented in Florida), most of this turnover takes place in an event that occurred between 2.0 and 2.5 Ma. North of Cape Hatteras, mollusk diversity declines during a turnover event that occurred between 3.0 and 3.5 Ma. Despite all the extinction that has occurred in the Western Atlantic, the species richness of the Recent gastropod fauna of the low- to mid-latitude Western Atlantic is not demonstrably different from that of that of the low- to mid-latitude Eastern Pacific (Allmon et al. 1993, 1996b).

Studies of individual gastropod clades confirm these overall patterns. Gastropods of the *Strombina* group (family Columbellidae) are prominent paciphiles (Woodring 1966). Extinction in both Eastern Pacific and Western Atlantic was concentrated in a brief interval near the Plio-Pleistocene boundary, but significant origination occurred at this time only in the Pacific (Jackson et al. 1993, 1996). In the gastropod family Turritellidae, Western Atlantic species show a sharp decline in diversity in the late Neogene, but Eastern Pacific turritellids do not (Allmon 1992b).

Fewer data have been published for bivalves. Those data that are available also point to substantial origination accompanying extinction in the Late Pliocene, although at a lower rate than in gastropods. In Florida, for example, between the Late Pliocene Pinecrest Formation and the Late Pliocene–Early Pleistocene Caloosahatchee Formation, data presented by Stanley (1986a) indicate extinction in bivalves of 47.9% (versus 62.4% for gastropods) and origination of 26.7% (versus 55.2% for gastropods) (Allmon et al. 1996b). Among chionine bivalves (Roopnarine 1996), extinction of Western Atlantic species (82.6%) exceeds origination during the entire Pliocene; origination decreases in the Pacific during the same interval but is never exceeded by extinction (38.5%) (Roopnarine 1996).

CORALS. Reef corals in the Caribbean show a pattern of origination and extinction somewhat similar to that shown by mollusks throughout the Western Atlantic (Johnson, Budd, and Stemann 1995; Budd, Johnson, and Stemann 1996). The major difference is that molluscan turnover appears to begin in the Caribbean and Coastal Plain around 2.4 Ma, whereas coral turnover may begin as much as 1.0 m.y. earlier. Between 4.0 and 1.0 Ma, extinction and origination rates in reef corals increase roughly simultaneously (Budd, Johnson, and Stemann 1996), although in species with smaller colonies the peak in extinction is preceded by a high level of origination (Johnson, Budd, and Stemann 1995). Approximately 64% of the Early Pliocene coral fauna becomes extinct; smaller, free-living colonies living in sea grasses are most affected by the turnover, with extinction rates of 30–50% per million years; other ecological assemblages average less than 30% extinction per million years. Species richness is relatively low (30–50%) throughout much of the Early to Middle Miocene (22–9 Ma), high (80–100%) from the Late Miocene to Early Pleistocene (9–1 Ma), and intermediate since the Early Pleistocene (1–0 Ma). Both extinction and origination are accelerated between 4.0 and 3.0 Ma. Extinction rates are also high between 2.0 and 1.0 Ma. (In the Eastern Pacific, reef corals suffered nearly complete extinction during the Pliocene. The present depauperate fauna includes species of Indo-Pacific as well as Caribbean affinities [Johnson, Budd, and Stemann 1995; Budd, Johnson, and Stemann 1996; Jackson and Budd 1996].) In southern Florida, reef corals show some decline after the Pliocene (Allmon et al. 1996a; Budd, Johnson, and Stemann 1996); extensive reefs evidently occurred farther north during the Pliocene than they do in Florida today.

MARINE VERTEBRATES. The late Cenozoic marine vertebrate record is particularly well known in Florida (see Allmon et al. 1996a, and references therein) and forms a basis for comparison with the invertebrate record.

In Florida, Lower Pliocene seabirds are well known (Emslie and Morgan 1994). Particularly noteworthy in this avifauna is the diversity of taxa with modern counterparts that are associated with cold water and upwelling marine systems, such as alcids, gannets, boobies, and cormorants (Allmon et al. 1996a). Breeding (or formerly breeding) seabirds in Florida and the Dry Tortugas are limited today to only eight species, or 47% fewer than in the Early Pliocene.

Among marine mammals, compared with Late Miocene, Pleistocene, and Recent faunas in Florida, abundance and diversity in the Early Pliocene are notably high. Lower Pliocene marine mammals include 10 species of cetaceans (whales and dolphins), dominated by baleen whales, and 4 species of pinnipeds (seals and walruses) (Allmon et al. 1996a). This is in striking contrast to the present marine mammal fauna of Florida. Although over 25 species of cetaceans have been recorded from Recent Florida waters, many of these species are very rare (Layne 1965). Baleen whales in particular are now virtually unknown in Florida (see discussion in Allmon et al. 1996a).

Late Cenozoic Change in the Western Atlantic: Environments

THE CENTRAL AMERICAN ISTHMUS. The emergence of the CAI may have begun to affect ocean circulation between the Atlantic and Pacific in the Middle to Late Miocene; deep water circulation was blocked no later than 3.6 Ma and shallow water circulation was blocked no later than about 3.0 Ma (Coates et al. 1992; Coates and Obando 1996; Collins et al. 1996). The emergence of the CAI led to separation and differentiation of Atlantic and Pacific water masses (Woodruff and Savin 1989; Wright, Miller, and Fairbanks 1991) and to changes in circulation in the Western Atlantic (Allmon et al. 1996a) and perhaps also in the Eastern Pacific (Weaver 1990). Present-day intermediate-depth Atlantic water is younger, more estuarine-influenced, and relatively nutrient-poor, whereas Pacific water is older, more lagoon-influenced, and relatively nutrient-rich. Present-day circulation in the Caribbean, Gulf of Mexico, and along the east coast of the United States is stronger, and oceanic upwelling may be weaker, than before formation of the CAI, whereas upwelling may be stronger on the west coast of Central America. (See Allmon et al. 1996a, for further discussion.)

TEMPERATURE. There is evidence of significant Northern Hemisphere ice at 3.0–3.4 Ma, but almost all available data indicate that a relatively small ice volume at this time compared to the ice buildup began around 2.5 Ma (references in Allmon et al. 1996b). A growing body of data suggests that sea surface temperatures in the North Atlantic between 3.5 and 3.0 Ma were higher than at present (e.g. Cronin and Dowsett 1996 and references therein). Although the pulse of Northern Hemisphere glaciation that began around 2.5 Ma appears to have been much more significant than earlier Neogene events, it remains not clear whether low latitude sea surface temperatures declined at this time. Specifically, on the Atlantic coast of North America, there is good evidence for cooling at around 2.5 Ma north of Cape Hatteras, but no unequivocal evidence for significant cooling south of there (Allmon et al. 1996a,b).

NUTRIENT SUPPLY. Several authors (e.g., Woodring 1966; Vermeij 1980, 1987a,b, 1989, 1990; Stanley 1986a) have suggested changes in the nutrient regimes (and resulting productivity) in the Western Atlantic as an explanation for the extinction of Neogene mollusks in the region. Good circumstantial evidence exists that (at least) local coastal upwelling and associated high productivity existed prior to around 3.0 Ma in the low-latitude Western Atlantic and then declined (Allmon et al. 1996a). In contrast, coastal upwelling and productivity appear to have changed little in the low-latitude Eastern Pacific. Evidence for upwelling in the Western Atlantic prior to 3.0 Ma includes (1) local areas of cooler temperatures in an otherwise warm Late Pliocene (Cronin and Dowsett 1990, 1996; Cronin 1991); (2) vertebrate and invertebrate faunal indicators of cool waters or high productivity, or both, amidst otherwise subtropical temperatures in the Late Pliocene Pinecrest Beds of Florida (Allmon et al. 1996a); (3) carbon and oxygen isotopic evidence (Jones and Allmon 1995; Allmon et al. 1996a); (4) widespread phosphogenesis in Florida and the Carolinas throughout the Miocene and into the Early Pliocene (Riggs 1984; Allmon et al. 1996a); (5) much more common accumulation of biogenic silica in the Atlantic prior to 10–11 Ma than at present, indicating higher productivity (Keller and Barron 1983).

A Model for Faunal Evolution in the Late Cenozoic Western Atlantic

If we focus on changes in temperature and nutrient supply and ask whether and how these environmental variables might have affected origination and extinction in late Cenozoic marine faunas of the Western Atlantic, we can use the three-stage framework to construct a simple model connecting environmental and evolutionary change (Allmon, in press; figure 7.3).

Faunal data clearly indicate that both speciation and extinction were ongoing processes in the Late Pliocene Western Atlantic. It is therefore reasonable to search for causal factors that might be connected to both processes. The model presented in figure 7.3 suggests a framework for such a search. Changes in nutrient conditions in the region at this time may have increased rates of isolate formation, and thus speciation, in some taxa, while increasing rates of extinction in other taxa. Some mollusk and coral clades show both enhanced origination and extinction, although it may never be possible to determine which occurred first. If temperature did decrease in mid-low latitudes, this may also have contributed to faunal changes in much the same way. The lack of compelling evidence for temperature change, however, implicates change in nutrient conditions as the environmental factor most closely connected to these faunal patterns.

As more and more data on particular clades become available, they can be analyzed using this model to determine exactly what kinds of processes may have occurred and when they occurred. This is a very different kind of analysis than simply attributing a "regional mass extinction" to temperature decrease and subsequent or approximately synchronous speciation to a black box called "diversification."

Conclusions

There is perhaps no issue more central to an understanding of the relationship between ecology and evolution than the problem of scale. Do processes and phenomena that occur and are observable on ecological timescales have the same kinds of effects when applied over geological timescales? If not, what are the effects that emerge at these longer timescales that would not be expected from examining the shorter timescales alone? What processes led to these faunal changes? We believe that these and many other similar cases can be usefully approached using an explicit analysis of the stages of the speciation process, and that when they are it will be clear that changes in nutrient conditions have important effects on the evolutionary process; that they are, in fact, one means by which ecological processes scale up to qualitatively different macroevolutionary processes.

Nutrients do not provide just a passive "backdrop" for evolution; they can affect the evolutionary process directly by affecting, or even controlling, the processes of speciation and extinction. They can affect speciation in separate but related ways; nutrient supply can affect the probability of the formation of isolated populations, the persistence of those populations, and the genetic differentiation of those populations. A given taxon may respond differently at different times, depending on the environmental conditions. Different taxa may respond differently based on their food requirements, among other factors. The point is that we have at least the potential to analyze episodes of

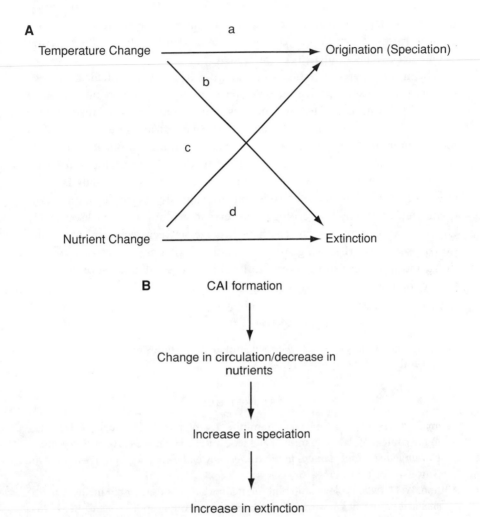

FIGURE 7.3. (A) Simple model of possible effects of change in temperature and nutrient supply on origination and extinction in the late Cenozoic of the Western Atlantic. (a) Possible effects of temperature change on species origination (e.g., creation of isolates by local extirpation and fragmentation of geographic range, especially of thermophilic taxa); (b) Possible effects of temperature change on species extinction (e.g., through cooling and selective killing off of thermophiles); (c) Possible effects of nutrient change on species origination (e.g., creation of isolates by local extirpation and fragmentation of geographic range, especially of nutriphilic taxa); (d) Possible effects of nutrient change on species extinction (e.g., through selective killing off of nutriphiles) (Allmon, in press). (B) Simple model for effects of formation of the Central American Isthmus (CAI) on origination and extinction in the late Cenozoic of the Western Atlantic (Allmon, in press).

diversification in much greater depth than we have before by breaking the speciation process down into its component stages and analyzing the potential and actual effects of nutrients on each stage.

The nutrient paradox results from the differing scales of ecological and evolutionary processes. Speciation, in general, does not happen on ecological timescales. It can be affected by processes that do, but it may be misleading to extrapolate smoothly from one to the other. High nutrient levels may produce locally lower taxonomic diversity in ecological communities, but at least among some taxa, may also lead to higher taxonomic diversity in evolutionary time.

Because energy flow is so important in ecological systems, analysis of the effects of energy flow in evolution is an important step in forging a more adequate view of exactly how ecology "matters" in evolution (cf., Jackson 1988). Unless we apply an explicit mechanistic linkage between ecological and evolutionary scales, we risk being stuck in making broad and untestable correlations, and in not exploring more thoroughly this essential question in evolutionary biology.

ACKNOWLEDGMENTS

We are grateful to J. Sepkoski, G. Vermeij, and an anonymous reviewer for comments on earlier drafts of the manuscript.

REFERENCES

Adams, C. G. 1990. Neogene larger foraminifera, evolutionary and geological events in the context of datum planes. In N. Ikebe and R. Tsuchi, eds., *Pacific Neogene Datum Planes: Contributions to Biostratigraphy and Chronology,* pp. 47–67. Tokyo: University of Tokyo Press.

Allmon, W. D. 1988. Ecology of Recent turritelline gastropods (Prosobranchia, Turritellidae): Current knowledge and paleontological implications. *Palaios* 3:259–284.

Allmon, W. D. 1992a. A causal analysis of stages in allopatric speciation. *Oxford Surveys in Evolutionary Biology* 8:219–257.

Allmon, W. D. 1992b. Role of temperature and nutrients in extinction of turritelline gastropods: Cenozoic of the northwestern Atlantic and northeastern Pacific. *Palaeogeography, Palaeoclimatology, Palaeoecology* 92:41–54.

Allmon, W. D. 1994. Taxic evolutionary paleoecology and the ecological context of macroevolutionary change. *Evolutionary Ecology* 8:95–112.

Allmon, W. D. In press. Nutrients, temperature, disturbance, and evolution: a model for the late Cenozoic marine record of the Western Atlantic. Palaeogeography, Palaeoclimatology, Palaeoecology.

Allmon, W. D., J. G. Carter, P. H. Kelley, and J. Schneider. 1991. Evolutionary dynamics of the Oligocene-Miocene molluscan radiation event in the western Atlantic region. *Geological Society of America Abstracts with Programs* 23(5):163

Allmon, W. D., D. H. Erwin, R. M. Linsley, and P. J. Morris. 1992. *Trophic level and evolution in Paleozoic gastropods*, p. 3. Fifth North American Paleontological Convention, Abstracts and Program. The Paleontological Society Special Publication No. 6. Knoxville TN: The Paleontological Society.

Allmon, W. D., G. Rosenberg, R. W. Portell, and K. Schindler. 1993. Diversity of Pliocene to Recent Atlantic coastal plain mollusks. *Science* 260:1626–1628.

Allmon, W. D., S. D. Emslie, D. S. Jones, and G. S. Morgan. 1996a. Late Neogene oceanographic change along Florida's west coast: Evidence and mechanisms. *Journal of Geology* 104:143–162.

Allmon, W. D., G. Rosenberg, R. W. Portell, and K. Schindler. 1996b. Diversity of Pliocene-Recent mollusks in the Western Atlantic: Extinction, origination, and environmental change. In J. B. C. Jackson, A. F. Budd, and A. G. Coates, eds. *Evolution and Environment in Tropical America*, pp. 271–302. Chicago: University of Chicago Press.

Allmon, W. D., P. J. Morris, and M. L. McKinney. 1998. An intermediate disturbance hypothesis of maximal speciation. In M. L. McKinney and J. A. Drake, eds., *Biodiversity Dynamics: Turnover of Populations, Taxa, and Communities*, pp. 349–376. New York: Columbia University Press.

Alongi, D. M. 1990. Abundances of benthic microfauna in relation to outwelling of mangrove detritus in a tropical coastal region. *Marine Ecology Progress Series* 63:53–63.

Andrews, J. C. 1983. Thermal waves on the Queensland shelf. *Australian Journal of Marine and Freshwater Research* 34:81–96.

Andrews, J. C. and P. Gentien. 1982. Upwelling as a source of nutrients for the Great Barrier Reef ecosystems: A solution to Darwin's questions? *Marine Ecology Progress Series* 8:257–269.

Andrie, C., I. Bouloubassi, H. Cornu, R. Fichez, C. Pierre, and F. Rougerie. 1992. Chemical and tracer studies in coral reef interstitial waters (French Polynesia): Implications for endo-upwelling circulation. *Proceedings of the Seventh International Coral Reefs Symposium*, Guam, pp. 1–12.

Aronson, R. B. 1994. Scale-independent biological processes in the marine environment. *Annual Review of Oceanography and Marine Biology* 32:435–460.

Arthur, M. A., J. C. Zachos, and D. S. Jones. 1987. Primary productivity and the Cretaceous/Tertiary boundary event in the oceans. *Cretaceous Research* 8:43–45.

Arx, W. S. von 1954. *Circulation systems of Bikini and Rongelap lagoons*, pp. 265–273. U.S. Geol. Surv. Prof. Paper 260B. Washington DC: U.S. Government Printing Office.

Atkinson, M. J., S. V. Smith, and E. D. Stroup. 1981. Circulation in Enewetak Atoll lagoon. *Limnology and Oceanography* 26:1074–1083.

Bambach, R. K. 1993. Seafood through time: Changes in biomass, energetics, and productivity in the marine ecosystem. *Paleobiology* 19(3):372–397.

Begon, M., J. L. Harper, and C. R. Townsend. 1996. *Ecology: Individuals, Populations, and Communities.* Boston: Blackwell Scientific Publications.

Bennett, K. D. 1990. Milankovitch cycles and their effects on species in ecological and evolutionary time. *Paleobiology* 16(1):11–21

Birkeland, C. 1987 Nutrient availability as a major determinant of differences among coastal hard-substratum communities in different regions of the tropics. *UNESCO Reports in Marine Science* 46:43–98.

Boucher, G. and J. Clavier. 1990. Contribution of benthic biomass to overall metabolism in New Caledonia sediments. *Marine Ecology Progress Series* 64:271–280.

Bramlette, M. N. 1965. Massive extinctions in biota at the end of Mesozoic time. *Science* 148:1696–1699.

Brasier, M. D. 1991. Nutrient flux and the evolutionary explosion across the Precambrian-Cambrian boundary interval. *Historical Biology* 5:85–93.

Brasier, M. D. 1992a. Nutrient-enriched waters and the early skeletal fossil record. *Journal of the Geological Society of London* 149:621–629.

Brasier, M. D. 1992b. Paleoceanography and changes in the biological cycling of phosphorous across the Precambrian-Cambrian boundary. In J. H. Lipps and P. W. Bengston, eds., *Origin and Early Evolution of the Metazoa,* pp. 483–523. New York: Plenum Press.

Brasier, M. D. 1995. Fossil indicators of nutrient levels: II. Evolution and extinction in relation to oligotrophy. In D. W J. Bosence and P. A. Allison, eds., *Marine Paleoenvironmental Analysis from Fossils,* pp. 133–150. Geological Society Special Publication No. 83. London: The Geological Society.

Brasier, M. D., R. M. Corfield, L. A. Derry, A. Y. Rozanov, and A. Y. Zhuravlev. 1994. Multiple 13C excursions spanning the Cambrian explosion to the Botomian crisis in Siberia. *Geology* 22:455–458.

Budd, A. F., K. G. Johnson, and T. A. Stemann. 1996. Plio-Pleistocene turnover and extinctions in the Caribbean reef-coral fauna. In J. B. C. Jackson, A. F. Budd, and A. G. Coates, eds., *Evolution and Environment in Tropical America,* pp. 168–204. Chicago: University of Chicago Press.

Butterfield, N. J. 1997. Plankton ecology and the Proterozoic-Phanerozoic transition. *Paleobiology* 23(2):247–262

Caldeira, K. and M. R. Rampino. 1993. Aftermath of the end-Cretaceous mass extinction: Possible biogeochemical stabilization of the carbon cycle and climate. *Paleoceanography* 8(4):515–525.

Carter, J. G. and P. H. Kelley. 1989. Biotic provinciality, nutrient levels, and the Miocene molluscan radiation event. *Geological Society of America Abstracts with Programs* 21(6):A111–112.

Charpy, L. and C. J. Charpy-Roubaud. 1990. Trophic structure and productivity of the lagoonal communities of Tikehau atoll (Tuamotu Archipelago, French Polynesia). *Hydrobiologia* 207:43–52.

Chevalier, J.-P. 1977. Apercu sur la faune corallienne recifale du Neogene. *Proceedings of the Second International Symposium on Corals and Fossil Coral Reefs*, pp. 359–366.

Clarke, A. 1993. Temperature and extinction in the sea: A physiologist's view. *Paleobiology* 19:499–518.

Coates, A. G. and J. A. Obando. 1996. The geologic evolution of the Central American Isthmus. In J. B. C. Jackson, A. F. Budd, and A. G. Coates, eds., *Evolution and Environment in Tropical America*, pp. 21–56. Chicago: University of Chicago Press.

Coates, A. G., J. B. C. Jackson, L. S. Collins, T. M. Cronin, H. J. Dowsett, L. M. Bybell, P. Jung, and J. A. Obando. 1992. Closure of the Isthmus of Panama: The nearshore marine record of Costa Rica and western Panama. *Geological Society of America Bulletin* 104:814–828.

Collins, L. S., A. G. Coates, W. A. Berggren, M. P. Aubry, and J. Zhang. 1996. The late Miocene Panama isthmian strait. *Geology* 24(8):687–690.

Cook, P. J. 1992. Phosphogenesis around the Proterozoic-Phanerozoic transition. *Journal of the Geological Society of London* 149:615–620.

Cook, P. J. and J. H. Shergold. 1984. Phosphorus, phosphorites and skeletal evolution at the Precambrian-Cambrian boundary. *Nature* 308:231–236.

Corfield, R. M. and N. J. Shackleton. 1988. Productivity change as a control on planktonic foraminiferal evolution after the Cretaceous/Tertiary boundary. *Historical Biology* 1(4):323–344.

Cronin, T. M. 1985. Speciation and stasis in marine Ostracoda: Climatic modulation of evolution. *Science* 227:60–63.

Cronin, T. M. 1987. Speciation and cyclic climate change. In M. R. Rampino, J. E. Sanders, W. S. Newman, and C. K. Kšnigsson, eds., *Climate: History, Periodicity, and Predictability*, pp. 333–342. New York: Van Nostrand Publishing Company.

Cronin, T. M. 1991. Pliocene shallow water paleoceanography of the North Atlantic Ocean based on marine ostracodes. *Quaternary Science Reviews* 10:175–188.

Cronin, T. M. and H. J. Dowsett. 1990. A quantitative micropaleontologic method for shallow marine paleoclimatology: application to Pliocene deposits of the western North Atlantic Ocean. *Marine Micropaleontology* 16:117–147.

Cronin, T. M. and N. Ikeya. 1990. Tectonic events and climatic change: Opportunities for speciation in Cenozoic marine Ostracoda. In R. Ross and W. Allmon, eds., *Causes of Evolution: A Paleontological Perspective*, pp. 210–248. Chicago: University of Chicago Press.

Cronin, T. M. and C. E. Schneider. 1990. Climatic influences on species: Evidence from the fossil record. *Trends in Ecology and Evolution* 5(9):275–279.

Cronin, T. M. and H. Dowsett. 1996. Biotic and oceanographic response to the Pliocene closing of the Central American Isthmus. In J. B. C. Jackson, A. F. Budd, and A. G. Coates, eds., *Evolution and Environment in Tropical America*, pp. 76–104. Chicago: University of Chicago Press.

Cronin, T. M., L. M. Bybell, E. M. Brouwers, T. G. Gibson, R. Margerum, and R. Z. Poore. 1991. Neogene biostratigraphy and paleoenvironments of Enewetak Atoll, equatorial Pacific Ocean. *Marine Micropaleontology* 18:101–114.

Crosslands, C. J., B. G. Hatcher, and S. V. Smith. 1991. Role of coral reefs in global ocean production. *Coral Reefs* 10:55–64.

Dall, W. H. 1892. Contributions to the Tertiary fauna of Florida. *Wagner Free Institute of Science Transactions* 3(2):201–473.

DeAngelis, D. L. 1992. *Dynamics of Nutrient Cycling and Food Webs.* London: Chapman and Hall.

Delaney, M. L. 1990. Miocene benthic foraminiferal Cd/Ca records: South Atlantic and western equatorial Pacific. *Paleoceanography* 5:743–760.

Delaney, M. L. and G. M. Filippelli. 1994. An apparent contradiction in the role of phosphorus in Cenozoic chemical mass balances for the world ocean. *Paleoceanography* 9:512–526.

Delgado, O. and B. E. LaPointe. 1994. Nutrient-limited productivity of calcareous versus fleshy macroalgae in a eutrophic, carbonate-rich tropical marine environment. *Coral Reefs* 13:151–159.

D'Elia, C. F. and W. J. Wiebe. 1990. Biogeochemical nutrient cycles in coral reef ecosystems. In S. Dubinsky, ed., *Coral Reefs*, pp. 49–74. Amsterdam: Elsevier.

D'Elia, C. F., K. L. Webb, and J. W. Porter. 1981. Nitrate-rich groundwater inouts to Discovery Bay, Jamaica: A significant source of N to local coral reefs? *Bulletin of Marine Science* 31:903–910.

D'Hondt, S., P. Donaghay, J. C. Zachos, D. Luttenberg, and M. Lindinger. 1998. Organic carbon fluxes and ecological recovery from Cretaceous–Tertiary mass extinction. *Science* 282:276–279.

Domning, D. P. 1981. Sea cows and sea grasses. *Paleobiology* 7(4):417–420.

Domning, D. P. 1982. Evolution of manatees: A speculative history. *Journal of Paleontology* 56(3):599–620.

Doty, M. S. and M. Oguri. 1956. The island mass effect. *Journal du Conseil, Conseil pour International pour l'Exploration de la Mer* 22:33–37.

Edinger, E. N. and M. J. Risk. 1994. Oligocene-Miocene extinction and geographic restriction of Caribbean corals: Roles of turbidity, temperature, and nutrients. *Palaios* 9:576–598.

Eldredge, N. 1982. Phenomenological levels and evolutionary rates. *Systematic Zoology* 31:338–347.

Elton, C. 1927. *Animal Ecology.* New York: Macmillan.

Emslie, S. D. and G. S. Morgan. 1994. A catastrophic death assemblage and paleoclimatic implications of Pliocene seabirds of Florida. *Science* 264:684–685.

Fagerstrom, J. A. 1987. *The Evolution of Reef Communities.* New York: John Wiley and Sons.

Falkowski, P. G., Z. Dubinsky, L. Muscatine, and L. McCloskey. 1993. Population controls in symbiotic corals. *BioScience* 43:606–611.

Fourqurean, J. W., J. C. Zieman, and G. V. N. Powell. 1992. Phosphorus limitation of primary production in Florida Bay: Evidence from C:N:P ratios of the dominant seagrass *Thalassia testudinum. Limnology and Oceanography* 37:162–171.

Furnas, M. J., A. W. Mitchell, M. Gilmartin, and N. Revelante. 1990. Phytoplankton biomass and primary production in semi-enclosed reef lagoons of the central Great Barrier Reef, Australia. *Coral Reefs* 9:1–10.

Glynn, P. W. and R. H. Stewart. 1973. Distribution of coral reefs in the Pearl Islands (Gulf of Panama) in relation to thermal conditions. *Limnology and Oceanography* 18:367–378.

Grassle, J. F. 1989. Species diversity in deep-sea communities. *TREE* 4:12–15.

Grassle, J. F. 1991. Deep-sea benthic biodiversity. *BioScience* 41:464–469.

Grassle, J. F., and N. J. Maciolek. 1992. Deep-sea species richness: Regional and local diversity estimates from quantitative bottom samples. *American Naturalist* 139:313–341.

Hallock, P. 1987. Fluctuations in the trophic resource continuum: A factor in global diversity cycles? *Paleoceanography* 2(5):457–471.

Hallock, P. 1988. Interoceanic differences in Foraminifera with symbiotic algae: A result of nutrient supplies? *Proceedings of the Sixth International Coral Reef Symposium*, Townsville, Australia, v. 3, pp. 251–255.

Hallock, P. and W. Schlager. 1986. Nutrient excess and the demise of coral reefs and carbonate platforms. *Palaios* 1:389–398.

Hamner, W. M. and I. R. Hauri. 1981. Effects of island mass: Water flow and plankton pattern around a reef in the Great Barrier Reef lagoon, Australia. *Limnology and Oceanography* 26:1084–1102.

Hargrave, B. T. 1980. Factors affecting the flux of organic matter to sediments in a marine bay. In K. R. Tenore and B. C. Coull, eds., *Marine Benthic Dynamics*. The Belle W. Baruch Library in Marine Science No. 11. Columbia SC: University of South Carolina Press.

Hiatt, R. W. and D. W. Strasburg. 1960. Ecological relationships of the fish fauna on coral reefs of the Marshall Islands. *Ecological Monographs* 30:65–127.

Holden, J. C. 1976. *Late Cenozoic Ostracoda from Midway Island drill holes*. U.S. Geol. Surv. Prof. Paper 680-F. Washington DC: U.S. Government Printing Office.

Hollander, D. J., J. A. McKenzie, and K. J. Hsü. 1993. Carbon isotope evidence for unusual plankton blooms and fluctuations of surface water CO_2 in "Strangelove Ocean" after terminal Cretaceous event. *Palaeogeography, Palaeoclimatology, Palaeoecology* 104:229–237.

Howarth, R. W. 1988. Nutrient limitation of net primary productivity in marine ecosystems. *Annual Review of Ecology and Systematics* 19:90–110.

Howarth, R. W. 1993. The role of nutrients in coastal waters. In *Managing Wastewater in Coastal Urban Areas*. Report from the National Research Committee on Wastewater Management for Coastal Urban Areas. Washington DC: National Academy Press.

Jablonski, D. and D. J. Bottjer. 1990. Onshore-offshore trends in marine invertebrate evolution. In R. M. Ross and W. D. Allmon, eds., *Causes of Evolution: A Paleontological Perspective*, pp. 21–75. Chicago: University of Chicago Press.

Jackscn, J. B. C. 1988. Does ecology matter? Review of Evolution and escalation, by G. J. Vermeij. *Paleobiology* 14(3):307–312.

Jackson, J. B. C. and A. F. Budd. 1996. Evolution and environment: Introduction and overview. In J. B. C. Jackson, A. F. Budd, and A. G. Coates, eds., *Evolution and Environment in Tropical America*, pp. 1–20. Chicago: University of Chicago Press.

Jackson, J. B. C., P. Jung, A. G. Coates, and L. S. Collins. 1993. Diversity and extinction of tropical American mollusks and emergence of the Isthmus of Panama. *Science* 260:1624–1626.

Jackson, J. B. C., P. Jung, and H. Fortunato. 1996. Paciphilia revisited: Transisthmian evolution of the *Strombina* group (Gastropoda: Columbellidae). In J. B. C. Jackson, A. F. Budd, and A. G. Coates, eds., *Evolution and Environment in Tropical America*, pp. 234–260. Chicago: University of Chicago Press.

Jeppson, L. 1990. An oceanic model for lithological and faunal changes tested on the Silurian record. *Journal of the Geological Society of London* 147:663–674.

Johnson, K. G., A. F. Budd, and T. A. Stemann. 1995. Extinction selectivity and ecology of Neogene Caribbean reef corals. *Paleobiology* 21(1):52–73.

Jones, D. S. and W. D. Allmon. 1995. Records of upwelling, seasonality, and growth in stable-isotope profiles of Pliocene mollusk shells from Florida. *Lethaia* 28:61–74.

Kauffman, E. G. 1972. *Evolutionary Rates and Patterns of North American Cretaceous Mollusca*, pp. 174–189. 24th International Geological Congress, Sect. 7.

Kauffman, E. G. 1977. Evolutionary rates and biostratigraphy. In E. G. Kauffman and J. E. Hazel, eds., *Concepts and Methods of Biostratigraphy*, pp. 109–142. Stroudsburg PA: Dowden, Hutchinson, and Ross.

Kay, E. A. 1980. Little worlds of the Pacific: An essay on Pacific basin biogeography. Lyon Arboretum Lecture 9, University of Hawaii, Honolulu.

Keen, A. M. 1971. *Sea Shells of Tropical West America*. Stanford CA: Stanford University Press.

Keller, G. and J. A. Barron. 1983. Paleoceanographic implications of Miocene deep-sea hiatuses. *Geological Society of America Bulletin* 94:590–613.

Kimmerer, W. J. and T. W. Walsh. 1981. Tawara Atoll Lagoon: Circulation, nutrient fluxes and the impact of human waste. *Micronesia* 17:161–179.

Kohn, A. J. 1985. Evolutionary ecology of Conus on Indo-Pacific coral reefs. *Proceedings of the Fifth International Coral Reef Congress*, Tahiti, v. 4, pp. 139–144.

Kohout, F. A. and M. C. Kolipinski. 1967. Biological zonation related to groundwater discharge along the shore of Biscayne Bay, Miami, Florida. In G. Lauff, ed., *Estuaries*, pp. 488–499. Washington DC: American Association for the Advancement of Science.

Ladd, H. S. 1960. Origin of the Pacific Island molluscan fauna. *American Journal of Science* 258-A:137–150.

Ladd, H. S. 1982. *Cenozoic fossil mollusks from western Pacific islands; Gastropods (Eulimidae and Volutidae through Terebridae)*, pp. 1–100. U.S. Geol. Surv. Prof. Paper 1171.

Layne, J. N. 1965. Observations on marine mammals in Florida waters. *Bulletin of the Florida State Museum, Biological Sciences* 9(4):131–181.

Levinton, J. S. 1974. Trophic group and evolution in bivalve mollusks. *Palaeontology* 17(3):579–585.

Levinton, J. S. 1996. Trophic group and the end-Cretaceous extinction: Did deposit feeders have it made in the shade? *Paleobiology* 22(1):104–112.

Lewis, J. B. 1987. Measurements of groundwater seepage flux onto coral reefs: Spatial and temporal variations. *Limnology and Oceanography* 32:1165–1169.

Lewis, J. B. 1989. Nutrients and the productivity of coral reef ecosystems. In M. M. Denis, ed., *Oceanologie: Actualite et Prospective*, pp. 367–387. Marseille: Centre d'Oceanologie de Marseille.

Lipps, J. H. and E. Mitchell. 1976. Trophic model for the adaptive radiations and extinctions of pelagic marine mammals. *Paleobiology* 2:147–155.

Littler, M. M. and D. S. Littler. 1984. Models of tropical reef biogenesis: The contribution of algae. In D. J. Round, ed., *Progress in Phycological Research*, pp. 323–364. Bristol: Biopress.

Littler, M. M., D. S. Littler, and E. A. Titlyanov. 1991. Comparison of N- and P-l imited productivity between high granitic islands versus low carbonate atolls in the Seychelles Archipelago: A test of the relative-dominance paradigm. *Coral Reefs* 10:199–209.

Marsh, J. A. 1977. Terrestrial inputs of nitrogen and phosphorus on fringing reefs of Guam. *Proceedings of the Fifth International Coral Reef Symposium*, Tahiti, v. 1, pp. 331–336.

Martin, R. E. 1995. Cyclic and secular variation in microfossil biomineralization: Clues to the biogeochemical evolution of Phanerozoic oceans. *Global and Planetary Change* 11:1–23

Martin, R. E. 1996. Secular increase in nutrient levels through the Phanerozoic: Implications for productivity, biomass, and diversity of the marine biosphere. *Palaios* 11:209–220.

Martin, R. E. 1998a. Catastrophic fluctuations in nutrient levels as an agent of mass extinction: Upward scaling of biological processes? In M. L. McKinney and J. A. Drake, eds., *Biodiversity Dynamics, Turnover of Populations, Taxa, and Communities*, pp. 405–429. New York: Columbia University Press.

Martin, R. E. 1998b. *One Long Experiment: Scale and Process in Earth History.* New York: Columbia University Press.

Matson, E. A. 1989. Biogeochemistry of Marianas Islands coastal sediments: Terrestrial influence of 13C, ash, $CaCO_3$, Al, Fe, Si, and P. *Coral Reefs* 7:153–160.

McKinney, M. L. and W. D. Allmon. 1995. Metapopulations and disturbance: From patch dynamics to biodiversity dynamics. In D. Erwin and R. Anstey, eds., *New Approaches to Speciation in the Fossil Record*, pp. 123–183. New York: Columbia University Press.

Muscatine, L., P. G. Falkowski, Z. Dubinsky, P. A. Cook, and L. R. McCloskey. 1989. The effect of external nutrient resources on the population dynamics of zooxanthellae in a reef coral. *Proceedings of the Royal Society of London*, B 236:311–326.

Nelson, G. and N. Platnick. 1981. *Systematics and Biogeography: Cladistics and Vicariance.* New York: Columbia University Press.

Newman, W. A. 1986. Origin of the Hawaiian marine fauna: Dispersal and vicariance as indicated by barnacles and other organisms. In K. L. Gore, ed., *Crustacean Biogeography*, pp. 21–49. Boston: Balkema.

Ogden, J. C. 1986. Comparison of the tropical western Atlantic (Caribbean) and the Indo-Pacific: Herbivore-plant interactions. *UNESCO Reports in Marine Science* 46:167–169.

Olsson, A. A. 1961. *Panamic-Pacific Pelecypoda*. Ithaca NY: Paleontological Research Institution.

Palumbi, S. R. 1992. Marine speciation on a small planet. *Trends in Ecology and Evolution* 7:114–118.

Parrish, J. D. 1986. Characteristics of fish communities on coral reefs and in potentially interacting shallow habitats in tropical oceans of the world. *UNESCO Reports in Marine Science* 46:171–218.

Paul, C. R. C. and S. F. Mitchell. 1994. Is famine a common factor in marine mass extinctions? *Geology* 22:679–682.

Paulay, G. 1990. Effects of late Cenozoic sea-level fluctuations on the bivalve faunas of tropical oceanic islands. *Paleobiology* 16:415–434.

Petuch, E. J. 1981. A relict Neogene caenogastropod fauna from northern South America. *Malacologia* 20:307–348.

Potts, D. C. 1984. Generation times and Quaternary evolution of reef-building corals. *Paleobiology* 10:48–58.

Potts, D. C. 1985. Sea level fluctuations and speciation in Scleractinia. *Proceedings of the Fifth Coral Reef Congress*, Tahiti, v. 4, pp. 127–132.

Pujos, M. and O. Javelaud. 1992. Late Quaternary carbonate sedimentation on the lower Guajira shelf, Colombia. *Marine Geology* 104:153–164.

Purton, L. and M. Brasier. 1997. Gastropod carbonate d18O and d13C values record strong seasonal productivity and stratification shifts during the late Eocene in England. *Geology* 25(10):871–874.

Rea, D. K., J. C. Zachos, R. M. Owen, and P. D. Gingerich. 1990. Global change at the Paleocene-Eocene boundary: Climatic and evolutionary consequences of tectonic events. *Palaeogeography, Palaeoclimatology, Palaeoecology* 79:117–128.

Rex, M. A., C. T. Stuart, R. R. Hessler, A. A. Allen, H. L. Sanders, and G. D. F. Wilson. 1993. Global-scale latitudinal patterns of species diversity in the deep-sea benthos. *Nature* 365:636–639.

Rex, M. A., R. J. Etter, and C. T. Stuart. 1997. Large-scale patterns of species diversity in the deep-sea benthos. In R. F. G. Ormond, J. D. Gage, and M. V. Angel, eds., *Marine Biodiversity: Patterns and Processes*, pp. 94–121. New York: Cambridge University Press.

Ricklefs, R. E. 1990. *Ecology*, 2nd ed. New York: W. H. Freeman.

Riebesell, U., D. A. Wolf-Gladrow, and V. Smetacek. 1993. Carbon dioxide limitation of marine phytoplankton growth rates. *Nature* 361:249–251.

Riggs, S. R. 1984. Paleoceanographic model of Neogene phosphorite deposition, U.S. Atlantic continental margin. *Science* 223:123–131.

Risk, M. J. and P. W. Sammarco. 1991. Cross-shelf trends in skeletal density of the massive coral *Porites lobata* from the Great Barrier Reef. *Marine Ecology Progress Series* 69:195–200.

Risk, M. J., P. W. Sammarco, and H. P. Schwarcz. 1994. Cross-continental shelf trends in 13C in coral on the Great Barrier Reef. *Marine Ecology Progress Series* 106:121–130.

Roopnarine, P. D. 1996. Systematics, biogeography and extinction of chionine bivalves (Bivalvia: Veneridae) in tropical America: Early Oligocene-Recent. *Malacologia* 38:103–142.

Rosenzweig, M. L. and Z. Abramsky. 1993. How are diversity and productivity related? In R. Ricklefs and D. Schluter, eds., *Species Diversity in Ecological Communities: Historical and Geographical Perspectives*, pp. 52–65. Chicago: University of Chicago Press.

Ross, R. M. 1990. The evolution and biogeography of Micronesian ostracodes. Ph.D. thesis, Harvard University, Cambridge MA.

Rougerie, R. and B. Wauthy. 1986. Le concept d'endo-upwelling dansle fonctionnement des atoll-oasis. *Oceanologica Acta* 9:133–148.

Rougerie, F., B. Wauthy, and C. Andrie. 1990. Geothermal endo-upwelling model testing for an atoll and high island barrier reef. *Proceedings of the International Society of Reef Studies*, Nouméa, New Caledonia, pp. 197–202.

Roy, J. M., M. A. S. McMenamin, and S. E. Alderman. 1990. Trophic differences, originations and extinctions during the Cenomanian and Maastrichtian stages of the Cretaceous. In E. G. Kauffman and O. H. Walliser, eds., *Extinction Events in Earth History*, pp. 299–303. Berlin: Springer-Verlag.

Ryther, J. H. 1969. Photosynthesis and fish production in the sea. *Science* 166:72–76.

Ryther, J. H. and W. M. Dunstan. 1971. Nitrogen, phosphorus, and eutrophication in the coastal marine environment. *Science* 171:1008–1013.

Sammarco, P. W. and M. J. Risk. 1990. Large-scale patterns in internal bioerosion of *Porites*: Cross continental shelf trends on the Great Barrier Reef. *Marine Ecology Progress Series* 59:145–156.

Sanders, F. 1981. A preliminary assessment of the main causative mechanisms of the "island mass" effect of Barbados. *Marine Biology* 64:199–205.

Sheehan, P. M. and T. A. Hansen. 1986. Detritus feeding as a buffer to extinction at the end of the Cretaceous. *Geology* 14:868–870.

Short, F. T., W. C. Dennison, and D. G. Capone. 1990. Phosphorus limited growth of the tropical seagrass *Syringodium filiforme* in carbonate sediments. *Marine Ecology Progress Series* 62:169–174.

Smayda, T. J. 1989. Primary production and the global epidemic of phytoplankton blooms in the sea: A linkage? In E. M. Cosper, E. J. Carpenter, and V. M. Bricelk, eds., *Novel Phytoplankton Blooms: Causes and Impacts of Recurrent Brown Tides and Other Unusual Blooms*, pp. 449–483. Lecture Notes on Coastal and Estuarine Studies. Berlin: Springer-Verlag.

Smith, A. B. and C. H. Jeffery. 1998. Selectivity of extinction among sea urchins at the end of the Cretaceous period. *Nature* 392:69–71.

Smith, S. V. and M. J. Atkinson. 1984. Phosphorus limitation of net production in a confined aquatic ecosystem. *Nature* 207:626–627.

Smith, S. V., W. J. Kimmerer, E. A. Laws, R. E. Brock, and T. W. Walsh. 1981. Kaneohe Bay sewage diversion experiment: Perspectives on ecosystem responses to nutritional perturbation. *Pacific Science* 35:279–395.

Stanley, S. M. 1986a. Anatomy of a regional mass extinction: Plio-Pleistocene decimation of the western Atlantic bivalve fauna. *Palaios* 1:17–36.

Stanley, S. M. 1986b. Population size, extinction, and speciation: The fission effect in Neogene Bivalvia. *Paleobiology* 12:89–110.

Stanley, S. M. 1990. The general correlation between rate of speciation and rate of extinction: Fortuitous causal linkages. In R. M. Ross and W. D. Allmon, eds., *Causes of Evolution. A Paleontological Perspective*, pp. 103–127. Chicago: University of Chicago Press.

Stanley, S. M. and L. D. Campbell. 1981. Neogene mass extinction of western Atlantic molluscs. *Nature* 293:457–459.

Stanley, S. M. and W. F. Ruddiman. 1995. Neogene ice age in the North Atlantic region: Climatic changes, biotic effects, and forcing factors. In A. H. Knoll and S. M. Stanley, eds., *Effects of Past Global Change on Life*, pp. 118–133. Washington DC: National Academy Press.

Suzuki, R. 1997. Distribution of shallow marine epiphytal ostracodes from northern Australia and Okinawa, Japan, and implications for dispersal. Thesis, Shizuoka University, Japan.

Tappan, H. 1968. Primary production, isotopes, extinctions, and the atmosphere. *Palaeogeography, Palaeoclimatology, Palaeoecology* 4:187–210.

Tappan, H. 1986. Phytoplankton: Below the salt at the global table. *Journal of Paleontology* 60(3):545–554.

Taylor, J. D. 1997. Diversity and structure of tropical Indo-Pacific benthic communities: Relation to regimes of nutrient input. In R. F. G. Ormond, J. D. Gage, and M. V. Angel, eds., *Marine Biodiversity: Patterns and Processes*, pp. 178–200. New York: Cambridge University Press.

Teeter, J. W. 1973. Geographic distribution and dispersal of some shallow-water marine Ostracoda. *Ohio Journal of Science* 73(1):46–54.

Tenore, K. R. and D. L. Rice. 1980. A review of trophic factors affecting secondary production of deposit-feeders. In K. R. Tenore and B. C. Coull, eds., *Marine Benthic Dynamics*, pp. 325–340. The Belle W. Baruch Library in Marine Science No. 11. Columbia SC: University of South Carolina Press.

Tucker, M. E. 1992. The Precambrian-Cambrian boundary: Seawater chemistry, ocean circulation and nutrient supply in metazoan evolution, extinction and biomineralization. *Journal of the Geological Society of London* 149:655–668.

Valentine, J. W. 1971. Resource supply and species diversity patterns. *Lethaia* 4:51–61.

Valentine, J. W. 1973. *Evolutionary Paleoecology of the Marine Biosphere*. Englewood Cliffs NJ: Prentice Hall.

Valentine, J. W. 1984. Climate and evolution in the shallow sea. In P. Brenchley, ed., *Fossils and Climate*, pp. 265–277. Chichester: John Wiley and Sons.

Valentine, J. W. and D. Jablonski. 1983. Speciation in the shallow sea: General patterns and biogeographic controls. In R. W. Sims, J. H. Price, and P. E. S. Whalley, eds., *Evolution, Time, and Space: The Emergence of the Biosphere*, pp. 201–226. London: Academic Press.

Valiela, I. 1984. *Marine Ecological Processes*. Berlin: Springer-Verlag.

Vermeij, G. J. 1978. *Biogeography and Adaptation*. Cambridge MA: Harvard University Press.

Vermeij, G. J. 1987a. Interoceanic differences in architecture and ecology: The effects of history and productivity. *UNESCO Reports in Marine Science* 46:105–125.

Vermeij, G. J. 1987b. *Evolution and Escalation*. Princeton NJ: Princeton University Press.

Vermeij, G. J. 1989. Interoceanic differences in adaptation: Effects of history and productivity. *Marine Ecology Progress Series* 57:293–305.

Vermeij, G. J. 1990. Tropical Pacific pelecypods and productivity: A hypothesis. *Bulletin of Marine Science* 47(1):62–67.

Vermeij, G. J. 1991a. When biotas meet: Understanding biotic interchange. *Science* 253:1099–1104.

Vermeij, G. J. 1991b. Anatomy of an invasion: The Trans-Arctic interchange. *Paleobiology* 17(3):281–307.

Vermeij, G. J. 1995. Economics, volcanoes, and Phanerozoic revolutions. *Paleobiology* 21(2):125–152.

Vermeij, G. J. and E. J. Petuch. 1986. Differential extinction in tropical American molluscs: Endemism, architecture, and the Panama land bridge. *Malacologia* 27:29–41.

Veron, J. E. N. and R. Kelley. 1988. Species stability in reef corals of Papua New Guinea and Indo-Pacific. *Memoirs of the Association of Australasian Palaeontologists* 6:1–69.

Vitousek, P. M. and R. W. Howarth. 1991. Nitrogen limitation on land and in the sea: How can it occur? *Biogeochemistry* 13:87–115.

Wang, K., H. H. J. Geldsetzer, and H. R. Krouse. 1994. Permian–Triassic extinction: Organic δ^{13}C evidence from British Columbia, Canada. *Geology* 22:580–584.

Weaver, A. J. 1990. Ocean currents and climate. *Nature* 347:432.

Weissleader, L. S., N. L. Gilinsky, R. M. Ross, and T. M. Cronin. 1989. Biogeography of marine podocopid Ostracoda from the western Equatorial Pacific. *Journal of Biogeography* 16:103–114.

Wiebe, W. J. 1987. Nutrient pools and dynamics in tropical, marine, coastal environments, with special reference to the Caribbean and Indo-West Pacific regions. *UNESCO Reports in Marine Science* 46:19–42.

Wilkinson, C. R. 1994. *The Nutritional Spectrum of Coral Reef Benthos*. Woods Hole MA: Woods Hole Oceanographic Institute.

Woodring, W. P. 1966. The Panama land bridge as a sea barrier. *Proceedings of the American Philosophical Society* 110:425–433.

Woodruff, F. and S. M. Savin. 1989. Miocene deepwater oceanography. *Paleoceanography* 4:87–140.

Wright, J. D., K. G. Miller, and R. G. Fairbanks. 1991. Evolution of modern deepwater circulation: Evidence from the Late Miocene Southern Ocean. *Paleoceanography* 6:275–290.

Zachos, J. C., M. A. Arthur, and W. E. Dean. 1989. Geochemical evidence for suppression of pelagic marine productivity at the Cretaceous/Tertiary boundary. *Nature* 337:61–64.

8

The Role of Ecological Interactions in the Evolution of Naticid Gastropods and Their Molluscan Prey

Patricia H. Kelley and Thor A. Hansen

DURING THE 25 YEARS since Eldredge and Gould (1972) proposed the hypothesis of punctuated equilibrium, many case studies have examined the tempo and mode of evolution of particular taxonomic groups. Although results vary among groups, stasis and punctuational change have been documented within the history of many benthic marine invertebrates (Gould and Eldredge 1993). The dominance of stasis in some lineages has raised the question of whether ecological interactions play a significant role in the evolutionary process (Gould 1985, 1990; Allmon 1992, 1994). In addition, the view that mass extinctions undo trends that may accumulate at lower "tiers" further questions the role of ecology in evolution (Gould 1985; Allmon 1992).

Predator–prey systems provide an ideal context in which to address the role of ecological interactions in evolution. Such systems have been used as evidence for hypotheses of coevolution and escalation, both of which assume a major role for ecological interactions in evolution. Coevolution and escalation are related, but different, concepts (see Vermeij 1994, for further clarification).

In this chapter, we define coevolution as the evolution of two or more species in response to one another (reciprocal adaptation). Such coevolution has been proposed mainly for terrestrial ecosystems, for instance, those involving plant–animal interactions (Ehrlich and Raven 1964; Thompson 1994), but it has also been suggested that marine predator–prey systems should coevolve (Kitchell 1982). Prey are expected to respond to predators by antipredatory adaptation, to which predators should respond by increasing their predatory capabilities.

How might such predator–prey coevolution, if it occurs, be documented by the fossil record? What tempo and mode should be expected? Schaffer and Rosenzweig (1978) and Dawkins and Krebs (1979) proposed that coevolution should produce a continuous arms race involving gradual improvement in prey defenses and in predator offensive capabilities. According to this view, the fossil record of such predator–prey systems should thus exhibit evolution in the gradual mode involving continuous accumulation of adaptations within species. Futuyma and Slatkin (1983:9) stated, "The ideal paleontological evidence would be a continuous deposit of strata in which each of two species shows gradual change in characters that reflect their interaction." DeAngelis, Kitchell, and Post (1985) agreed that coevolution should characterize predator–prey systems but suggested that the situation is more complex. Mathematical modeling of the naticid gastropod-bivalve prey system showed that gradual change, punctuation, or stasis are possible depending on intensity of predation, size of the predator, and prey defensive strategies (early reproduction, rapid growth to large size, increased shell thickness). The broad range of possible outcomes of coevolution makes the model proposed by DeAngelis, Kitchell, and Post difficult to test.

Vermeij (1987, 1994) argued that predator–prey systems escalate but that they need not coevolve. Escalation involves adaptation to enemies; Vermeij hypothesized that biologic hazards have become more intense through geologic time, and adaptations to those hazards have become better expressed. Although escalation may involve coevolution, Vermeij (1983, 1987, 1994) considered it more likely that predators respond evolutionarily to their own enemies (for instance, their own predators or competitors) rather than to their prey. Thus adaptation is not necessarily reciprocal.

Vermeij (1987) provided a lengthy argument for escalation, based on numerous lines of evidence. For instance, he reported temporal increases in various modes of predation, including drilling by gastropods, and increased development of shell armor by prey. Vermeij also considered the pace at which escalation might occur. At the level of faunas, Vermeij (1987, 1994) claimed that escalation has been episodic; extrinsic events such as transgression, climatic warming, and increases in productivity fostered episodes of escalation. At the scale of lineages, however, Vermeij (1987:376) stated that tempo and mode of escalation are uncertain: "the important question . . . is whether adaptive changes in the competitive and defensive attributes of species come only when enemies appear and disappear, or whether they can also occur when the populations of species and their enemies are persistent without great fluctuation. The evidence from living species favors the former viewpoint, but in the fossil record the question remains an unresolved and important issue." Thus it is unclear whether escalation is confined to speciation, or whether adaptation is possible within lineages once a species originates.

Gould (1990:22) criticized the hypothesis of escalation, expressing puzzlement over the mechanism involved: "If 'arms races' exist in a world of biotically driven trends . . . then how do they work in the speciational mode? . . . the anagenetic mode is easier to visualise, but cannot possibly apply—for any continuous escalation in such positive feedback would drive the trend to its furthest point in a geological instant, while the actual events span tens of millions of years. Yet if the trend must occur more episodically by occasional frog-hops of speciation . . . , then how is the locking of biotic interaction maintained?"

Several questions thus exist regarding evolution of predator–prey systems. What is the role of ecological interactions in the evolution of such systems? Are they characterized by coevolution or escalation, both of which claim a significant role for ecological interactions? What processes are important in such evolution, and what tempo and mode result?

The naticid gastropod predator–prey system provides the opportunity to address these questions. Naticids have been important shell-drilling molluscan predators since the Cretaceous (Sohl 1969); Vermeij (1987) cited temporal increases in naticid predation as one line of evidence supporting his hypothesis of escalation. The dynamics of the Recent naticid predator–prey system are well understood. Studies of extant naticids have elucidated the mechanics of drilling (Ziegelmeier 1954; Carriker 1981) and determined the "rules" of selective predation (Kitchell et al. 1981). Studies of Neogene fossil assemblages (Kitchell et al. 1981; Kitchell 1982; Kelley 1988, 1991a) indicate that similar rules characterized naticid prey selection in the past (although prey preferences appear to be less predictable in the Paleogene; Kelley and Hansen 1996b; Sickler, Kelley, and Hansen 1996). Because predators choose prey based on morphology, it is likely that prey may have evolved antipredatory morphological adaptations. Thus ecology may have played a role in the evolution of this system.

Our work on the naticid gastropod predator–prey system has proceeded along two fronts.

1. As an explicit evaluation of Vermeij's hypothesis of escalation, we have studied large-scale patterns of predation within Gulf and Atlantic Coastal Plain mollusk faunas. Our initial work focused on Cretaceous through Paleogene faunas (Kelley and Hansen 1993, 1996a, 1996b; Hansen

and Kelley 1995a), while current efforts span the Neogene. This approach traces changes in the predator–prey system observable at the faunal level.

2. In order to determine the microevolutionary patterns that underlie the large-scale faunal patterns, we have examined morphologic change in a variety of characters within individual genera of naticids and their bivalve prey, including taxa from the Eocene, Oligocene, and Miocene (Kelley 1979, 1983a, 1983b, 1984, 1989, 1991a,b; Kelley and Hansen, in revision). In particular, we are interested in whether the large-scale patterns resulted from within-species adaptation to predator–prey interactions.

Large-Scale Patterns of Predation

Materials and Methods

Vermeij (1987) suggested that the Cretaceous through Paleogene was an important period of escalation in predation by drilling gastropods, with very low predation levels in the Cretaceous and attainment of modern levels of drilling by the Eocene. His conclusion was based on data from half a dozen assemblages of Cretaceous and Eocene age. In order to provide a more detailed test of this hypothesis, ongoing work by Kelley and Hansen (1993, 1996a,b; Hansen and Kelley 1995a; Hansen, Graham, and Kelley 1996) surveys the history of naticid predation in the Gulf and Atlantic Coastal Plains. We compiled a database, including more than 46,000 specimens for 14 stratigraphic levels (17 formations) of Cretaceous through Oligocene age. Bulk, hand-picked samples provided data on frequency of complete and incomplete drillholes on all preserved taxa within these units (see Kelley and Hansen 1993 for details). We are currently continuing the study through the Pleistocene; the database now includes 150,000 specimens.

Large-Scale Patterns in Drilling Frequency

Our data show an episodic pattern of drilling frequencies during the Cretaceous and Paleogene. Drilling frequencies for bivalves were moderate in the Cretaceous (table 8.1) and dropped to very low levels (3–5%) crossing the Cretaceous–Tertiary boundary. Drilling rose dramatically in the Early Paleocene (to 33% in the Brightseat Formation), then stabilized and eventually declined to a low of 7–8% in the Late Eocene. A rebound of drilling frequencies occurred in the Oligocene, though the high levels of drilling seen in the Paleocene and Eocene were not reached. Drilling of gastropods followed a similar episodic pattern, though predation levels were lower than for bivalves

TABLE 8.1. Results of Comprehensive Survey of Naticid Predation for Cretaceous, Paleocene, Eocene, and Oligocene Bivalves

LEVEL	AGE	# SPEC.[a]	DR. FREQ.[b]	C.D.[c]	P.E.[d]	MULT.[e]
Ripley	L. Cret.	2629	0.132	174	0.054	0
Providence	L. Cret.	639	0.194	62	0.016	0
Corsicana	L. Cret.	291	0.055	8	0.273	0
Kincaid	E. Paleo.	1054	0.034	18	0.053	0.105
Brightseat	E. Paleo.	1743	0.327	285	0.087	0.026
Matthews Landing	M. Paleo.	235	0.111	13	0.235	0.118
Bells Landing	L. Paleo.	62	0	0	1	0
Bashi	E. Eo.	1124	0.415	233	0.045	0.066
Cook Mountain	L. M. Eo.	5035	0.287	721	0.062	0.039
Gosport	L. M. Eo.	468	0.073	17	0.105	0.105
Moodys Branch	L. Eo.	16209	0.079	641	0.181	0.097
Red Bluff	E. Olig.	2197	0.235	258	0.189	0.154
Mint Spring	E. Olig.	1596	0.204	163	0.133	0.154
Byram	E. Olig.	662	0.139	46	0.148	0.074

[a] # SPEC = number of specimens.
[b] DR. FREQ. = drilling frequency (2D/N where D = number of valves with a complete drillhole and N = total number of valves, because drilling only one valve is sufficient to kill a bivalved individual).
[c] C.D. = number of complete drillholes.
[d] P.E. = prey effectiveness, or number of incomplete drillholes : total number of attempted holes.
[e] MULT. = number of holes in multiply bored shells : total number of attempted holes.

in the Cretaceous, and the Oligocene rebound was less dramatic (table 8.2). Work in progress (Hansen, Graham, and Kelley 1996) indicates that drilling declined again through the Early Miocene, rose abruptly in the Middle Miocene, and declined into the Pliocene and Pleistocene.

Data on incomplete drillholes and multiple holes are also shown in tables 8.1 and 8.2. Both represent failed predation attempts (Kitchell et al. 1986; Vermeij, Dudley, and Zipser 1989) and indicate the relative effectiveness of predator offenses and prey defenses. Kelley and Hansen (1993) demonstrated a statistically significant temporal increase in incomplete and multiple holes in both bivalve and gastropod prey. The frequency of incomplete and multiple drilling generally was low in the Cretaceous through Early Eocene, with the exception of samples containing very few drillholes (e.g., the Corsicana Formation); the Middle Paleocene Matthews Landing Formation contains an anomalously high frequency of multiple drillholes, primarily in the gastropod *Turritella aldrichi*. Incomplete and multiple drilling increased significantly in the Late Eocene and Oligocene for both bivalve and gastropod prey.

TABLE 8.2. Results of Comprehensive Survey of Naticid Predation for Cretaceous, Paleocene, Eocene, and Oligocene Gastropods

LEVEL	AGE	# SPEC.[a]	DR. FREQ.[b]	C.D.[c]	P.E.[d]	MULT.[e]
Ripley	L. Cret.	2426	0.037	90	0.011	0.044
Providence	L. Cret.	516	0.060	31	0	0
Brightseat	E. Paleo.	414	0.377	156	0	0
Matthews Landing	M. Paleo.	297	0.465	138	0	0.341
Bells Landing	L. Paleo.	742	0.352	261	0.011	0.015
Bashi	E. Eo.	271	0.203	55	0.018	0.036
Cook Mountain	L. M. Eo.	2761	0.162	447	0.049	0.102
Gosport	L. M. Eo.	200	0.075	15	0.118	0.118
Moodys Branch	L. Eo.	1222	0.095	116	0.252	0.161
Red Bluff	E. Olig.	329	0.119	39	0.025	0.225
Mint Spring	E. Olig.	737	0.145	107	0.053	0.159
Byram	E. Olig.	178	0.174	31	0.139	0.194

[a] # SPEC = number of specimens.
[b] DR. FREQ. = drilling frequency calculated as drilled/undrilled individuals.
[c] C.D. = number of complete drillholes.
[d] P.E. = prey effectiveness, or number of incomplete drillholes : total number of attempted holes.
[e] MULT. = number of holes in multiply bored shells : total number of attempted holes.

Interpretation

Patterns of drilling frequencies indicate significant changes in the predator–prey system through time, as predicted by Vermeij's hypothesis of escalation and supporting an important role for ecology in evolution. The pattern is more complex, however, than implied by Vermeij's initial data that suggested limited Cretaceous drilling and escalation to modern levels by the Eocene. Vermeij (1987) predicted that warming, transgression, and high primary productivity should foster escalation and that mass extinctions involving a drop in productivity or cooling should interrupt it. Our data support a significant role for mass extinctions in escalation. We have proposed (Kelley and Hansen 1996a) that escalation in the naticid predator–prey system occurred in cycles controlled by mass extinctions. Following Vermeij (1987), we have suggested that highly escalated species with antipredatory adaptations would be lost selectively at mass extinctions (particularly those associated with cooling or a decline in productivity) due to the higher energy requirements for maintaining those adaptations. Loss of such highly escalated prey allows drilling frequencies to rise abruptly after mass extinctions; as prey escalate their defenses, drilling frequencies subsequently stabilize or decline until the next mass extinction. This model seems to fit the changes in drilling frequencies for the K/T and E/O extinctions; work in progress (Hansen, Graham, and Kelley 1996) indicates consistency between the model and drilling frequencies across the Middle Miocene extinction. We are in the process of testing the proposed mechanism for episodic escalation by examining the nature of pre-extinction and postextinction faunas in greater detail, particularly with respect to the occurrence of antipredatory traits (Hansen and Kelley 1995b; Melland, Kelley, and Hansen 1996). Prey effectiveness or multiple drilling frequency, or both, were low in the Paleocene Brightseat Formation but high in the Oligocene Red Bluff Formation (tables 8.1 and 8.2). Those results suggest that highly escalated prey were more likely to have been eliminated at the K/T than at the E/O boundary.

What evolutionary mechanisms were involved in the apparent escalation observed in our comprehensive survey of naticid predation? Was coevolution involved in the changes in the predator–prey system indicated by the patterns in drilling frequencies? Was adaptation of predator and prey reciprocal? The eventual decline in drilling frequencies after the postextinction highs suggests that adaptation by the prey outpaced that by the predator. Data on incomplete and multiple drillholes from the Cretaceous and Paleogene also suggest a lack of reciprocal adaptation. Superimposed on the pattern of episodic escalation of drilling frequencies is an increase in the frequency of failed predation attempts, especially in the later Eocene and Oligocene. This relative increase in prey defensive abilities suggests that, during the Paleogene, adaptation of predator and prey was not reciprocal, or at least not balanced. Escalation appears to have occurred, but the coevolutionary component is not significant.

Evolutionary Patterns Within Genera

Large-scale changes in the predator–prey system revealed by this comprehensive survey of predation suggest an important role for ecological interactions in macroevolution. To determine how these large-scale faunal patterns are linked to evolution within lineages, we have reexamined a database established in a series of studies by Kelley (1979, 1983a,b, 1984, 1989, 1991b, 1992) and Kelley and Hansen (in revision). We are interested in examining the following questions: What microevolutionary processes were responsible for the large-scale changes? Did adaptation of prey to predators, or vice versa, occur within species or only during speciation events?

Materials

Kelley and Hansen (in revision) presented data on populations of species through continuously sampled sections from the Eocene and Oligocene of

Mississippi, including samples from the Moodys Branch, Red Bluff, Mint Spring, and Byram Formations. Seven species were analyzed: *Hilgardia multilineata, Glycymeris idonea, Caestocorbula wailesiana,* and *Spisula jacksonensis* of the Moodys Branch Formation, *Corbula rufaripa* of the Red Bluff and Mint Spring Formation, *Corbula laqueata* (Mint Spring and Byram), and *Scapharca lesueuri* (Byram Formation). Details concerning stratigraphy and sampling are included in Kelley and Hansen (in revision).

Miocene material previously studied by Kelley was collected from 20 localities on the western shore of Chesapeake Bay and the Patuxent and St. Marys Rivers. Samples were taken at intervals of a few centimeters to a meter from the Calvert Formation [zones 10 and 14, Shattuck (1904)], Choptank Formation (zones 16, 17, and 19 = Calvert Beach, Drumcliff, and Boston Cliffs Members, respectively, of Gernant 1970), and St. Marys Formation (zones 22 and 24). Details regarding sampling are provided in Kelley (1983a). The interval studied had been estimated to represent about three to five million years (Ward and Blackwelder 1980; Wright and Eshelman 1987); recent dates from strontium ratios (D. Jones, personal communication, 1996) indicate an interval of about eight million years.

In this chapter, we reexamine morphometric data for two naticid and five bivalve genera from the Miocene fauna. (The Miocene fauna has recently undergone systematic revision by Ward 1992; his taxonomy is followed here.) The taxa analyzed include four long-ranging species, the naticids *Neverita duplicata* and *Euspira heros,* and the bivalves *Bicorbula idonea* and *Stewartia anodonta,* which commonly were preyed upon by naticids. Three bivalve prey lineages consisting of multiple species were also studied: *Marvacrassatella melina* (Calvert Formation); *M. turgidula* (Choptank zone 16 and 17); *M. marylandica* (Choptank zone 19); *Dallarca subrostrata* (Calvert Formation); *D. elnia* (Choptank zone 17); *D. elevata* (Choptank zone 19); *D. idonea* (St. Marys Formation); and *Astarte cuneiformis* (Calvert Formation); *A. thisphila* (Choptank Formation); *A. perplana* (St. Marys Formation). These lineages were proposed by B. Blackwelder (personal communication, 1977) using stratophenetic information and modified by Kelley (1983a) based on morphometric analyses. Kelley (1984) demonstrated that morphologic change in these taxa was evolutionary and not an ecophenotypic response to environmental change.

Characters Measured

Measurements were taken on a variety of characters for each taxon. For naticids, characters were selected because they affected naticid efficiency as predators, their ability to escape their own predators, or both. Characters measured on naticids represented shell size, globosity, aperture characteristics, and shell thickness. Shell size and globosity (linked by Kitchell 1986 to locomotory efficiency) affect both offensive and defensive capabilities of naticids. Aperture dimensions reflect foot size and thus the predator's ability to manipulate its prey. Increased shell thickness is a defense against predation, including cannibalism.

For the bivalve prey taxa, we examined two characters that are particularly significant in determining prey choice by the predator: thickness (TH), which controls the drilling time needed by a naticid predator to penetrate the prey shell, and internal volume (IV) of the shell, which is highly correlated with prey biomass and thus determines the benefit gained by the predator. A third character (CB) represents the cost-benefit of a particular prey item and is calculated as the ratio of TH:IV. Cost-benefit ratios have been used successfully to predict naticid prey selectivity; naticids generally select prey items with the lowest CB ratio in the size range they can handle (Kitchell et al. 1981; Kitchell 1982; Kelley 1988, 1991a).

In addition to the predation-related characters, for each Miocene bivalve taxon, characters were measured in the categories of overall size and shape, hinge line, ornamentation, and internal anatomy. These characters were chosen by Kelley (1979, 1983a) to quantify a range of aspects of morphology during a tabulation of the relative frequencies of stasis and gradualism within the Miocene fauna. Some characters had been employed previously by systematists (Glenn 1904; Schoonover 1941) to differentiate species in the lineages being studied. Several additional characters were identified by Kelley (1984) as being important in species discrimination; these characters are indicated in table 8.5 with an asterisk. We were thus able to compare evolutionary patterns for characters used in species discrimination (i.e., those that changed in association with the origin of species) versus those not specifically used to differentiate species.

Previous work (Kelley 1983a) documented evolutionary patterns among Chesapeake Group mollusks without regard to the nature of the characters examined. In contrast, in this chapter we categorize characters with respect to two criteria:

1. Significance to predator–prey interactions;
2. Importance in species discrimination.

By comparing evolutionary patterns of characters in different categories, we explore the role of ecology and the significance of morphologic trends in the evolutionary process.

Methods

To determine how ontogenetic growth patterns changed through time, bivariate methods were used. Regressions of each variable on length were calculated for each stratigraphic sample of a taxon. Then the value of each variable at a standard length (the mean of all sample means for that taxon) was predicted from the regression for each stratigraphic level.

Spearman's rank correlation coefficient, C, was used to determine whether there was a correlation between predicted magnitude of the variable at the standard length and stratigraphic position. (Although the Spearman coefficient is somewhat less powerful than the parametric Pearson correlation coefficient, absolute dates were not available for each sampling level, thus precluding calculation of parametric correlations between morphology and time.) A significant Spearman's C suggested directional change within a taxon for that variable; conversely, a nonsignificant correlation indicated stasis or nondirectional fluctuation. Variables for which Spearman's C was significant were examined further to determine whether the significant rank correlation was caused by gradual evolution. Gradualism was not indicated if any of the following were true.

1. The net change during an interval was less than geographic variation at a stratigraphic level
2. The direction of the trend was not oriented toward the morphology of the next species in a lineage
3. An across-lineage trend consisted of "plateaus" between species with no gradation between them.

For each character within a taxon, we present the following data: Spearman's rank correlation coefficient and its significance (indicating possible trends), the net morphologic change during the interval studied (as a percent of the magnitude of the variable at the start of the interval), and the geographic variation where data exist for multiple localities (calculated as the range of values divided by their mean). For the Miocene bivalve taxa, characters used in species differentiation are compared with those that have not been used; special attention is paid to the characters relating to predation.

Results

Tables 8.3–8.5 present the results of the analysis of evolutionary patterns within genera. Among the naticid species (table 8.3), most characters thought to affect predator efficiency exhibited stasis or nondirectional fluctuation. Of

TABLE 8.3. Evolutionary Patterns Exhibited by Offensive and Defensive Predation-Related Variables Measured for Chesapeake Group Naticid Gastropod Taxa.[a]

TAXON	VAR	C-VALUE	SIGNIF	NET CHGE	GEOG VAR	GRAD
Euspira	TH	0.4000	ns	0.627	0.136	
	H	0.3697	ns	0.194	0.506	
	AH	−0.6485	0.05	-0.089	0.115	
	G	−0.4061	ns	-0.057	0.079	
	AL	−0.2606	ns	-0.055	0.088	
	AA	−0.2970	ns	-0.055	0.208	
	L	−0.2727	ns	-0.046	0.095	
Neverita	H	1.0000	—	0.598	—	y?
	TH	−1.0000	—	-0.373	—	y?
	AL	−0.8000	ns	-0.073	—	
	L	−0.8000	ns	-0.021	—	
	AH	−0.4000	ns	-0.028	—	
	AA	0	ns	0.012	—	
	G	0	ns	0.005	—	

[a] Abbreviations for variables are as follows: AA = aperture area, AH = aperture height, AL = aperture length, G = globosity, H = height, TH = thickness. Shown are Spearman's rank correlation coefficient (C-VALUE) and its significance (SIGNIF), the net change (NET CHGE) and geographic variation (GEOG VAR). Trends exhibiting gradualism are indicated by "y" in "GRAD" column.

14 cases, only *Euspira* aperture height (AH) exhibited a significant rank correlation coefficient, decreasing by 9% through the section (i.e., in the opposite direction from increased predator efficiency). This decrease is comparable to the geographic variation in AH at a stratigraphic level (11.5%) and thus not an example of gradualism. Possible morphologic trends occurred in shell thickness relative to height and in shell height. *Euspira* increased in TH by 63% despite a reversal of the morphologic trend in the final sample. *Neverita* increased mean height by 60% while simultaneously decreasing shell thickness by 37%.

For the Eocene and Oligocene species, only patterns in the predation-related characters (TH, IV, and TH:IV) were examined (table 8.4). Directional trends were apparent in the species *Hilgardia multilineata* and *Glycymeris idonea*. *Hilgardia* TH increased monotonically, while IV decreased, resulting in a net increase in CB ratio of 122%. Simultaneous increases in TH and decreases in IV of *Glycymeris* yielded a 95% increase in CB ratio through the Moodys Branch Formation. *Caestocorbula wailesiana* also showed a monotonic increase in TH, a decrease in IV, and increase in CB, but the stratigraphic sequence was too short to calculate meaningful C-values. Finally, *Scapharca lesueuri* of the Byram Formation also showed significant rank correlation

TABLE 8. 4. Evolutionary Patterns Exhibited by Predation-Related Variables Measured for Eocene and Oligocene Bivalve Prey Taxa.[a]

TAXON	VAR	C-VALUE	SIGNIF	NET CHGE	GRAD
Hilgardia multilineata	TH	1.0000	<0.001	0.216	y
	IV	−0.6000	ns	−0.453	
	CB	0.8000	ns	1.221	
Glycymeris idonea	TH	0.8857	<0.01	0.104	y
	IV	−1.000	<0.001	−0.435	y
	CB	1.000	<0.001	0.953	y
Caestocorbula wailesiana	TH	−	−	0.075	
	IV	−	−	−0.047	
	CB	−	−	0.128	
Spisula jacksonensis	TH	−0.1429	ns	−0.199	
	IV	0.5238	ns	0.847	
	CB	−0.5238	ns	−0.567	
Corbula rufaripa	TH	0.3571	ns	0.462	
	IV	−0.2143	ns	−0.432	
	CB	0.4286	ns	1.574	
Corbula laqueata	TH	−0.7000	ns	−0.470	
	IV	0.2500	ns	1.019	
	CB	−0.1000	ns	−0.527	
Scapharca lesueuri	TH	0.3000	ns	0.0025	
	IV	1.000	<0.001	0.452	y
	CB	−1.000	<0.001	−0.309	y

[a] Abbreviations for variables are as follows: TH = thickness, IV = internal volume, CB = cost-benefit ratio (TH:IV). Shown are Spearman's rank correlation coefficient (C-VALUE), its significance (SIGNIF) and the net change (NET CHGE). Trends exhibiting gradualism are indicated by "y" in "GRAD" column.

coefficients for IV and CB ratio, but trends are in the direction opposite that expected in response to predation. None of the remaining species showed significant directional change in any characters.

For the Miocene bivalve prey, we have data on a wider variety of characters (table 8.5). Most characters measured for *Stewartia anodonta* exhibited stasis or nondirectional fluctuation (as indicated by nonsignificant C-values), producing net changes through the section of 1–15%. The two characters useful in discriminating this species from the associated *S. foremani*, convexity (CON) and shell width (W), increased by 17% and 25%, respectively, but through a pattern of nondirectional fluctuation that produced nonsignificant C-values. In contrast, TH and the related variable CB increased by 157% and 145%, respectively, compared to the geographic variation of 7% and 2%. Significant rank correlation coefficients suggest that these substantial increases were accomplished through gradual trends.

TABLE 8.5. Evolutionary Patterns Exhibited by Variables Measured for Chesapeake Group Bivalve Taxa.[a]

TAXON	VAR	USED	CATEG	C-VALUE	SIGNIF	NET CHGE	GEOG VAR	GRAD
Stewartia	TH		P	0.5879	<0.05	1.573	0.068	y
	CB		P	0.6485	<0.05	1.447	0.019	y
	W	*	Sh	−0.2545	ns	0.253	0.001	
	CON	*	Sh	−0.1182	ns	0.169	0.041	
	LAA		I	0.2091	ns	0.145	0.025	
	LPA		I	−0.2000	ns	0.123	0.007	
	DDM		I	0.2545	ns	0.113	0.015	
	DBA		I	0.3545	ns	0.047	0.011	
	DPM		I	0.1455	ns	0.047	0.070	
	IV		P	−0.1455	ns	0.038	0.050	
	H		Sh	−0.4818	ns	−0.013	0.007	
Bicorbula	CB		P	0.6121	<0.05	0.142	0.091	y
	TT	*	H	−0.4571	ns	0.123	0.055	
	TH		P	0.5636	<0.05	0.078	0.033	y
	DDM		I	−0.4607	ns	0.028	0.064	
	H	*	Sh	−0.3857	ns	0.005	0.018	
	W	*	Sh	−0.7000	<0.01	0.004	0.027	
	DPS	*	I	−0.4077	ns	−0.001	0.026	
	DPM		I	−0.6500	<0.01	−0.009	0.054	
	CON	*	Sh	−0.6536	<0.01	−0.039	0.022	y
	IV		P	−0.3455	ns	−0.057	0.124	
	LPA		I	−0.8572	<0.001	−0.169	0.012	y
	O		Sh	−0.7036	<0.01	−0.217	0.061	y
Dallarca	TH	*	P	0.2179	ns	0.945	0.109	
	CB		P	−0.0681	ns	0.491	0.157	
	W	*	Sh	0.4571	ns	0.317	0.036	
	IV		P	0.7275	<0.01	0.304	0.100	y
	CON	*	Sh	0.4125	ns	0.288	0.043	
	H	*	Sh	0.6893	<0.01	0.255	0.018	
	LAA		I	0.7821	<0.001	0.251	0.017	y
	DDM		I	0.5500	<0.05	0.231	0.019	y
	DBAM	*	Sh	0.2821	ns	0.205	0.028	
	HCA	*	H	0.3393	ns	0.053	0.089	
	LCA	*	H	−0.1304	ns	−0.039	0.034	
	NR	*	O	−0.3607	ns	−0.073	0.036	
D. elevata	CB		P	0.7000	ns	0.407	0.062	
	HCA	*	H	0.9857	<0.01	0.217	0.048	
	TH	*	P	0.5000	ns	0.170	0.059	
	LCA	*	H	0.6714	ns	0.041	0.024	
	DDM		I	0.5000	ns	0.033	0.012	
	LAA		I	0.4714	ns	0.025	0.008	
	NR	*	O	0.5571	ns	0.012	0.031	
	W	*	Sh	−0.0429	ns	−0.004	0.027	
	H	*	Sh	0.3000	ns	−0.005	0.036	
	CON	*	Sh	−0.2134	ns	−0.010	0.038	
	DBAM	*	Sh	−0.3000	ns	−0.059	0.022	
	IV		P	−0.3857	ns	−0.168	0.068	
Marvacrassatella	TH	*	P	0.7818	<0.01	0.721	0.008	y
	IV		P	0.1455	ns	0.543	0.055	
	DRB	*	O	−0.0545	ns	0.413	0.063	

(Continued on next page)

TABLE 8.5. (*continued*)

TAXON	VAR	USED	CATEG	C-VALUE	SIGNIF	NET CHGE	GEOG VAR	GRAD
	TL	*	H	0.7091	<0.02	0.336	0.028	
	CON	*	Sh	0.1182	ns	0.239	0.014	
	W	*	Sh	0.0682	ns	0.203	0.012	
	DBAM	*	Sh	0.8727	<0.001	0.125	0.022	
	CB		P	0.6273	<0.05	0.115	0.047	y
	H		Sh	0.1455	ns	0.099	0.038	
	LPA		I	−0.3091	ns	0.058	0.049	
	SHPA	*	Sh	−0.4364	ns	0.038	0.058	
	DPM	*	Sh	−0.1091	ns	0.031	0.038	
Astarte	IV		P	0.8571	<0.001	0.497	0.049	y
	LAA	*	I	0.8250	<0.001	0.160	0.030	
	TL		H	0.1964	ns	0.155	0.021	
	CON	*	Sh	0.5429	<0.05	0.132	0.076	
	H	*	Sh	0.7714	<0.001	0.087	0.016	y
	W	*	Sh	0.5000	ns	0.043	0.086	
	DDM		I	0.7571	<0.01	0.038	0.033	y
	LL		H	0.6036	<0.02	−0.027	0.032	y
	DAM		I	0.3607	ns	−0.155	0.098	
	TH		P	0.3893	ns	−0.339	0.128	
	CB		P	0.0964	ns	−0.558	0.147	
	DRB	*	O	0.0250	ns			
A. cuneiformis	TL		H	0.7143	ns	0.258	0.021	
	H	*	Sh	0.3714	ns	0.041	0.016	
	LL		H	0.2000	ns	0.035	0.032	
	DDM		I	0.4286	ns	0.019	0.033	
	CON	*	Sh	0.3714	ns	−0.096·	0.076	
	DAM		I	0.4857	ns	−0.114	0.098	
	W	*	Sh	−0.4857	ns	−0.185	0.086	
	LAA	*	I	−0.4857	ns	−0.197	0.030	
	IV		P	−0.7751	0.05	−0.236	0.049	
	CB		P	0.0286	ns	−0.295	0.147	
	TH		P	0.0286	ns	−0.461	0.128	
A. thisphila	IV		P	0.08095	0.01	0.240	0.049	y
	DRB	*	O	0.1429	ns	0.087		
	DDM		I	0.4762	ns	0.038	0.033	
	LAA	*	I	0.4524	ns	0.005	0.030	
	H	*	Sh	0.5952	ns	0.004	0.016	
	LL		H	0.0952	ns	−0.012	0.032	
	TL		H	−0.4524	ns	−0.036	0.021	
	CON	*	Sh	−0.6190	ns	−0.075	0.076	
	DAM		I	−0.1273	ns	−0.114	0.098	
	W	*	Sh	−0.8571	<0.01	−0.164	0.086	y
	TH		P	−0.3809	ns	−0.210	0.128	
	CB		P	−0.7381	<0.05	−0.363	0.147	y

[a] Abbreviations for most variables given in text; remainder defined in appendix of Kelley (1983a). Characters useful in species discrimination are indicated by an asterisk in "USED" column. CATEG. = category of variable, where P = predation-related, Sh = shape, H = hinge, I = internal anatomy, and O = ornamentation. Also shown are Spearman's rank correlation coefficient (C-VALUE) and its significance (SIGNIF), the net change (NET CHGE) and geographic variation (GEOG VAR). Trends exhibiting gradualism are indicated by "y" in "GRAD" column.

Three of the five variables useful in discriminating *Bicorbula idonea* from co-occurring corbulids possessed nonsignificant rank correlation coefficients. Width and the related variable CON showed a statistically significant rank correlation but minimal net change (0.4% for W, −4% for CON). Seven characters are not generally used in species discrimination; four appear to show directional changes of greater magnitude than the geographic variation. Of the predation-related variables, IV exhibited nondirectional fluctuation and TH showed directional change that resulted in an 8% increase. In combination, these changes produced an increase in CB of 14%; the C-value for CB is significant ($p < 0.05$).

Eight of twelve *Marvacrassatella* characters are useful in species discrimination. Most showed no across-lineage trends. Tooth length (TL) exhibited a significant rank correlation and major changes between species (a net change of 34%) but little intraspecific temporal variation (Kelley 1983a). The position of the beak (DBAM) showed a net change of 12%, also with most change occurring between, rather than within, species (Kelley 1983a). Thickness, however, increased both within and between species (Kelley 1989), resulting in a net change of 72%. The increase in CB was smaller (12%) but also involved directional change within and between species.

For the genus *Dallarca*, patterns were examined across all three species and within the long-ranging *D. elevata*. Eight of twelve characters were useful in species discrimination. In the across-lineage analysis, only one of those eight variables showed a significant C-value. Height (H) increased by 26%, in contrast to its geographic variation of 2%, but through a series of increasingly higher individual species' plateaus. Three additional variables [IV; length of the anterior adductor (LAA); and its distance from the dorsal margin (DDM)], not used in species discrimination, also had significant rank correlations and net changes of 23–30%. Both LAA and DDM were reported by Kelley (1983a) to exhibit gradual interspecific trends. All other variables possessed nonsignificant across-lineage rank correlations. Nevertheless, some variables showed very large net changes (TH, 95%; W, 32%; CON, 29%).

Results for *Dallarca elevata* yielded only one significant rank correlation coefficient. Height of the cardinal area (HCA), a character useful in species discrimination, increased by 22%. However, the trend is oriented away from the direction of the next species in the lineage and thus not in the gradual mode.

In *Astarte*, six variables yielded significant across-lineage rank correlations. Three are useful in differentiating species: CON, LAA, and H (13%, 16%, and 9% net change, respectively). Of these, only H exhibited a trend between species of the lineage. The other three variables with significant C-values have not been used to discriminate species. DDM and LL (lunule length) exhibited minimal net change comparable to the geographic variation, but IV increased by 50% across the lineage. All other variables showed nonsignificant rank correlations. (TH showed a major increase between A. *cuneiformis* and A. *thisphila* but then decreased for A. *perplana,* producing a nonsignificant C; see Kelley 1989).

Rank correlation coefficients were calculated separately for the long-ranging species A. *cuneiformis* and A. *thisphila.* Only IV exhibited a significant correlation for A. *cuneiformis,* decreasing by 24% in comparison to the geographic variation of 5%, but the trend is not in the direction of the next species of the lineage. IV increased by 24% within A. *thisphila,* for a significant rank correlation; as a consequence, CB decreased by 36%, also gradually. Width, a variable useful in species discrimination, showed a significant trend of decrease (16%).

For the Miocene bivalve prey, table 8.6 summarizes tempo and mode of evolution with respect to utility of characters in species discrimination and with respect to their relationship to predation. Of 94 characters examined, 19 (20%) are characterized by gradualism. Only four of those gradualistic characters were useful in species discrimination: *Bicorbula* CON, *Marvacrassatella* TH, *Astarte* H, and *Astarte thisphila* W. With the exception of TH, all are shape variables. There is a significant association between mode of evolution and utility in species discrimination; for the 2 x 2 contingency table shown in Table 8.6, chi square = 6.86 ($p < 0.01$).

Table 8.6 also shows results for tempo and mode versus ecological significance of variables for the Miocene bivalve prey. Ten of 24 (42%) predation-related characters evolved gradually, in contrast to 9 of 70 (13%) characters unrelated to predation. The association between predation and mode of evolution is significant. Chi square = 9.20 for the 2 x 2 contingency table ($p < 0.005$).

TABLE 8.6. Number of Characters of Miocene Bivalves Exhibiting Gradual and Punctuational Evolution versus Their Utility in Species Discrimination and Their Relation to Predation

	Gradual evolution	Punctuational evolution
Characters useful in species discrimination	4	41
Nonuseful characters	15	34
Predation-related characters	10	14
Nonrelated characters	9	61

Implications

Analyses at the level of individual taxa yield results consistent with the hypothesis that within-species antipredatory adaptation by prey occurred in response to predation by naticids (and possibly by a variety of durophagous predators). The evidence is particularly strong for the Miocene prey, but examples of directional change in predation-related characters occurred within Eocene species also (*Hilgardia multilineata, Glycymeris idonea,* and possibly *Caestocorbula wailesiana; Scapharca lesueuri* also showed directional trends but opposite those predicted as antipredatory adaptation).

Among Miocene bivalve prey, characters closely related to predation (TH, IV, and CB ratio) exhibited patterns different from characters less closely tied to predator selectivity. Predation-related characters showed a greater frequency of directional within-species change than did other characters, and the net change that occurred (whether through gradual or punctuational change) was usually large compared to that for other characters. Thus prey evolved antipredatory adaptations, most often in the gradual mode. This situation is in apparent agreement with the suggestion by Schaffer and Rosenzweig (1978) and Dawkins and Krebs (1979) that predator–prey systems should exhibit gradual coevolution. (Conversely, the results do not support the coevolutionary model of DeAngelis, Kitchell, and Post 1985, in which the prey strategy of extreme thickness increase was considered attainable only by speciation.)

Coevolution, however, involves reciprocal evolution of predator and prey. Such reciprocal adaptation is not supported by patterns of evolution within the common Chesapeake Group naticids, *Euspira heros* and *Neverita duplicata.* Characters affecting naticid predatory efficiency exhibited nondirectional fluctuation resulting in minimal net change (a statistically significant trend in *Euspira heros* aperture height produced net change less than the geographic variation at a stratigraphic level and was directed opposite to increasing predator efficiency). The lack of predatory adaptation within the naticids is consistent with Kelley's (1989) observation that, in the Miocene, as thickness of prey increased through time, predation intensity decreased. In addition, predators of a given size preyed on smaller, less profitable individuals of *Marvacrassatella* and *Astarte* in the stratigraphically higher Choptank Formation than in the Calvert (Kelley 1989).

Although offensive capabilities of the predator did not keep pace with increases in defensive abilities of the prey, two characters showed substantial directional change (37–63%) within naticid species: relative TH of both species and mean height of *Neverita.* Kelley (1992) interpreted these changes as increases in naticid defenses against their own enemies (other naticids) rather than as an increase in offense in response to evolution of their prey.

Discussion

What was the role of ecological interactions in the evolution of naticid gastropods and their prey? Did the naticid gastropod predator–prey system exhibit either coevolution or escalation, and if so, at what pace?

At the scale both of faunas and lineages, escalation involving increased adaptation to biological hazards is apparent. At the scale of faunas, escalation was episodic and involved cycles apparently initiated by mass extinctions. Within individual lineages, both punctuational and gradual change occurred; characters related to predation tended to evolve in the gradual mode more frequently and involved larger changes than did characters unrelated to predation. At neither scale did the predator–prey system exhibit coevolution in the sense of reciprocal adaptation. Within predator species, no increases in predatory capabilities occurred, and drilling frequencies declined as prey lineages increased their defenses (for instance, through increased shell thickness). At the faunal scale, the frequency of failed predation attempts suggests that prey antipredatory adaptation outpaced the predator's response to evolution of its prey.

Despite the lack of coevolution, predator–prey interactions provided the primary pathway for anagenetic change, and some very substantial within-lineage changes were produced. This observation is in apparent contradiction to Gould's (1990:22) suggestion that the anagenetic mode "cannot possibly apply" to long-term biotically driven trends (though in this case the trends span a period of eight million years at most, rather than the "tens of millions of years" mentioned by Gould).

Although predation-mediated gradual change occurred within lineages, such gradual adaptation was generally unimportant in the evolution of new species. (Gradual change occurred in 27% of the characters that were not used in species differentiation, whereas only 9% of the characters useful in species discrimination exhibited gradualism. The difference is statistically significant.) Gradual change was most likely to occur in characters incidental to species discrimination, especially if those characters were related to predation. Such characters included *Stewartia anodonta* and *Bicorbula idonea* TH, and IV of *Astarte thisphila* (as well as the across-lineage trend in *Astarte* IV; *Astarte cuneiformis* also displayed a trend in IV but it was not oriented in the direction of the next species of the lineage). Only one predation-related character useful in species discrimination displayed unequivocal gradual change; *Marvacrassatella* TH showed both inter- and intraspecific gradual trends.

Despite the relatively high incidence of gradualism within characters related to predation, antipredatory adaptation also occurred in the punctuational mode, during speciation (for instance, the TH increase between *Astarte cuneiformis* and *A. thisphila,* or the decrease in IV between *Marvacrassatella melina* and *M. turgidula*). Allmon (1994) stated that ecology could be important in evolution if ecological interactions were significant in speciation. For instance, he suggested that predation-resistant morphologies may increase the likelihood of speciation by enhancing the probability of persistence of allopatric isolates. This suggestion seems plausible; the same ecological interactions and selective pressures that promote gradual within-species change might also foster "directional" speciation of predation-resistant morphologies. The changes in antipredatory characters that occurred during speciation in the Miocene fauna could have resulted from the mechanism envisioned by Allmon.

Gould (1990:22) viewed the occurrence of escalation in the speciational mode as problematic because it required "locking of biotic interactions" over the duration of the trend. Such "locking," however, may not be necessary. Because there appears to be no strong coevolutionary component to escalation, maintenance of specific predator–prey interactions is not required. Prey respond to their enemies, which respond to their own enemies; evolutionary responses of predator and prey need not be reciprocal.

Evidence for escalation within faunas and individual lineages indicates the importance of ecological interactions in evolution. The episodic nature of escalation at the level of faunas may indicate that the processes that regulate large-scale escalation may differ from those that produce within-lineage change. Whereas ecological interactions may play a significant role in the evolution of individual lineages (in both the gradual and the speciational mode), mass extinctions appear to impose an episodic pattern on the escalation of faunas. This hypothesis is in accord with Gould's (1990:23) view that "the increasing importance awarded to episodes of mass extinction . . . suggests that abiotic struggle may be more important than previously thought by palaeontologists." Our ongoing research addresses the link between mass extinctions and escalation by examining the effect of the Cretaceous-Tertiary, Eocene-Oligocene, middle Miocene, and Plio-Pleistocene extinctions on survivorship of escalated prey.

ACKNOWLEDGMENTS

We thank the following individuals for assistance in the field: J. Kelley, J. Olivier, R. A. Dickerson, D. Bohaska, D. Haasl, and D. Dockery. R. A. Dickerson, J. Olivier, C. R. McMullen, D. Haasl, B. Farrell, E. Akins, V. Melland, R. Sickler, K. Bradbury, and A. Huntoon assisted with various phases of data tabulation. W. Allmon, R. Aronson, N. Gilinsky, G. Vermeij, D. Miller, and J. Sepkoski reviewed versions of this manuscript and contributed useful discussion. This research has been supported by National Science Foundation grants EAR-8507293 (to Kelley), EAR-8915725 (to Kelley and Hansen), and collaborative grants EAR-9405104 (to Kelley) and EAR-9406479 (to Hansen). This chapter is CMSR Contribution Number 179.

REFERENCES

Allmon, W. D. 1992. What, if anything, should evolutionary paleoecology be? *Palaios* 7:557–558.

Allmon, W. D. 1994. Taxic evolutionary paleoecology and the ecological context of macroevolutionary change. *Evolutionary Ecology* 8:95–112.

Carriker, M. R. 1981. Shell penetration and feeding by naticacean and muricacean predatory gastropods: A synthesis. *Malacologia* 20:403–422.

Dawkins, R. and J. R. Krebs. 1979. Arms races within and between species. *Proceedings of the Royal Society of London* B205:489–511.

DeAngelis, D. L., J. A. Kitchell, and W. M. Post. 1985. The influence of naticid predation on evolutionary strategies of bivalve prey: Conclusions from a model. *American Naturalist* 126:817–842.

Ehrlich, P. R. and P. H. Raven. 1964. Butterflies and plants: A study in coevolution. *Evolution* 18:586–608.

Eldredge, N. and S. J. Gould. 1972. Punctuated equilibria: An alternative to phyletic gradualism. In T. J. M. Schopf, ed., *Models in Paleobiology,* pp. 82–115. San Francisco: Freeman, Cooper, and Co.

Futuyma, D.J. and M. Slatkin. 1983. Introduction. In D. J. Futuyma and M. Slatkin, eds., *Coevolution,* pp. 1–13. Sunderland MA: Sinauer Associates.

Gernant, R. E. 1970. Paleoecology of the Choptank Formation (Miocene) of Maryland and Virginia. *Maryland Geological Survey Report of Investigations* 12:1–90.

Glenn, L. C. 1904. Systematic paleontology, Miocene Pelecypoda. In W. B. Clark, G. B. Shattuck, and W. H. Dall, eds., *The Miocene Deposits of Maryland,* pp. 274–401. Maryland Geological Survey, Miocene Volume. Baltimore: Johns Hopkins Press.

Gould, S. J. 1985. The paradox of the first tier: An agenda for paleobiology. *Paleobiology* 11:2–12.

Gould, S. J. 1990. Speciation and sorting as the source of evolutionary trends, or 'Things are seldom what they seem'. In K. McNamara, ed., *Evolutionary Trends,* pp. 3–27. London: Belhaven Press.

Gould, S. J. and N. Eldredge. 1993. Punctuated equilibrium comes of age. *Nature* 366:223–227.

Hansen, T. A. and P. H. Kelley. 1995a. Spatial variation of naticid gastropod predation in the Eocene of North America. *Palaios* 10:268–278.

Hansen, T. A. and P. H. Kelley. 1995b. The effect of mass extinctions on escalation in molluscs: A test at the Cretaceous-Tertiary boundary. *Geological Society of America Abstracts with Programs* 27(6):A164.

Hansen, T. A., S. E. Graham, and P. H. Kelley. 1996. Does escalation occur in cycles? Evidence from the Neogene. *Sixth North American Paleontological Convention Abstracts* 160.

Kelley, P. H. 1979. Mollusc lineages of the Chesapeake Group (Miocene). Ph.D. thesis, Harvard University, Cambridge MA.

Kelley, P. H. 1983a. Evolutionary patterns of eight Chesapeake Group molluscs: Evidence for the model of punctuated equilibria. *Journal of Paleontology* 57:581–598.

Kelley, P. H. 1983b. The role of within-species differentiation in macroevolution of Chesapeake Group bivalves. *Paleobiology* 9:261–268.

Kelley, P. H. 1984. Multivariate analysis of evolutionary patterns of seven Miocene Chesapeake Group molluscs. *Journal of Paleontology* 58:1235–1250.

Kelley, P. H. 1988. Predation by Miocene gastropods of the Chesapeake Group: Stereotyped and predictable. *Palaios* 3:436–448.

Kelley, P. H. 1989. Evolutionary trends within bivalve prey of Chesapeake Group naticid gastropods. *Historical Biology* 2:139–156.

Kelley, P. H. 1991a. Cannibalism by Chesapeake Group naticid gastropods: A predictable result of stereotyped predation. *Journal of Paleontology* 65:75–79.

Kelley, P. H. 1991b. The effect of predation intensity on rate of evolution of five Miocene bivalves. *Historical Biology* 5:65–78.

Kelley, P. H. 1992. Evolutionary patterns of naticid gastropods of the Chesapeake Group: An example of coevolution? *Journal of Paleontology* 66:794–800.

Kelley, P. H. and T. A. Hansen. 1993. Evolution of the naticid gastropod predator-prey system: An evaluation of the hypothesis of escalation. *Palaios* 8:358–375.

Kelley, P. H. and T. A. Hansen. 1996a. Recovery of the naticid gastropod predator–prey system from the Cretaceous-Tertiary and Eocene-Oligocene extinction. *Geological Society Special Publication* 102:373–386.

Kelley, P. H. and T. A. Hansen. 1996b. Naticid gastropod prey selectivity through time and the hypothesis of escalation. *Palaios* 11:437–445.

Kelley, P. H. and T. A. Hansen. In revision. The role of within-species adaptation in escalation of Neogene bivalves. *Journal of Paleontology*.

Kitchell, J. A. 1982. Coevolution in a predator–prey system. *Third North American Paleontological Convention Proceedings* 2:301–305.

Kitchell, J. A. 1986. The evolution of predator–prey behavior: naticid gastropods and their molluscan prey. In M. Nitecki and J. A. Kitchell, eds., *Evolution of Animal Behavior: Paleontological and Field Approaches,* pp. 88–110. Oxford, U.K.: Oxford University Press.

Kitchell, J. A., C. H. Boggs, J. F. Kitchell, and J. A. Rice. 1981. Prey selection by naticid gastropods: Experimental tests and application to the fossil record. *Paleobiology* 7:533–552.

Kitchell, J. A., C. H. Boggs, J. A. Rice, J. F. Kitchell, A. Hoffman, and J. Martinell. 1986. Anomalies in naticid predatory behavior: A critique and experimental observations. *Malacologia* 27:291–298.

Melland, V. D., P. H. Kelley, and T. A. Hansen. 1996. Long-term effects of the Cretaceous-Tertiary and Eocene/Oligocene mass extinctions on escalation in bivalves. *Geological Society of America Abstracts with Programs* 28(6):56–57.

Schaffer, W. M. and M. L. Rosenzweig. 1978. Homage to the Red Queen: I. Coevolution of predators and their victims. *Theoretical Population Biology* 14:135–157.

Schoonover, L. M. 1941. A stratigraphic study of the molluscs of the Calvert and Choptank Formations of Southern Maryland. *Bulletins of American Paleontology* 25:169–299.

Shattuck, G. B. 1904. Geological and paleontological relations, with a review of earlier investigations. In W. B. Clark, G. B. Shattuck, and W. H. Dall, eds., *The Miocene Deposits of Maryland*, pp. 33–137. Maryland Geological Survey, Miocene Volume. Baltimore: Johns Hopkins Press.

Sickler, R. N., P. H. Kelley, and T. A. Hansen. 1996. Prey selectivity by naticid gastropods from Tertiary sediments of the United States Coastal Plain. *Geological Society of America Abstracts with Programs* 28(6):64.

Sohl, N. F. 1969. The fossil record of shell boring by snails. *American Zoologist* 9:725–734.

Thompson, J. N. 1994. *The Coevolutionary Process.* Chicago: University of Chicago Press.

Vermeij, G. J. 1983. Intimate associations and coevolution in the sea. In D.J. Futuyma and M. Slatkin, eds., *Coevolution*, pp. 311–324. Sunderland MA: Sinauer Associates.

Vermeij, G. J. 1987. *Evolution and Escalation: An Ecological History of Life.* Princeton NJ: Princeton University Press.

Vermeij, G. J. 1994. The evolutionary interaction among species: Selection, escalation, and coevolution. *Annual Reviews of Ecology and Systematics* 25:219–236.

Vermeij, G. J., E. C. Dudley, and E. Zipser. 1989. Successful and unsuccessful drilling predation in Recent pelecypods. *Veliger* 32:266–273.

Ward, L. W. 1992. Molluscan biostratigraphy of the Miocene, Middle Atlantic Coastal Plain of North America. Virginia Museum of Natural History Memoir 2. Martinsville VA: Virginia Museum of Natural History.

Ward, L. W. and B. W. Blackwelder. 1980. Stratigraphic revision of upper Miocene and lower Pliocene beds of the Chesapeake Group, middle Atlantic Coastal Plain. *U.S. Geological Survey Bulletin* 1482D:1–61.

Wright, D. B. and R. E. Eshelman. 1987. Miocene Tyassuidae (Mammalia) from the Chesapeake Group of the mid-Atlantic coast and their bearing on marine-nonmarine correlation. *Journal of Paleontology* 61:604–618.

Ziegelmeier, E. 1954. Beobachtungen uber den Nahrungserwerb bei der Naticide *Lunatia nitida* Donovan (Gastropoda Prosobranchia). *Helgolander Wissenschaftliche Meeresforschung* 5:1–33.

9

Evolutionary Paleoecology
of Caribbean Coral Reefs

Richard B. Aronson and William F. Precht

ECOLOGISTS HAVE RADICALLY ALTERED their thinking about coral reefs, and those views continue to evolve. Coral reefs, formerly viewed as stable, equilibrial systems (Newell 1971), are now interpreted as nonequilibrial on ecologically relevant scales of time (years to decades) and space (landscapes to reef systems) (Grigg and Dollar 1990; Karlson and Hurd 1993; Edmunds and Bruno 1996; Brown 1997). Increasing awareness of this variability is motivating a strategic shift in coral reef research. Paleontologists are taking up the quest for predictability, searching for large-scale patterns in the fossil record (Jackson, Budd, and Pandolfi 1996).

The nonequilibrial view of reef ecology gained initial support when the intermediate disturbance hypothesis was formulated for coral reefs in the 1970s (Grassle 1973; Connell 1978). This hypothesis predicts maximum diversity at intermediate levels of disturbance. At low disturbance levels the competitive dominants exclude other species, whereas at high disturbance levels good colonizers, which are poor competitors, are the only ones able to persist. Intermediate disturbance levels strike a balance between the slowly recruiting competitive dominants and the rapidly recruiting competitive subordinates, leading to coexistence and higher diversity than at the extremes of the disturbance continuum. In 1980, Hurricane Allen further promoted the idea of nonequilibrium on coral reefs by devastating the best-studied reef in the Caribbean: Discovery Bay, on the north coast of Jamaica (Woodley et al. 1981). Attempts to fit the effects of Hurricane Allen into an intermediate disturbance scenario of increasing coral diversity (Porter et al. 1981) helped convince many ecologists that hurricanes are an important structuring force in coral reef communities.

Although intermediate disturbance effects occur on coral reefs under certain conditions (Rogers 1993a; Aronson and Precht 1995), serious doubts persist over the explanatory value of the hypothesis (Huston 1985a; Jackson 1991). Hurricane Allen and its indirect effects (Knowlton et al. 1981; Knowlton, Lang, and Keller 1990) caused such massive destruction at Discovery Bay that diversity ultimately decreased, a result that neither supports nor falsifies the intermediate disturbance hypothesis (Hughes 1994; Hughes and Connell 1999). Despite the enormity of its consequences in Jamaica, even Hurricane Allen, which is arguably the most famous disturbance in the literature on coral reefs, could not have prepared ecologists for what followed throughout the Caribbean region.

Many coral reefs worldwide have changed dramatically since the 1970s. Coral cover has typically declined, and the cover of noncoralline, fleshy and filamentous macroalgae (henceforth "macroalgae") has increased (Done 1992a; Wilkinson 1993; Ginsburg 1994; Cortés 1997). This transition has been particularly pronounced in the Caribbean region, including the Florida Keys and the Bahamas (Rogers 1985; Porter and Meier 1992; Ginsburg 1994; Hughes 1994; Connell 1997).

Corals are, of course, the principal framework builders of coral reefs. Bioeroders, including fish and invertebrates, break down that framework (Hutchings 1986). Crustose coralline algae are thought to play a role in reef cementation, but their importance has recently been questioned (Macintyre 1997). From the 1950s through the 1970s, living communities on the open surfaces of Caribbean reefs were dominated by corals, crustose coralline algae, and algal turfs. Today, macroalgae are the dominant space-occupiers on most Caribbean reefs.

Disturbances of various types have been invoked to explain the changing face of Caribbean reefs. These include hurricanes, coral bleaching, diseases of corals and sea urchins, overfishing, nutrient loading, sedimentation, and pollution (Tomascik and Sander 1987; Lessios 1988; Bythell et al. 1989; Hatcher, Johannes, and Robertson 1989; Rogers 1990; Glynn 1993; Liddell and Ohlhorst 1993; Hughes 1994; Shulman and Robertson 1996; Peters et al. 1997; and many others). Whether the coral-to-macroalgal transition is due primarily to natural or anthropogenic impacts remains controversial (Grigg and Dollar 1990).

Macroalgal dominance of Caribbean reef communities may not be unusual. On the contrary, Discovery Bay prior to 1980, from which the ecology of Caribbean reefs was originally described, may represent an atypical condition (Precht 1990; Woodley 1992). On average, hurricanes pass close enough to Discovery Bay to do significant damage more often than once per decade (Kjerfve et al. 1986; Woodley 1992). It just happened that the now-classic studies of zonation on Jamaican reefs (Goreau 1959; Goreau and Wells 1967; Goreau and Goreau 1973; Kinzie 1973) were conducted following several decades without a hurricane. This is a problem of scale and variance; the unusual lack of physical disturbance prior to Hurricane Allen may explain the "typically" high coral cover observed earlier.

Before the 1980s, living bank/barrier reefs of the Caribbean displayed a generalized zonation pattern of three common species that were the primary builders of reef framework (Goreau 1959; Graus and Macintyre 1989). The thickly branching elkhorn coral, *Acropora palmata,* was dominant at the reef crest and the shallowest depths of the fore reef (0–5 m depth). The more thinly branching staghorn coral, *A. cervicornis,* was dominant at intermediate depths (~5–25 m) on exposed reefs, and it ranged into shallower habitats on more protected reefs (Adey and Burke 1977; Geister 1977; Hubbard 1988). The massive corals of the *Montastraea annularis* species complex (Knowlton et al. 1992, 1997) were common in reef habitats from approximately 5 m to greater than 30 m.

Woodley (1992) suggested that Discovery Bay, and by extension other Caribbean reefs, which currently have low coral cover and high macroalgal cover, may be that way much or most of the time. Because most study reefs were initially chosen for their luxuriant coral growth, it is perhaps not surprising that the predominant direction of change has been toward declining coral cover (Hughes 1992). The alternative interpretation is that high coral cover is the usual condition. By this reasoning, current dominance by macroalgae is the result of a novel combination of circumstances (Hughes, Reed, and Boyle 1987; Jackson 1991, 1992; Knowlton 1992; Liddell and Ohlhorst 1993; Hughes 1994).

The fossil record has the potential to discriminate these two possibilities. Reef zonation patterns similar to Jamaica before 1980 have been recognized in Pleistocene and Holocene reef deposits throughout the region (Mesolella 1967; Macintyre, Burke, and Stuckenrath 1977; Geister 1980, 1983; Lighty, Macintyre, and Stuckenrath 1982; James and Macintyre 1985; Boss and Liddell 1987; Macintyre 1988; Fairbanks 1989; Stemann and Johnson 1992). Thick deposits that display this zonation have been taken as prima facie evidence that (1) corals dominated Caribbean reef communities for most of Pleistocene-

Holocene time, and (2) macroalgae dominated rarely if at all (Jackson 1991, 1992; see also Stemann and Johnson 1995). Thus, despite nonequilibrial dynamics on small spatio-temporal scales, corals are thought to have persisted in interspecific associations that are predictable on larger scales, forming long-lived communities.

Knowlton (1992) hypothesized that the coral- and macroalgal-dominated situations represent alternative community states, each of which is resistant to conversion to the other (see also Lighty 1981; Hatcher 1984; Precht 1990; McClanahan et al. 1999). The conjunction of recent disturbances has disrupted associations of coral species, shifting many Caribbean reefs to the macroalgal state (Jackson 1991, 1992, 1994b). The implication is that reef communities display emergent properties based on direct or indirect interspecific interactions, which influence their response to disturbance at various scales. The structure of coral reef communities may therefore differ fundamentally from that of other Neogene marine communities, which are interpreted as assemblages of independently distributed species (Jablonski and Sepkoski 1996).

This chapter critically reviews our current understanding of changes to Caribbean reefs on ecological and geological–evolutionary scales. We first examine causes of the recent shift to macroalgal dominance. Reduced herbivory and nutrient loading each played a role. We argue, however, that coral mortality, especially from disease, was a major driving force in the transition. Rates of recovery on Caribbean reefs are strongly influenced by the life history strategies of the corals. Second, we review work on the historical forces that have shaped the Caribbean coral fauna since the early Miocene. The evolutionary history of the fauna explains a great deal about the current state of Caribbean reefs. Third, we discuss the implications of that history for the hypothesis of whole-community responses to disturbance. The "community integration" hypothesis, an explanation based on emergent properties of biological systems, is pitted against an individualistic, "independent species distribution" hypothesis. These alternative models of community structure can be tested by considering the variability of coral reefs at multiple scales, connecting living reef communities to their fossil record. Finally, we suggest avenues of future research on coral reefs in the emerging field of evolutionary paleoecology.

Much of our knowledge of the coral-to-macroalgal shift comes from research at Discovery Bay and environs over the past 30 years. It will come as no surprise that generalizing those results to the rest of the Caribbean has often proved erroneous. Nevertheless, Discovery Bay provides a convenient point of departure for our discussion.

Herbivory and Its Consequences

If they are present in sufficient numbers, herbivores limit the rapid accumulation of macroalgal biomass, thereby promoting the growth and recruitment of corals (Hay 1981; Sammarco 1982; Hay and Goertemiller 1983; Carpenter 1986; Lewis 1986; Steneck 1988; Knowlton 1992). The most influential herbivores on Caribbean reefs are parrotfish (Scaridae; placed in the Labridae by Kaufman and Liem 1982), surgeonfish (Acanthuridae), and sea urchins (Echinoidea) (Ogden and Lobel 1978; Glynn 1990a; Choat 1991; Carpenter 1997; Hixon 1997).

Damselfish (Pomacentridae) are also herbivores, but they are ecologically quite different from scarids, acanthurids, and echinoids. Some pomacentrids kill living coral tissue and cultivate algal lawns on the dead surfaces for feeding and breeding purposes. They have had significant negative effects on coral cover in Recent and Pleistocene reef communities of the Caribbean (Kaufman 1977, 1981; Potts 1977; Lobel 1980). Because they promote algal growth rather than inhibiting it, pomacentrids will not be considered functional herbivores in this discussion.

The blackspined urchin, *Diadema antillarum,* was an important herbivore on many Caribbean reefs, including Discovery Bay and nearby areas (Sammarco 1982; Hay 1984; Carpenter 1986, 1997; Liddell and Ohlhorst 1986; Hughes, Reed, and Boyle 1987). Where it was abundant, this sea urchin competed strongly with other herbivores. *Diadema* negatively affected population densities of herbivorous fishes, although there was no evidence of a converse effect (Hay and Taylor 1985; Carpenter 1988, 1990b; Morrison 1988; Robertson 1991). *Diadema* severely limited the growth of macroalgae on the north coast of Jamaica before and for several years after Hurricane Allen (Hughes, Reed, and Boyle 1987). At high densities, *Diadema* also suppressed coral recruitment by grazing small colonies (Sammarco 1980).

A water-borne pathogen of unknown origin caused mass mortality of *Diadema* throughout the region in 1983–1984 (Lessios, Robertson, and Cubit 1984; Lessios 1988; Peters 1997). In the absence of *Diadema,* macroalgae flourished on heavily fished reefs like the one at Discovery Bay, because scarids and acanthurids had been largely eliminated by humans over the previous decades (Liddell and Ohlhorst 1986; Hughes, Reed, and Boyle 1987). Macroalgae are capable of overgrowing and shading small, recently settled corals (Birkeland 1977; Sammarco 1980, 1982), and, in extreme situations like Discovery Bay, even adult corals, including massive colonies of the *M. annularis* complex, were overgrown (Hughes 1994). The cover of crustose coralline algae declined as well (Steneck 1997).

This scenario is widely accepted as a model for the recent dynamics of Caribbean reefs. It was inferred largely from events of the past several decades at Discovery Bay and nearby sites, and it emphasizes the role of herbivory by *Diadema* in community turnover (e.g., Glynn 1990a; Steneck 1994; Carpenter 1997). In contrast, Hay (1984) argued that Discovery Bay and many other reefs had artificially elevated densities of *Diadema* because of fishing pressure, whereas scarids and acanthurids were more important on less fished reefs (see also Lewis and Wainwright 1985; Lewis 1986; Rogers, Garrison, and Grober-Dunsmore 1997). *Diadema* populations were released from predation by human exploitation of their predators, particularly the queen triggerfish, *Balistes vetula,* and the hogfish, *Lachnolaimus maximus* (Munro 1983; Hay 1984; Reinthal, Kensley, and Lewis 1984; Hughes, Reed, and Boyle 1987; Roberts 1995). The negative influence of *Diadema* on scarids and acanthurids, combined with overfishing of those herbivorous fishes, made *Diadema* by far the most important herbivore, at least down to 10 m depth (Liddell and Ohlhorst 1986; Jackson 1991). The mass mortality of *Diadema* then resulted in high macroalgal cover, a condition that persisted for more than a decade (for Discovery Bay: Liddell and Ohlhorst 1993; Aronson et al. 1994; Hughes 1994; Steneck 1994; Andres and Witman 1995; Edmunds and Bruno 1996; Aronson and Precht 2000; figure 9.1). Similar effects of fishing pressure on sea urchin abundance, sea urchin grazing on macroalgal abundance, and macroalgal abundance on coral cover have been observed on coral reefs of the Indo-Pacific (Muthiga and McClanahan 1987; McClanahan and Muthiga 1988, 1989; Tanner 1995; McClanahan et al. 1999).

Levitan (1992) uncovered a positive relationship between *Diadema* population density in the Caribbean and inferred levels of human exploitation pressure through time. Increasing ratios of jaw size to test size in museum specimens collected over the century before 1983 implied that *Diadema* density increased as human population density increased. This finding supported Hay's (1984) contention that *Diadema* density was related to fishing pressure. Levitan pointed out, however, that this result may have been confounded by the proximity of many less populated, less fished sites to continental land-masses.

One problem with generalizing the herbivory scenario beyond Discovery Bay is that macroalgal cover has increased on protected and moderately fished reefs, as well as on heavily fished ones, although to a lesser degree (e.g., Levitan 1988 for St. John, U.S. Virgin Islands; Carpenter 1990a for St. Croix, U.S. Virgin Islands ; McClanahan and Muthiga 1998 for Belize). Carpenter (1986 for St. Croix; see also Foster 1987 for Panama) explained these observations with evidence that *Diadema* were the most important herbivores even on lightly fished reefs, contradicting Hay (1984). By Carpenter's reasoning, the 1983–

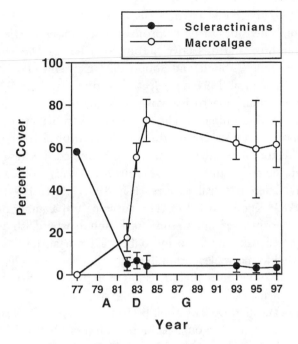

FIGURE 9.1. Changes in scleractinian coral and macroalgal cover at 5–6 m depth on the West Fore Reef at Discovery Bay, Jamaica. Letters beneath the *x*-axis mark significant events in the history of the site: A, Hurricane Allen (1980); D, *Diadema* mass mortality (1983); G, Hurricane Gilbert (1988). Data for 1977 are from visual transects by M. A. Huston, reported in Liddell and Ohlhorst (1987); data for 1982–84 are from visual transects by Liddell and Ohlhorst (1986); data for 1993–97 are from transects videotaped by the authors and analyzed by M. L. Kellogg, per Aronson et al. (1994). Error bars represent standard deviations, which are not available for 1977 data.

1984 *Diadema* mass mortality significantly decreased herbivory at both unfished and fished sites.

Carpenter's hypothesis, however, does not explain the response of Belizean reefs to damage from Hurricane Hattie in the early 1960s, two decades before the *Diadema* mass mortality. After Hurricane Hattie, macroalgae dominated for more than ten years before corals recovered on those unfished reefs (Stoddart 1963, 1969, 1974). Likewise, as coral cover declined in the 1980s, macroalgal cover increased from less than 5% to greater than 60% on the fore reef at Carrie Bow Cay, Belize (12–15 m depth; Littler et al. 1987; Aronson et al. 1994). Carrie Bow Cay was lightly fished at the time, and *Diadema* were virtually absent below 6 m depth before, as well as after, their mass mortality (Lewis and Wainwright 1985).

Knowlton (1992) proposed a hypothesis to explain macroalgal dominance in areas with high densities of herbivorous fishes, such as Carrie Bow Cay. Scarids and acanthurids are selective herbivores, whereas *Diadema* are less selective (but see Ogden, Abbott, and Abbott 1973; Carpenter 1981, 1997; and Hay, Kappel, and Fenical 1994 on selective feeding by *Diadema*). Where *Diadema* were abundant, macroalgal cover was low because the urchins ate algae more or less indiscriminately. The *Diadema* mass mortality removed the single most important nonselective herbivore from Caribbean reefs. Following space-clearing disturbances, macroalgal cover increased, because the algae grew faster than the corals and because herbivorous fishes avoided physically and chemically defended algal species (Hay 1985, 1991, 1997; Hay and Goertemiller 1983; Knowlton 1992). The cover of unpalatable macroalgae increased even on some reefs and in some reef habitats where fish density was high and *Diadema* density was low prior to the mass mortality. In those places, coral mortality led to macroalgal dominance precisely because *Diadema* were historically rare. In other cases, high densities of herbivorous fishes were able to maintain low algal cover.

In St. Croix, microherbivores (amphipods, tanaids, polychaetes, and gastropods) had a negligible effect on algal biomass and productivity (Carpenter 1986). On the other hand, microherbivores (gastropods and hermit crabs) were important in preventing macroalgae from dominating a site in Bermuda disturbed by a ship grounding (Smith 1988). From this and the foregoing observations, we can only conclude that although herbivory influences the distribution, abundance, and productivity of algae, it is sometimes difficult to predict specific effects.

Arguments about the ecological effects of herbivores on "pristine" versus "disturbed" reefs may well be moot, because pristine reefs probably have not existed in the Caribbean for a very long time. Megaherbivores such as sea turtles and manatees most likely had important effects on algal dynamics until their near-extermination in the eighteenth century (Jackson 1997). The loss of megaherbivores, along with the subsequent exploitation of reef fishes, is evidence that Caribbean reefs have been disturbed for centuries.

Of course, it is still important to understand the dynamics of change in modern reef ecosystems, and many ambiguities remain. Jackson (1997), citing historical descriptions, argued that *Diadema* were naturally abundant on at least some Caribbean reefs. Historically, *Diadema* may have exerted a generally greater influence on the (oceanically influenced) islands of the eastern Caribbean than along the (terrestrially influenced) Central American coast (Levitan 1992; Jackson 1997). Perhaps this biogeographical pattern was erroneously attributed to recent differences in fishing pressure; from the literature cited previously, *Diadema* appear to have been less important in Belize (less fished) and more important in the Antilles (more heavily fished).

Nutrient loading has increased macroalgal cover on some Caribbean reefs (Tomascik and Sander 1987; Tomascik 1991; Knowlton 1992). In addition to increasing algal growth rates, dissolved nutrients disrupt the coral-zooxanthellae symbiosis, poison the calcification process, and promote bioerosion (Highsmith 1980; Barnes and Chalker 1990; Jokiel, Dubinsky, and Stambler 1994; Dubinsky and Stambler 1996; Marubini and Davies 1996). Consequently, reef development is poor in areas influenced by upwelling, terrestrial runoff, or other nutrient sources (Neumann and Macintyre 1985; Hallock and Schlager 1986; Hallock 1988; Wood 1993; Burke 1994; Ginsburg and Shinn 1994). Temporary, nutrient-induced algal blooms often occur in fore-reef habitats following storms (e.g., Hughes 1994; Rogers, Garrison, and Grober-Dunsmore 1997). These blooms probably result from the sudden resuspension of nutrients stored in sediments, as well as from the outflow of nutrient-rich lagoonal waters and terrestrial runoff (Szmant and Forrester 1996; Szmant 1997). Debate continues over the importance of natural and anthropogenic nutrient input to the balance between corals and macroalgae on coral reefs (Lewis 1984; Hunte and Wittenberg 1992; Rougerie, Fagerstrom, and Andrie 1992; Wittenberg and Hunte 1992; Hunter and Evans 1995; McCook 1996; Szmant and Forrester 1996; Lapointe 1997; McCook, Price, and Klumpp 1997; McClanahan and Muthiga 1998; McCook 1999; Miller et al. 1999).

Szmant (1997) suggested that the level of topographic complexity on a reef indirectly influences the effects of nutrient input and uptake. In the absence of overfishing or mass mortalities of herbivores, topographic complexity determines the availability of shelter for herbivores, thereby mediating rates of consumption of algae in the face of pulsed or chronic nutrification. Nutrients and herbivory clearly interact in complex ways to determine the outcome of competition between corals and macroalgae (Meyer, Schultz, and Helfman 1983; Littler and Littler 1985; Littler, Littler, and Titlyanov 1991; McClanahan 1997; Miller 1998).

Attention to herbivory and nutrient loading de-emphasizes the pivotal role of coral mortality in the transition to macroalgal dominance. Some authors have explicitly acknowledged the importance of coral mortality (Adey et al. 1977; Hughes, Reed, and Boyle 1987; Done 1992a; Knowlton 1992; Rogers, Garrison, and Grober-Dunsmore 1997; Hughes and Connell 1999; McCook 1999), but others advocate a primary role for herbivory (Steneck 1994; McClanahan 1995; Hay 1997) or nutrient enrichment (Lapointe 1997). Although algae have higher growth rates and are capable of outcompeting corals, they must recruit to the reef in order to do so (Umar, McCook, and Price 1998). Discovery Bay had been heavily fished for decades to centuries (Munro 1983; Jackson 1997) but supported high coral cover prior to 1980, even in intermediate to deeper habitats in which *Diadema* were less common (Morrison 1988; Jackson 1991). Before 1980, the corals apparently pre-empted space (Ohlhorst 1984; Littler and Littler 1985). Had Hurricane Allen not occurred and had coral cover remained high at Discovery Bay, the *Diadema* die-off probably would not have had such a rapid and dramatic effect, because there would have been far less free space available on which *Diadema* grazing was suddenly alleviated.

Even in situations where herbivores are abundant and nutrient concentrations are low, coral mortality can result in macroalgal dominance if the amount of empty space created exceeds the feeding capacity of consumer populations. The potential of herbivores to respond numerically to the availability of algae will be limited by the loss of topographic heterogeneity that accompanies coral mortality from hurricanes and other disturbances (e.g., Adey et al. 1977; Kaufman 1983; Szmant 1997). Furthermore, in the absence of *Diadema*, the scarids and acanthurids, which are selective feeders, may be unable to control physically and chemically defended macroalgae (Knowlton 1992).

Differential Coral Mortality

Observations of interspecific differences in coral mortality support the contention that mortality is a precondition for macroalgal dominance. Branching corals, including the two *Acropora* species, are particularly susceptible to hurricane damage. Both species were devastated in Discovery Bay by Hurricane Allen (Woodley et al. 1981).

Edmunds and Bruno (1996) compared the West Fore Reef at Discovery Bay to other reef sites along the north coast of Jamaica in 1995. Fishing pressure was historically heavy throughout the area studied, and *Diadema* had been absent for more than a decade. One comparison site was located at Dairy Bull, a few kilometers east of Discovery Bay. At the intermediate depth of 10 m, coral cover was considerably higher at Dairy Bull than at Discovery Bay (table 9.1), and macroalgal cover was lower (40% macroalgal cover at Dairy Bull versus 62% at Discovery Bay; Edmunds and Bruno 1996). A principal reason for the difference between sites was that the cover of *M. annularis* complex was much higher at Dairy Bull than at Discovery Bay. In fact, the cover of *M. annularis* was higher at Dairy Bull in 1995 than it was at Discovery Bay prior to Hurricane Allen (table 9.1).

The maximum skeletal growth rate measured for *M. annularis* is ~1.5 cm linear extension per year (Hudson 1981a,b). The average growth rate is lower: a 50 cm tall colony is approximately 50 yr old under good conditions (Dodge,

TABLE 9.1. Comparison of Coral Cover at 10 m Depth Between the West Fore Reef at Discovery Bay (WFR) and Dairy Bull, Jamaica.[a]

Category	WFR 1977	WFR 1995	Dairy Bull 1995
Total coral cover	42%	2%	23%
M. annularis complex	7%	1%[b]	13%
A. cervicornis	12%	0%	0%

[a] The 1977 data for Discovery Bay are from Huston (1985b). The 1995 data for Discovery Bay and Dairy Bull are from Edmunds and Bruno (1996 and unpubl. data).
[b] Value for *M. annularis* cover at Discovery Bay in 1995 agrees with values for the same depth in 1993 reported by Hughes (1994).

Aller, and Thomson 1974; Dustan 1975; Graus and Macintyre 1982; Highsmith, Lueptow, and Schonberg 1983; Shinn et al. 1989). Because most of the *M. annularis* colonies at Discovery Bay and Dairy Bull were considerably taller than 50 cm in 1995, they must have predated Hurricane Allen and the *Diadema* mass mortality. This suggests that much of the difference between the sites in 1995 was due to the mortality and subsequent macroalgal overgrowth of corals more susceptible to hurricane damage than *M. annularis.*

Prior to 1980, the most common coral at the two sites other than *M. annularis* was *A. cervicornis,* judging from (1) surveys conducted at Discovery Bay in the 1970s (Porter et al. 1981; Huston 1985b), and (2) the enormous quantity of *A. cervicornis* rubble at 10 m depth at both sites in the 1990s. Hurricane Allen had the main effect of subtracting *A. cervicornis* from Discovery Bay and (presumably) Dairy Bull, leaving *M. annularis* more or less intact (Woodley et al. 1981; Hughes 1989; Steneck 1994; see also Bak and Luckhurst 1980). Hurricane Gilbert in 1988 had less dramatic effects on corals at Discovery Bay and (presumably) Dairy Bull, because *A. cervicornis* and other corals had only begun to recover from Hurricane Allen (Woodley 1989). *M. annularis* colonies again were largely unaffected (Hughes and Connell 1999).

This pattern was repeated elsewhere. Stoddart (1963) reported the nearly complete destruction of *A. cervicornis* but only moderate to minimal destruction of *M. annularis* following Hurricane Hattie in Belize (see also Glynn, Almodóvar, and González 1964 on Hurricane Edith's impact in Puerto Rico). Shinn (1976; Shinn et al. 1989) commented on the volatility of *A. cervicornis* populations in the Florida Keys, and Curran et al. (1994) noted that *A. cervicornis* died but *M. annularis* remained intact in the Bahamas. Macroalgal overgrowth subsequent to the *Diadema* mass mortality reduced the cover of *M. annularis* in Jamaica (Hughes 1994), but this did not happen in Belize, the Florida Keys, or the Bahamas.

In extreme cases of herbivore removal, macroalgae will overgrow massive corals (Hay and Taylor 1985; Lewis 1986; but see de Ruyter van Steveninck, van Mulekom, and Breeman 1988). Although the absence of herbivores may have been largely responsible, the mortality of *M. annularis* at Discovery Bay can also be linked to the activities of pomacentrids. L. S. Kaufman (personal communication, 1998) noted an increase in the number of three-spot damselfish (*Stegastes planifrons*) territories on *M. annularis* after Hurricane Allen. *S. planifrons* apparently moved their territories to head corals after the hurricane eliminated their preferred microhabitat, *A. cervicornis* thickets. *M. annularis* heads were less preferred before the demise of *A. cervicornis* (Kaufman 1977; see also Williams 1978; Ebersole 1985), but those coral heads are currently the microhabitat of choice for *S. planifrons* in St. Croix (Tolimieri 1998). Eakin (1989) also observed that, in the absence of live branching corals, juvenile *S. planifrons* occupied living *M. annularis* colonies in Florida. By partially killing head corals, *S. planifrons* probably contributed to macroalgal overgrowth. Qualitative observations over the past few years indicate increasing numbers of *S. planifrons* territories on massive corals in Belize and the Florida Keys, again causing partial mortality (W. F. Precht, unpublished observations, 1998; S. L. Miller and A. M. Szmant, personal communications, 1998).

Differential coral mortality has also occurred between reef zones. For example, fore-reef zones at Carrie Bow Cay, Belize, experienced differential mortality in the 1980s. The shallow spur-and-groove zone (3–6 m depth) had high coral cover in the 1970s to early 1980s, dominated by the blade-shaped lettuce coral, *Agaricia tenuifolia*. Even though *Diadema* were present in this shallow zone at that time, scarids and acanthurids were the most important herbivores (Lewis and Wainwright 1985). The deeper spur-and-groove zone at intermediate depths (9–15 m) also had high coral cover, dominated by *A. cervicornis* (Burke 1982; Rützler and Macintyre 1982). *Diadema* were rare in this deeper zone, herbivorous fishes were more abundant in the deeper zone, and herbivory was intense and macroalgal cover was low in both zones (Lewis and Wainwright 1985; Littler et al. 1987). In the 1990s, *Diadema* were absent from both zones, and scarids and acanthurids were abundant and not subject to fishing pressure. Coral cover remained variable but high in the shallow spur-and-groove (Aronson and Precht 1995), while the deeper spur-and-groove was largely covered by macroalgae (Aronson et al. 1994). The difference in macroalgal cover between zones in the 1990s arose because most of the *A. cervicornis* in the deeper zone died in the early 1980s.

These observations highlight the importance of coral mortality, specifically the mortality of *A. cervicornis*, as a prerequisite to macroalgal dominance at intermediate depths. Presumably the mortality of *A. palmata* is equally necessary for coral-to-macroalgal transitions in reef crest and shallow fore-reef habitats. What are the important causes of coral mortality in the Caribbean?

Causes of Coral Mortality

Like the herbivory scenario, much of what reef ecologists believe about coral mortality is influenced by the experience at Discovery Bay. The devastation of *A. cervicornis* from Hurricane Allen and collateral mortality from pomacentrids and corallivorous invertebrates (Porter et al. 1981; Woodley et al. 1981; Knowlton et al. 1981; Knowlton, Lang, and Keller 1990) was followed by some damage from Hurricane Gilbert in 1988 (Woodley 1989; Hughes 1994). In 1989 Hurricane Hugo damaged reefs in the U.S. Virgin Islands (Edmunds and Witman 1991; Hubbard et al. 1991; Rogers, McLain, and Tobias 1991; Bythell, Bythell, and Gladfelter 1993; Aronson, Ebersole, and Sebens 1994). These studies and other ecological and geological work (e.g., Ball, Shinn, and Stockman 1967; Perkins and Enos 1968; Stoddart 1974; Rogers, Suchanek, and Pecora 1982; Mah and Stearn 1986; Fenner 1991; Kobluk and Lysenko 1992; Blair, McIntosh, and Mostkoff 1994; Blanchon 1997; Blanchon, Jones, and Kalbfleisch 1997) led to the prevailing opinion that hurricanes are a primary cause of present and past coral mortality in the Caribbean.

Hurricanes have been important at some localities, but they do not explain recent patterns of coral mortality in much of the region. Damage from hurricanes is patchy on many spatial scales (Hubbard et al. 1991; Rogers 1993b; Steneck 1994). Some areas of the Caribbean, such as Trinidad, Costa Rica, and Panama, receive virtually no hurricanes, while others, including Jamaica, are struck frequently (Neumann et al. 1987; Woodley 1992; Tremel, Colgan, and Keevican 1997). Coral populations often suffer more extensive damage from chronic disturbances (Rogers 1993b). In Belize, for example, the deeper, *A. cervicornis*-dominated spur-and-groove zone lost cover beginning in the 1980s, as described previously. Hurricane Greta is known to have damaged the fore reef in 1978 (Rützler and Macintyre 1982; Kjerfve and Dinnel 1983); however, Hurricane Greta was not responsible for the mass destruction of *A. cervicornis* populations in the 1980s. White-band disease (WBD) was the principal cause of *A. cervicornis* mortality.

WBD is a presumed bacterial infection that is specific to *Acropora* spp. (Antonius 1981; Gladfelter 1982; Peters 1993). Cases of WBD are recognizable as segments of bare skeleton, sometimes bordered by narrow bands of disintegrating, necrotic coral tissue, on otherwise healthy-looking, brown *Acropora* branches. The "bands" of disease spread along the branches, generally from base to tip, and eventually kill entire colonies. The etiology of WBD and the causes of outbreaks are poorly understood, and recent reports suggest that there are several varieties of the disease with different symptoms (Peters 1993, 1997; Antonius 1995; Peters and McCarty 1996; Santavy and Peters 1997; Richardson 1998).

Despite the difference in appearance upon careful examination, WBD in Caribbean *Acropora* spp. has often been mistaken for coral bleaching, which has received considerably more attention in the literature. Dramatic bleaching events at times of elevated sea temperature and ultraviolet radiation have caused mass mortalities of corals in the Indo-Pacific, although bleaching-related mortality has until recently been more limited in the Caribbean (Jaap 1979; Brown 1987, 1997; Ogden and Wicklund 1988; Glynn 1990b, 1993; Williams and Bunkley-Williams 1990; D'Elia, Buddemeier, and Smith 1991; Glynn and Colgan 1992; Lang et al. 1992; Fitt et al. 1993; Shick, Lesser, and Jokiel 1996; Aronson et al. 2000). An important reason for the current interest in bleaching is that global warming may increase its frequency and extent (Smith and Buddemeier 1992; Glynn 1996).

Acropora spp. will bleach under thermal stress (Cortés 1994). Other sources of *Acropora* mortality include predation by corallivores (Knowlton et al. 1981; Knowlton, Lang, and Keller 1990; Tunnicliffe 1983), nutrient loading (Weiss and Goddard 1977; Lewis 1984; Bell and Tomascik 1994), sedimentation (Rogers 1990; Cortés 1994), and, in Florida and the Bahamas, cold water stress (Davis 1982; Porter, Battey, and Smith 1982; Roberts et al. 1982; Burns 1985). Although quantitative data are generally lacking, it is becoming apparent that WBD epizootics have been the primary cause of the recent mortality of *A. cervicornis* and *A. palmata* over wide areas of the Caribbean (Rogers 1985; Wells 1988; Sheppard 1993; table 9.2).

Stands of *A. cervicornis* killed by WBD generally collapse due to weakening of the skeletons by bioerosion, and the result is large fields of *A. cervicornis* rubble. *A. palmata* is more robust, and stands of this species remain in growth position for longer periods after they have been killed, as has been observed in Anguilla, Belize, the Florida Keys, St. Croix, and elsewhere. Dead stands of *A. palmata* are then leveled by storms (Bythell et al. 1989; Hubbard et al. 1991). Hurricane damage has clearly been a more localized cause of *Acropora* mortality over the past few decades (table 9.3).

Compared with *Acropora* spp., the mortality of massive corals has been more variable. Colonies of *M. annularis* complex have been affected by hurricanes, bleaching, and disease (Rützler, Santavy, and Antonius 1983; Porter et al. 1989; Edmunds 1991; Edmunds and Witman 1991; Fitt et al. 1993; Meesters and Bak 1993; Kuta and Richardson 1996), but large stands of *M. annularis* can still be observed throughout the region (e.g., Shinn et al. 1989; Precht 1993; Curran et al. 1994; Edmunds and Bruno 1996; Burke et al. 1998; McClanahan and Muthiga 1998).

TABLE 9.2. Reports of White-Band Disease as the Primary Cause or an Important Cause of *Acropora* Mortality on Caribbean Reefs over the Past Few Decades[a]

Location (habitat)	Species Affected	Time Period	Source
Anguilla			
(rc)	A. palmata[b]	1980s–90s	Sheppard et al. 1995
Bahamas			
Andros Barrier Reef			
(br)	A. cervicornis	1980s	S. Cove, personal communication
(rc)	A. palmata[c]	1980s	S. Cove, personal communication
(fr)	A. cervicornis	1980s	S. Cove, personal communication
New Providence Island			
(br)	A. cervicornis	1980s	S. Cove, personal communication
(rc)	A. palmata[c]	1980s	S. Cove, personal communication
(fr)	A. cervicornis	1980s	S. Cove, personal communication
San Salvador Island			
(rc)	A. palmata	1980s	Shinn 1989
(pr)	A. cervicornis	1980s	Shinn 1989; Curran et al. 1994
Lee Stocking Island, Exumas			
(pr)	A. cervicornis[c]	1980s–90s	J. C. Lang, personal communication
Belize			
Ambergris Cay			
(pr)	A. cervicornis	1980s–90s	Precht, unpublished observation
(rc)	A. palmata	1980s–90s	Precht, unpublished observation
(fr)	A. cervicornis	1980s–90s	Precht, unpublished observation
Carrie Bow Cay			
(pr)	A. cervicornis	1980s–90s	Precht, unpublished observation
(rc)	A. palmata[c]	1980s–90s	Aronson and Precht, unpublished observation
(fr)	A. cervicornis	1980s	Precht, unpublished observation
Central Shelf Lagoon			
(l)	A. cervicornis	1986–90	Aronson and Precht 1997
Ranguana Cay			
(rc)	A. palmata	1980s–90s	Precht, unpublished observation
(fr)	A. cervicornis	1980s–90s	Precht, unpublished observation
British Virgin Islands			
(rc/fr)	A. palmata	1980s–90s	Bythell and Sheppard 1993
Cayman Islands			
Grand Cayman			
(fr)	A. cervicornis[d]	1980s	Woodley et al. 1997
Colombia			
(g)	Acropora ssp.[b]	1970s–90s	Garzón-Ferreira and Kielman 1994
San Andrés			
(l/pr)	A. palmata[b,d]	1970s–90s	Zea et al. 1998
(l)	A. cervicornis[b,d]	1970s–90s	Zea et al. 1998

(*Continued on next page*)

TABLE 9.2. (*continued*)

Location (habitat)	Species Affected	Time Period	Source
Cuba			
(rc/fr)	*A. palmata*[c]	1980s–90s	P.M. Alcolado, personal communication
(fr)	*A. cervicornis*[c]	1980s–90s	P.M. Alcolado, personal communication
Dominican Republic			
(fr)	*A. cervicornis*[b]	1980s–90s	M. Vega, personal communication
Florida Reef Tract			
Biscayne National Park (2 reefs), Northern Keys (1 reef), and Looe Key (1 reef)			
(rc/fr)	*A. palmata*[d]	1980s	Porter and Meier 1992; Porter et al. 1993
Northern Keys (4 reefs)			
(fr)	*A. cervicornis*	1980s	Jaap, Halas and Muller 1988; Shinn, et al. 1989
Jamaica			
Discovery Bay			
(fr)	*A. cervicornis*[d]	1980–88	Knowlton, Lang, and Keller 1990; Tunnicliffe 1983
Netherlands Antilles			
(rc/fr)	*A. palmata*[c]	1970s–80s	Bak and Criens 1981
(fr)	*A. cervicornis*	1970s–80s	Bak and Criens 1981; van Duyl 1982, 1985; Wells 1988
Puerto Rico			
La Parguera			
(rc/fr, l)	*A. palmata*[d]	1990s	Bruckner et al. 1997
(pr)	*A. cervicornis*[d]	1990s	Bruckner and Bruckner 1997
Panama			
San Blas Islands			
(pr)	*Acropora* spp.[b]	1970s–80s	Ogden and Ogden 1994
Trinidad and Tobago			
Tobago			
(rc/fr)	*A. palmata*	1980s	Laydoo 1984
U.S. Virgin Islands			
St. Croix: Buck Island			
(rc/fr)	*A. palmata*	1976–85	Gladfelter 1982; Bythell et al. 1989
(fr)	*A. cervicornis*	1976–85	Bythell et al. 1989
St. John			
(rc/fr)	*A. palmata*[d]	1980s–90s	Rogers 2000

[a] Habitat abbreviations: rc, reef crest; fr, fore reef; pr, patch reef; l, lagoon; g, reef habitats in general. No superscript, WBD as principal cause of widespread mortality. See Rogers (1985) and Wells (1988) for earlier reports of the presence of WBD around the Caribbean.

[b] WBD as probable cause of widespread mortality.

[c] WBD as cause of some mortality.

[d] WBD associated with other sources of coral mortality.

TABLE 9.3. Reports of Hurricane Damage as the Primary Cause or an Important Cause of *Acropora* Mortality on Caribbean Reefs over the Past Few Decades.[a]

Location (habitat)	Species Affected	Year (Storm)	Source
Belize			
Turneffe Islands			
(rc)	A. palmata	1961 (Hattie)	Stoddart 1963
(fr)	A. cervicornis	1961 (Hattie)	Stoddart 1963
Barrier Reef leeward of Turneffe Islands			
(rc)	A. palmata	1961 (Hattie)	Stoddart 1963
(fr)	A. cervicornis	1961 (Hattie)	Stoddart 1963
Peter Douglas Cay			
(l)	A. cervicornis[c]	1961 (Hattie)	Stoddart 1963
Carrie Bow Cay			
(rc)	A. palmata	1978 (Greta)	Highsmith et al. 1980; Kjerfve and Dinnel 1983
(fr)	A. cervicornis[c]	1978 (Greta)	Highsmith et al. 1980; Kjerfve and Dinnel 1983
Cayman Islands			
Grand Cayman			
(rc/fr)	A. palmata[c]	1988 (Gilbert)	Blanchon et al. 1997
Colombia			
San Andrés			
(l/pr)	A. palmata[b?]	1988 (Joan)	Zea et al. 1998
(l)	A. cervicornis[c?]	1961 (Hattie)	Zea et al. 1998
Florida Reef Tract			
Biscayne National Park			
(pr)	A. palmata[c]	1992 (Andrew)	Rogers 1994; Lirman and Fong 1996, 1997a
(pr)	A. palmata[c]	1993 (Storm of the Century)	Lirman and Fong 1997b
(pr)	A. palmata[c]	1994 (Gordon)	Lirman and Fong 1997b
Northern Keys, off Key Largo			
(rc/fr)	A. palmata[c]	1960 (Donna)	Ball et al. 1967
(rc/fr)	A. palmata	1965 (Betsy)	Perkins and Enos 1968
(fr)	A. cervicornis	1965 (Betsy)	Perkins and Enos 1968
(rc/fr)	A. palmata[c]	1992 (Andrew)	Precht et al. 1993
Jamaica			
Port Royal Cays			
(rc)	A. palmata[b]	1951 (Charlie)	Goreau 1959; Woodley 1992
Discovery Bay			
(rc)	A. palmata	1980 (Allen)	Porter et al. 1981; Woodley et al. 1981
(fr)	A. cervicornis	1980 (Allen)	Woodley et al. 1981; Knowlton et al. 1990
(rc)	A. palmata[c]	1988 (Gilbert)	J. D. Woodley, pers. comm.
(fr)	A. cervicornis[c]	1988 (Gilbert)	Woodley 1989

(*Continued on next page*)

TABLE 9.3. (*continued*)

Location (habitat)	Species Affected	Year (Storm)	Source
Netherlands Antilles			
Bonaire			
(rc/fr)	*A. palmata*[c]	1988 (Gilbert, Joan)	Kobluk and Lysenko 1992
(fr)	*A. cervicornis*[c]	1988 (Gilbert, Joan)	Kobluk and Lysenko 1992
Puerto Rico			
La Parguera			
(rc)	*A. palmata*	1963 (Edith)	Glynn et al. 1964
(fr)	*A. cervicornis*	1963 (Edith)	Glynn et al. 1964
(rc)	*A. palmata*	1979 (David)	Vicente 1994
(fr)	*A. cervicornis*	1979 (David)	Vicente 1994
U.S. Virgin Islands			
St. Croix: Buck Island			
(rc/fr)	*A. palmata*	1979 (David, Frederic)	Rogers et al. 1982
(rc/fr)	*A. palmata*[c,d]	1989 (Hugo)	Bythell et al. 1989; Rogers 1992

[a] Habitat abbreviations as in Table 2. No superscript, hurricanes as principal cause of widespread mortality. Note that the effects of Hurricane Gilbert (1988) in Jamaica were limited because the reef had not yet recovered from Hurricane Allen (1980). The same can be said for the effects of the Storm of the Century (1993) and Tropical Storm Gordon (1994) in Florida, both of which following Hurricane Andrew (1992).
[b] Hurricanes as probable cause of widespread mortality.
[c] Hurricanes as cause of some mortality.
[d] Hurricanes associated with other sources of coral mortality.

In summary, WBD, and to a lesser extent hurricanes, caused the mortality of *Acropora* throughout the Caribbean, reducing coral cover substantially on most reefs. Even at Discovery Bay, WBD was noted on *A. cervicornis* prior to Hurricane Allen, and the disease may have killed surviving fragments after the storm (Knowlton et al. 1981; Knowlton, Lang, and Keller 1990; Tunnicliffe 1983; Woodley et al. 1997). Hurricanes, diseases, bleaching, and probably nutrient loading have caused mortality in populations of nonacroporid corals in recent years, including massive *Montastraea* spp. and *Diploria* spp., foliose *Agaricia* spp., branching and head-forming *Porites* spp., and many others. This mortality has been highly variable: As an example, Hurricane Hugo severely damaged *M. annularis* populations in St. John, U.S. Virgin Islands (Edmunds and Witman 1991), but hurricanes had minimal impacts on *M. annularis* populations in Jamaica and Belize.

The effects of corallivorous invertebrates on Caribbean reefs remain poorly understood, but there may be a causal link between coral diseases and outbreaks of corallivores (Knowlton, Lang, and Keller 1990; Antonius and Riegl 1997; Bruckner, Bruckner, and Williams 1997). Also, despite earlier reports to the contrary, some parrotfish species eat corals at some localities in the Caribbean, but the importance of this trophic linkage to the ecology of Caribbean reefs requires further study (Bruckner and Bruckner 1998; Miller and Hay 1998). The prospects for recovery of coral populations depend to a large extent on the nature and recurrence of disturbance and on the life-history strategies of the corals.

Coral Reproductive Strategies

Corals reproduce asexually, primarily by fragmentation. They also employ two strategies of sexual reproduction: broadcast spawning of gametes and release of brooded planula larvae (Szmant 1986; Richmond and Hunter 1990; Smith 1992; Richmond 1997). Slow recovery times of populations of *A. palmata*, *A. cervicornis*, and (to a lesser extent) *M. annularis* complex can be tied to their reproductive strategies.

The key to success for *A. cervicornis* prior to 1980 was rapid growth coupled with reproduction by branch fragmentation (Shinn 1966; Gilmore and Hall 1976; Tunnicliffe 1981; Highsmith 1982). The dominance of *Acropora* spp. through asexual reproduction likely resulted in low genetic variation in at least some localities, possibly increasing the susceptibility of populations to disease (Bak 1983; Neigel and Avise 1983). However successful *A. cervicornis* was formerly, a nearly exclusive dependence on asexual reproduction, with limited potential for larval recruitment (Hughes 1985; Sammarco 1985; Knowlton, Lang, and Keller 1990; Hughes, Ayre, and Connell 1992), has slowed its recovery at most localities. *A. palmata* has shown higher rates of sexual recruitment than *A. cervicornis* in some situations (Rosesmyth 1984; Jordán-Dahlgren 1992), but *M. annularis* complex, like *A. cervicornis*, has recruited poorly (Bak and Engel 1979; Hughes 1985, 1989; Szmant 1986; Smith 1992).

A. palmata, *A. cervicornis*, and *M. annularis* complex are all broadcast spawners. Because they are now rare, *A. palmata* and *A. cervicornis* may be experiencing an Allee effect: colonies may be too far apart for high fertilization success (Knowlton 1992; Kojis and Quinn 1994; Levitan and Petersen 1995). The same may be true for *M. annularis* complex, although the mortality of adult colonies has not been nearly as dramatic or complete as in the *Acropora* spp. (see Bak and Meesters 1999).

At this point, brooding corals are recruiting more successfully than broadcast spawners in the Caribbean. *Ag. agaricites,* a morphologically plastic, foliose species; other *Agaricia* spp.; *Porites astreoides,* a head-forming species; and *P. porites,* a branching species, are among the first to appear on disturbed reef surfaces, including *Acropora* rubble fields (Bak and Engel 1979; Neese and Goldhammer 1981; Rylaarsdam 1983; Rogers et al. 1984; Hughes 1985, 1989; Sammarco 1985; Hunte and Wittenberg 1992; Smith 1992, 1997; Chiappone and Sullivan 1996; Aronson and Precht 1997; Edmunds et al. 1998). All Caribbean representatives of the families Agariciidae and Poritidae are brooders.

One explanation is that the flexibility of larval lifespan afforded by lecithotrophy enables brooded planulae to settle either near or far from the mother colony (Fadlallah 1983; Szmant-Froelich, Ruetter, and Riggs 1985; Richmond 1987; Sammarco and Andrews 1989; Harrison and Wallace 1990; Ward 1992; Edinger and Risk 1995; Sakai 1997). Another possibility is that brooders have some advantage over broadcasters in fertilization success. *Acropora* spp., *Montastraea* spp., *Ag. agaricites,* and *P. astreoides* are all hermaphroditic, although *P. porites* may be gonochoristic (Harrison and Wallace 1990; Szmant 1986; Richmond and Hunter 1990; Richmond 1997). Information on rates of self-fertilization in hermaphroditic brooders *versus* hermaphroditic broadcasters would be highly relevant to the issue of recruitment success; the brooding *P. astreoides* and *Favia fragum,* for example, can exhibit high rates of self-fertilization (Gleason and Brazeau 1997; Brazeau, Gleason, and Morgan 1998), whereas self-fertilization is probably rare in broadcasting *Montastraea* spp. that are hermaphroditic (Knowlton et al. 1997; Szmant et al. 1997) and in other hermaphroditic broadcasters (Hagman, Gittings, and Vize 1998).

In the Indo-Pacific, brooding and broadcasting acroporids and other brooding and broadcasting species rapidly recolonize hard substrata in the wake of crown-of-thorns starfish (*Acanthaster planci*) outbreaks, coral bleaching, and other disturbances. Post-disturbance periods of algal dominance are generally shorter than in the Caribbean (Pearson 1981; Sammarco 1985; Colgan 1987; D. Smith 1991; Done 1992b; S. Smith 1992; Kojis and Quinn 1994; Gleason 1996; Fabricius 1997; Connell, Hughes, and Wallace 1997; but see Hatcher 1984; Endean, Cameron, and DeVantier 1988). The difference is in part a function of the small size of the Caribbean, where disturbances are more likely to have regional-scale effects, limiting the availability of coral recruits from upstream reefs in the wake of coral mortality (Connell 1997; Roberts 1997). Interoceanic differences in life history strategies, including a greater emphasis on sexual reproduction in Indo-Pacific acroporids, are probably also a factor.

It may be that the successful colonists of the Caribbean, which are not important framework builders, just happen to be brooders because a high proportion of living Caribbean species are brooders (Richmond and Hunter 1990; Richmond 1997). In other words, even if recruitment success on Caribbean reefs is independent of reproductive strategy, most good recruiters could be brooders based on chance alone. The emphasis on brooding in the Caribbean is tied to the evolutionary history of the modern coral fauna, as we shall see in the next section.

Origin of the Modern Coral Fauna of the Caribbean

The present-day coral fauna of the Caribbean is very unlike that of the Indo-Pacific, although both originated from the same Eocene–Oligocene pantropical species pool. Many of these differences can be traced to two intervals of evolutionary change in the Caribbean: A regional extinction in the early Miocene and a period of accelerated turnover during the Plio-Pleistocene. These two formative episodes affected both the taxonomic composition and ecological attributes of the fauna.

Following an extinction interval in the late Eocene (Budd, Stemann, and Stewart 1992), coral reefs were well developed and the scleractinian fauna was diverse in the Caribbean by the late Oligocene (Frost and Langenheim 1974; Frost 1977b; Veron 1995). At the beginning of the Miocene, closure of the eastern end of the Tethyan Seaway and the opening of the Drake Passage increased upwelling in the Caribbean (references in Edinger and Risk 1994, 1995). Environmental changes related to upwelling, including increased nutrient loading and turbidity, and lowered sea temperatures, were apparently responsible for the regional extinction of half the genera living in the Caribbean at the time (Edinger and Risk 1994, 1995). The survivors were primarily eurytopic genera in terms of both habitat distribution and environmental tolerance (the latter surmised from the autecology of modern representatives). Genera that were tolerant of the cool, turbid water produced by upwelling survived in far greater proportion than genera tolerant of cool water only, turbid water only, or neither cool nor turbid water (Edinger and Risk 1994). Broad habitat distribution was also correlated with survival.

Taxonomically, the result of this extinction was a culling of genera in the Caribbean. Most of the genera that became extinct in the Caribbean are extant in the Indo-Pacific (Frost and Langenheim 1974; Frost 1977a). Ecologically, the early Miocene extinction may have fundamentally changed the distribution of reproductive strategies in Caribbean corals.

Edinger and Risk (1995) inferred the reproductive strategies of Miocene coral genera from the modern representatives of those genera. If such inferences are valid (and they require caution since coral reproductive modes are evolutionarily labile; Kinzie 1997), then most genera that survived the extinction contained all or at least some brooding species. This apparent bias in survival during the Miocene contributed to the greater emphasis on brooding reproduction in the Caribbean today; most modern Indo-Pacific species are broadcasters (Smith 1992; Richmond 1997).

Edinger and Risk (1995) suggested two possible explanations for the preferential survival of brooding genera in the early Miocene, one based on dispersal and the other on recruitment. The first hypothesis was that brooded, lecithotrophic planulae dispersed further than the planktotrophic planulae produced by broadcast spawning (Richmond 1987, 1988; Richmond and Hunter 1990). The observed geographic ranges of brooding and broadcasting corals in the Indo-Pacific do not support this dispersal hypothesis. Instead, Edinger and Risk (1995:212–213) preferred a recruitment-based hypothesis in which the high proportion of brooding genera surviving the Miocene extinction event was a consequence of species sorting. In this hypothesis, brooding was associated with other features that conferred recruitment success and resultant extinction resistance:

> Brooding . . . coral genera are mostly eurytopic, tolerant of both turbid and cold conditions. . . . Onset of upwelling in the Caribbean during the Early Miocene apparently favored brooding corals, but by virtue of the ecological correlates of brooding, rather than reproductive mode per se. Rather, brooding is correlated with other traits which helped corals survive in the deteriorating habitats of the Miocene Caribbean.

This argument is based on the observed correlations among brooding, eurytopy, and survival in the Miocene.

If brooding was a consequence rather than a cause of generic survival, then we can make a testable prediction: There should be no difference in generic survival between eurytopic brooders and eurytopic broadcasters. We used Edinger and Risk's (1995) data on eurytopic Miocene coral genera from Puerto Rico to examine the relationship between survival and reproductive mode. One null hypothesis was that survival among genera tolerant of both cold water and turbidity was independent of reproductive mode. We could not falsify this null hypothesis (one-tailed Fisher's exact test, $0.20 < P < 0.50$), a result that probably reflects low sample size (figure 9.2A). Brooding genera with a wide habitat distribution, however, were significantly more likely to survive than broadcasting genera with a wide habitat distribution ($P < 0.05$; figure 9.2B). Therefore, the possibility remains that brooding itself conferred survival advantage during the Miocene crisis.

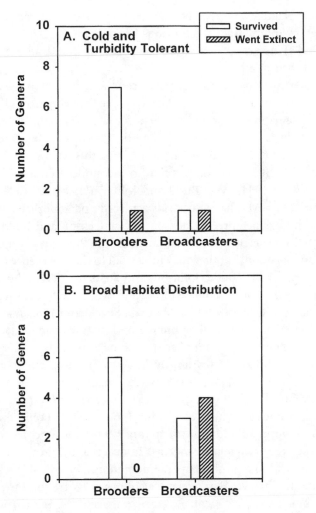

FIGURE 9.2. Relationships between generic survivorship through the early Miocene extinction interval and mode of sexual reproduction (genera in which some or all species are brooders versus genera in which all species are broadcasters) for (A) genera tolerant of cold water and turbidity, and (B) genera with a broad habitat distribution. Data are from a survey in Puerto Rico by Edinger and Risk (1995, their table 3).

Brooding genera may have survived preferentially in the Miocene because of the enhanced recruitment success of individual, brooded coral planulae, but not because of long-distance dispersal ability. The advantage of brooding in the Miocene, as well as today, may lie precisely in the potential for a short larval life. Short dispersal distances enable brooding colonies that survive a disturbance to disperse their planulae effectively to nearby patches of substratum that have been opened by mortality of the previous incumbents (Szmant 1986; Carlon and Olson 1993). As a result, local brooders may have the advantage of priority over distant broadcasters.

After the extinctions of the early Miocene and further origination and extinction activity in the middle and late Miocene (Kauffman and Fagerstrom 1993; Budd, Stemann, and Johnson 1994; Veron 1995), the Caribbean coral fauna changed little from ~8 Ma until the Plio-Pleistocene. Increased rates of origination from 4 to 3 Ma and extinction from 2 to 1 Ma resulted in a fauna virtually identical in species composition to the modern fauna (Budd, Johnson, and Stemann 1994, 1996; Budd and Johnson 1999). It was during the interval from 2 to 1 Ma that acroporids replaced pocilloporids as the dominant corals on Caribbean reefs (Jackson 1994a; Jackson and Budd 1996; see also Frost 1977a). The environmental causes of the Plio-Pleistocene turnover are not well understood, but they may be related to the emergence of the Isthmus of Panama 3.5 Ma (Budd, Johnson, and Stemann 1994; Johnson, Budd, and Stemann 1995). Global climate change associated with the onset of prolonged glacial eustasy from 2.5 to 1 Ma (e.g., Shackleton 1985; Stanley 1986) could have played a role as well. Subsequent glaciations and deglaciations in the Pleistocene, however, did not affect the fauna appreciably. After 1 Ma, the Caribbean coral fauna persisted largely intact until the present (Budd, Stemann, and Johnson 1994; Jackson 1994b; Hunter and Jones 1996).

Like the Miocene extinctions, the Plio-Pleistocene turnover had ecological as well as taxonomic consequences. Neither mode of sexual reproduction was favored, but Johnson, Budd, and Stemann (1995) documented extinction resistance and a resulting relative increase in taxa that (1) grow as large, long-lived colonies, and (2) reproduce by fragmentation. This Plio-Pleistocene bias underlies the dominance of modern Caribbean reefs by the three primary framework builders: *A. palmata, A. cervicornis,* and *M. annularis* complex (Budd, Johnson, and Stemann 1994; Johnson, Budd, and Stemann 1995).

In summary, after passing through environmental filters in the early Miocene and the Plio-Pleistocene, the Caribbean coral fauna, and the zonation patterns that the fauna created, survived the glacial cycles of the Pleistocene. Jackson (1994a; Jackson and Budd 1996) raised the possibility that species selection or species sorting accounts for the recent dominance and subsequent demise of the three primary species. Features such as high growth rates and an emphasis on fragmentation in *Acropora* spp., acquired before the Pleistocene, may have helped these species persist through the Pleistocene glaciations to become framework builders on modern reefs (Jackson 1994a;

Veron 1995; McNeill, Budd, and Borne 1997). *Acropora* spp. in particular have been decimated over the past two decades as some of those features have become less advantageous or even detrimental (Jackson and Budd 1996; Jackson, Budd, and Pandolfi 1996).

We have described a relatively static coral fauna interrupted by two historical intervals of accelerated turnover. The Recent episode of rapid transition could also have consequences beyond ecological time. We now turn to the community integration hypothesis posed in the Introduction. Does the long-term persistence of species composition on reefs imply that Caribbean corals form interactive, tightly integrated communities? The next section reviews the theory behind the community integration debate, and the two sections following apply those theoretical considerations to Caribbean reefs.

Coordinated Stasis and the Response to Disturbance

A number of studies have documented the persistence of terrestrial and marine paleocommunities through environmental fluctuations for as long as several million years. Community replacement then occurred during ecosystem reorganization in the face of more intense environmental change (e.g., Vrba 1985; Brett and Baird 1995; Morris et al. 1995; Brett, Ivany, and Schopf 1996; DiMichele, Pfefferkorn, and Phillips 1996; Holterhoff 1996; Tang and Bottjer 1996). This temporal pattern has been dubbed *coordinated stasis*, and it is viewed as the paleoecological analogue of phylogenetic punctuated equilibrium. Most observations of coordinated stasis come from Paleozoic sequences.

For paleobiologists, coordinated stasis revived a classical debate in ecology over the nature of community structure. In the Clementsian view, the community was interpreted as a superorganism. The component species were highly interactive and their distributions were strongly associated along environmental gradients (Clements 1916, 1936). The Gleasonian model rejected the idea of tight community integration. Instead, the community was seen as a collection of independently distributed species (Gleason 1926). The Gleasonian model does not exclude the possibility of succession, competition, niche partitioning, assembly rules, and other interspecific interactions. Rather, it denies interspecific interdependence as the cause of species distributions (Hoffman 1979; Allen and Hoekstra 1992; McIntosh 1995).

Few ecologists subscribe to the Clementsian superorganism concept at this point, the majority opinion being that communities conform to the Gleasonian "independent-but-interactive" model (Underwood and Denley 1984;

Roughgarden 1989; McIntosh 1995; Dayton et al. 1998). Nevertheless, some studies of terrestrial ecosystems suggest a causal connection between high species diversity and "superior" ecosystem function (high productivity, stability, resistance to invasion) (Bengtsson, Jones, and Setälä 1997). Whether diversity per se is important to the function of coral reef ecosystems is an unresolved question (Done et al. 1996).

Coordinated stasis in the fossil record has been interpreted as evidence in favor of the Clementsian model of community structure: Communities persisted for geologically significant periods as tightly integrated entities, and sets of mutually dependent component species appeared and disappeared as units (DiMichele 1994). Morris et al. (1995) suggested that interspecific interactions led to "ecological locking." According to this hypothesis, interspecific dependencies enabled the living communities to resist disturbance, up to some threshold disturbance intensity that caused whole-community reorganization. Brett, Ivany, and Schopf (1996:15) stated, "Some form of internal coherency appears to be necessary to maintain the stability of assemblages (derived from either traditional notions of incumbency, or ecosystem organization e.g. ecological locking. . .)" These "autogenic" models (Miller 1996) invoke interspecific interaction or dependence as an emergent property that stabilizes community composition for long periods. "Allogenic" explanations for the persistence of communities include persistently stable environments and faunal tracking of environments when conditions vary (Brett 1998).

Ironically, even as paleobiologists considered patterns of coordinated stasis, ecologists discovered the large body of paleontological literature that supports the Gleasonian model (Walter and Paterson 1994). In contrast to the coordinated stasis observed in Paleozoic paleocommunities, species in Cenozoic paleocommunities are in general distributed independently (e.g., Davis 1986; Paulay 1990; Valentine and Jablonski 1993; Buzas and Culver 1994, 1998; Roy, Jablonski, and Valentine 1995; Bennett 1997). Some studies of Paleozoic marine faunas also support the Gleasonian model (Westrop 1996; Patzkowsky and Holland 1997).

Brett, Ivany, and Schopf (1996) attributed the apparent Paleozoic–Cenozoic difference in community structure to better stratigraphic resolution in the Cenozoic. Because variance is high at the scales of the living community and preserved paleocommunity, biotic patterns occur at and should be sought at scales and organizational levels above the community and paleocommunity: the assemblage or community type of Boucot (1983, 1990), the paleocommunity type of Bennington and Bambach (1996), or the metacommunity of Jackson, Budd, and Pandolfi (1996). A metacommunity is a set of connected communities occupying a habitat type. Ecological-scale noise of the fluctuating component communities should be small compared to larger-scale, metacommunity signals. For example, patterns should emerge when coral reefs are studied in aggregate, over areas larger or through time intervals longer than the spans of individual reefs.

Searching for pattern at larger scales is entirely reasonable, but inferring process at those larger scales is problematic. In terms of the putative mechanisms underlying coordinated stasis, it seems unrealistic to try to reconcile individualistic distributions at ecological scales and community integration in time-averaged paleocommunities at larger scales. Species from different successional stages may interact in ecological time (through substratum conditioning, etc.), but it is far more difficult to argue for dependence among species that lived in a habitat thousands or tens of thousands of years apart (Aronson 1994). (The dynamics of coral reefs may actually be amenable to such multiscaled explanations under certain circumstances, and we will return to this possibility later.)

Community Structure of Coral Reefs

From a broad-scale, regional perspective, at least some assemblages of Cenozoic reef corals display distributional patterns that suggest coordinated stasis punctuated by community change. Does this mean that coral reefs are an exception to the Gleasonian dynamics of most Cenozoic communities (Brett, Ivany, and Schopf 1996; Jablonski and Sepkoski 1996)? While rejecting a strictly Clementsian interpretation, Jackson (1994a; Jackson, Budd, and Pandolfi 1996) proffered a version of the ecological locking hypothesis for Caribbean reefs, in which interspecific interactions or associations at some scale keep species bound up in stable communities or metacommunities.

To test for the pattern of coordinated stasis, Budd and colleagues examined the Plio-Pleistocene turnover at several Caribbean localities with good stratigraphic resolution. They found that the dynamics of faunal replacement varied with geographic location (Budd, Johnson, and Jackson 1994; Budd, Johnson, and Stemann 1996). In the Bahamas, pre- and post-turnover coral assemblages were discrete and did not overlap in species composition, favoring the coordinated stasis model. At other localities, however, the model was not supported. The turnover in Curaçao and Costa Rica was characterized by protracted, stepwise addition and removal of species rather than by whole-community collapse and reorganization (Budd and Johnson 1997; Budd, Petersen, and McNeill 1998).

What of the static coral fauna that followed during the Pleistocene–Holocene? Ecological locking, community integration, or mutual dependence of species within associations may account for the observed constancy of assemblage composition. On the other hand, individualistic interpretations may have equal or greater explanatory value while requiring fewer assumptions about higher-order processes.

There is a striking concordance in species composition between Pleistocene–Holocene fossil reefs and living reefs of the Caribbean. Most impressive are fossil coral assemblages that are compositionally similar to living communities occurring in the same place (e.g., Hubbard, Gladfelter, and Bythell 1994; Hunter and Jones 1996). Observations of this sort have been tendered as evidence for the community integration hypothesis.

In Jamaica, limestone outcrops spanning 0–6 m above present sea level represent coral reefs deposited ~125 Ka, during the last major interglacial high sea stand (Liddell, Ohlhorst, and Coates 1984; Boss and Liddell 1987). The 125 Ka Pleistocene bank/barrier reef exposed along the eastern margin of Rio Bueno Harbor, on the north coast of the island, is dominated by *A. palmata, A. cervicornis, M. annularis* complex, and *Porites porites* in the fore-reef facies (Liddell, Ohlhorst, and Coates 1984). The same species characterize(d) the living fore-reef community in the waters just below at Rio Bueno and at nearby Discovery Bay (Precht and Hoyt 1991).

The living community at Negril, on Jamaica's west coast, is quite different. Here the primary habitat type is flat limestone pavement. As is typical of hardground habitats in the Caribbean, Negril is populated primarily by gorgonians and massive corals. In this case *M. annularis* complex, *Siderastrea siderea, Diploria* spp., and the pillar coral *Dendrogyra cylindrus* are among the dominant scleractinians. Examination of the emergent, 125 Ka Pleistocene deposit just above reveals the same suite of massive coral species. Gorgonians are not preserved, but flamingo tongues, *Cyphoma gibbosum* (Gastropoda), are abundant (W. F. Precht, unpublished data). Since flamingo tongues prey significantly and exclusively on gorgonians in modern Caribbean communities (Harvell and Suchanek 1987; Lasker and Coffroth 1988), their abundance in the Pleistocene of Negril signifies that gorgonians were common at the time. Thus, the difference between locations is far greater than the difference between 125 Ka and the present.

These qualitative observations agree with a detailed, quantitative study of reef coral assemblages in the Pleistocene of New Guinea. Pandolfi (1996) examined paleocommunity composition based on species presence and absence on nine sequentially uplifted, Pleistocene reef terraces on the Huon Peninsula, Papua New Guinea. The interval examined spanned 95 k.y. and included nine glacial cycles. Variation in species composition was greater across space than through time. Although there was considerable variation through time even at a single locality, only a limited subset of the available, habitat-specific species pool constituted the metacommunity characteristic of a particular place. This suggested to Pandolfi (1996:152) that "local environmental parameters associated with riverine and terrestrial sources had a greater influence on reef coral composition than global climate and sea level changes." Pandolfi (1996), however, argued against the hypothesis that metacommunities were simply collections of species that tracked habitats and local environments. Instead he favored a community integration model to explain limited membership in coral assemblages.

Pandolfi's argument was based on the inference of interspecific competition in Caribbean corals, for which zonation patterns are better understood than Indo-Pacific corals. A few extant coral species, such as *A. palmata,* live only under a narrow range of environmental conditions. Most coral species occupy a broad range of environments, but they are common or dominant only within a subset of that range. This observation could imply present or past competitive interactions, which could in turn imply community integration.

As discussed previously, competition and other biological interactions such as niche partitioning (Hubbell 1997) can occur in reef communities even if they are collections of independently distributed species. More significant are the observations that (1) many Caribbean coral species are common over a broad range of environments (e.g., Porter et al. 1981; Aronson and Precht 1997; Bruno and Edmunds 1997), and (2) most Caribbean species are broadly distributed on a regional scale (Veron 1995; Pandolfi and Jackson 1997; Karlson and Cornell 1998). Both of these patterns are consistent with the independent distribution model. Given the clear influence of local physical conditions in Pandolfi's (1996) study, it is difficult to reject the hypothesis that the shared distributional patterns of Indo-Pacific coral species in New Guinea resulted from shared environmental preferences, which have persisted since the Pleistocene.

Coral diversity within reef communities depends on both the regional species pool and a suite of local factors (Karlson and Cornell 1998). The redundancy of ecologies observed in Indo-Pacific corals and other reef-dwelling taxa ensures that those reef communities function in essentially the same manner over a broad range of species richness values and through large variations in species composition (Roberts 1995; Paulay 1997). Because the Caribbean is dominated by only three coral species, the resilience of entire reef communities after disturbance may be more limited (Kojis and Quinn 1994; Birkeland 1996; see also Coral Reproductive Strategies). One might therefore predict that the community integration model is more likely to apply to Caribbean reefs.

Following Pandolfi's (1996) approach, Jackson, Budd, and Pandolfi (1996) presented evidence for limited membership in Caribbean fore-reef communities: Species composition varied geographically among fore-reef assemblages, but the fauna occurring in a particular environment through time was a limited subset of the available species pool. The same criticisms of proposed process apply. It is well established that Caribbean reef corals track environmental conditions through space and through changes in sea level (Goreau 1959, 1969; Lighty, Macintyre, and Stuckenrath 1982; Macintyre 1988; Fairbanks 1989; Pandolfi and Jackson 1997). In fact, one environmental signal that Jackson, Budd, and Pandolfi (1996) discovered was a difference in assemblage composition between the eastern/oceanic and western/continental areas of the Caribbean. Again, the persistence of species composition does not appear to be predicated on integrated species assemblages or interspecific associations within (meta)communities.

In many situations, small-scale variability masks large-scale pattern and higher-order process. In this case, we suggest that the small-scale variability detected among assemblages within environments is an important aspect of pattern: It reflects the independent distribution of species on coral reefs. The search for large-scale pattern has, we think, confused the issue somewhat through deliberate time-averaging.

Modern and "Postmodern" Reefs of the Caribbean

Even if community composition exhibits a pattern of stasis and punctuated change, paleontological data alone cannot eliminate one or another model of underlying causality (Miller 1997). The only hope for such a test is to combine paleontological observations of pattern with ecological observations of process. Is it possible to distinguish the independent distribution and community integration models with respect to the transition to macroalgal dominance on Caribbean reefs? Regardless of the availability of ecological data, the coral-to-macroalgal transition in the Caribbean presents a paleontological problem: Corals preserve in the fossil record, but fleshy macroalgae do not (Kauffman and Fagerstrom 1993).

Aronson and Precht (1997; Aronson, Precht, and Macintyre 1998) avoided this taphonomic pitfall when they documented a recent transition from *A. cervicornis* to *Agaricia* spp. (primarily *Ag. tenuifolia*) on reefs in the central shelf lagoon of the Belizean Barrier Reef. One lagoonal reef, Channel Cay, was studied intensively over a 10 yr period. Beginning in 1986, the *A. cervicornis* population at Channel Cay died off catastrophically from white-band disease. *Agaricia*, rather than macroalgae, recruited to the *A. cervicornis* rubble, and subsequent *Agaricia* growth produced a thick accumulation of dead skeletal material.

Both *A. cervicornis* and *Agaricia* have high preservation potential in the lagoonal environment in Belize. By 1995 the community shift had produced a clear signal in the sedimentary record: a thin layer of heavily eroded *A. cervicornis* rubble topped by a 22 cm thick layer of imbricated *Agaricia* plates. The uncemented, uncompacted nature of the subsurface sediments made it possible to core the reef in search of previous transitions of this sort.

Cores were extracted from Channel Cay and radiocarbon dated. The 7.6 cm diameter cores revealed no significant reworking of the reef sediments and no layers of *Agaricia* rubble below the post-1986 accumulation (figure 9.3). From the age determinations, Aronson and Precht (1997) concluded that prior to 1986, a transition to *Agaricia* had not occurred since at least 3–4 Ka. We have now taken cores from reefs throughout a 375 km² area of the central shelf lagoon. The new cores support the conclusion of a unique transition. The unique transition, however, did not result because the incumbent, *A. cervicornis*–dominated community was tightly integrated, nor did it result because that community as a whole somehow resisted less intense disturbances.

Point-count data along replicate transects at Channel Cay showed that *Agaricia* spp. were present but rare in 1986 (figure 9.4). As the *A. cervicornis* population crashed, *Agaricia* opportunistically recruited to the *A. cervicornis* skeletal rubble and increased in percent cover. Macroalgal cover remained low (<10%) throughout the period 1986–1995. *Agaricia* increased instead of macroalgae because of (1) the brooding reproductive strategy of *Agaricia* spp., which promoted local recruitment, combined with (2) intense herbivory by the abundant sea urchin *Echinometra viridis,* which suppressed macroalgal populations. Meanwhile, the 12 other coral species that were present maintained low cover (<10%) and did not fluctuate significantly. The trajectories of percent cover of coral species and genera were thus largely independent, at least on a decadal scale. Whether the independent distribution model will still explain the dynamics of Channel Cay several decades from now is unknown.

Agaricia spp. are not the only coral species that can take advantage of WBD outbreaks. *Porites porites* opportunistically replaced *A. cervicornis* on a patch reef at San Salvador Island, Bahamas (Curran et al. 1994). This replacement sequence was also noted in backreef habitats in Belize (Macintyre and Aronson 1997). Like the Caribbean *Agaricia* spp., *P. porites* is a brooder and a good colonizing species (see Coral Reproductive Strategies). If more than one coral species can be the replacement species, it is again difficult to argue for community integration.

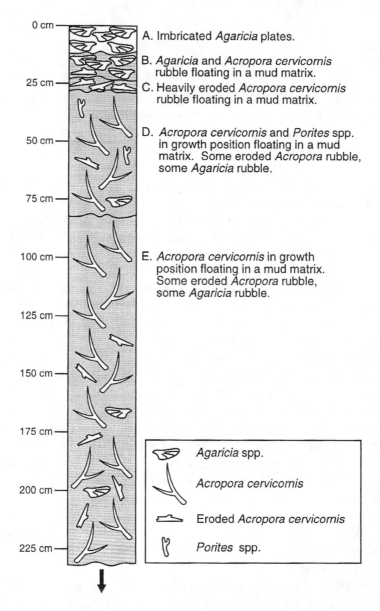

0 cm —

A. Imbricated *Agaricia* plates.

B. *Agaricia* and *Acropora cervicornis* rubble floating in a mud matrix.

25 cm —

C. Heavily eroded *Acropora cervicornis* rubble floating in a mud matrix.

50 cm —

D. *Acropora cervicornis* and *Porites* spp. in growth position floating in a mud matrix. Some eroded *Acropora* rubble, some *Agaricia* rubble.

75 cm —

100 cm —

E. *Acropora cervicornis* in growth position floating in a mud matrix. Some eroded *Acropora* rubble, some *Agaricia* rubble.

125 cm —

150 cm —

175 cm —

	Agaricia spp.
	Acropora cervicornis
	Eroded *Acropora cervicornis*
	Porites spp.

200 cm —

225 cm —

FIGURE 9.3. Schematic drawing of a generalized core from Channel Cay, a reef in the central lagoon of the Belizean Barrier Reef. The living community was removed prior to coring. Wavy, horizontal lines separate conformable variations in core composition, as a convenience to the reader. (A), (B), and (C) represent deposition during the period 1986–1995. All coral species other than those figured constituted <5% of the material. Centimeter scale shows depth within the core, and shading represents mud matrix. Redrawn from Aronson and Precht (1997).

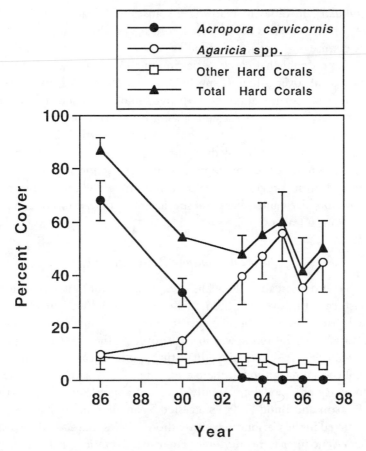

FIGURE 9.4. Changes in percent cover of living corals at Channel Cay over the period 1986–1997. Percent cover values were calculated from point counts along transects laid down the slope of the reef at two stations, perpendicular to depth contours. Data were pooled across the 3–15 m depth range of the transects. Error bars represent standard deviations; positive or negative errors are omitted for clarity.

Turning to the more widespread coral-to-macroalgal transition, even if it represents a rare or unique occurrence it would be premature to conclude that it was caused by the disruption of integrated reef communities. One could argue that the persistence of *M. annularis* depended indirectly on the presence of *A. cervicornis*. Macroalgal overgrowth of adult *M. annularis* colonies at Discovery Bay may have occurred because algal colonization was enhanced by the availability of *A. cervicornis* rubble or because pomacentrids switched their territories to *M. annularis*. In general, however, different things have happened to *M. annularis* in different places (see Causes of Coral Mortality). Where mortality of *M. annularis* has occurred, it has generally been a separate issue from the demise of *A. cervicornis* populations.

There is no doubt that coral species interact on Caribbean reefs, particularly when coral cover is high (Lang 1973; Porter 1974; Lang and Chornesky 1990). There is also no doubt that strong biological associations exist on Caribbean reefs; corals and zooxanthellae are mutually dependent, and flamingo tongues cannot exist without gorgonians. Nevertheless, the transition to macroalgal dominance is only the summation of coral species' independent responses to a number of contemporaneous perturbations. The excision of *Acropora* spp. from previously stable Caribbean reef communities has not dragged *M. annularis* and other coral species into a maelstrom of community disintegration, at least not yet.

Stasis and Change Reconsidered

Potts (1984) proposed an individualistic explanation for faunal stasis after the Plio-Pleistocene turnover. For Indo-Pacific corals with long generation times, continual disturbance in the Plio-Pleistocene maintained high intraspecific genetic variation. The result was low turnover of the coral fauna (see also Veron 1985, 1995). A similar individualistic hypothesis can be constructed for the Caribbean fauna, although there are some species with short generation times to which the model apparently does not apply (Potts and Garthwaite 1991). Jackson and Budd (1996) suggested as an alternative that Pleistocene stasis occurred in the Caribbean because those species that passed through the Plio-Pleistocene filter were eurytopic: Their preadaptations made them resistant to subsequent fluctuations.

The two explanations are really quite similar. In each hypothesis, the mechanisms maintaining stasis before and after the turnover event were different. Before the turnover, a long period of stasis occurred in a relatively benign environment. That stasis was shattered by environmental change during the Plio-Pleistocene. Then, after the turnover, stasis occurred despite continued environmental change because of the eurytopic (Jackson and Budd 1996) or plastic and variable (Potts 1984) nature of the coral species. Most recently, new combinations of coral diseases and other factors have overwhelmed the capacity of the eurytopic or plastic species to adjust to new conditions, hence the sudden biotic shift of the last 20 years. Ecological and evolutionary stasis and change of the Caribbean coral fauna thus becomes largely an observation of species' independent responses to the tempo and amplitude of environmental change.

Future Research

The recent transition to macroalgal dominance is a regional phenomenon. Widespread mortality of incumbent coral populations was a prerequisite, and white-band disease was probably the most important cause of *Acropora* mortality over the past decades. Paleontological data suggest that, although populations of *Acropora* spp. have died off locally in the past (Shinn et al. 1981; Lewis 1984; Shinn et al. 1989), a Caribbean-wide mass mortality of these corals has not occurred previously during the Holocene. Core data from St. Croix suggest that the WBD-induced mortality event was unique in the Holocene (Hubbard, Gladfelter, and Bythell 1994). Our studies in the shelf lagoon in Belize support the hypothesis of a unique event, if the WBD-mediated, *Acropora*-to-*Agaricia* transition can be considered a preservable proxy for the more widespread *Acropora*-to-macroalgal transition. Likewise, Greenstein, Curran, and Pandolfi (1998) argued that the WBD-mediated *Acropora*-to-*Porites* transition observed at San Salvador had not occurred previously. Mann et al. (1984; Taylor et al. 1985) described a lagoonal reef in the Dominican Republic that dated to 9–5 Ka. Judging from the excellent preservation of the corals, the shallow outer slope of this fossil reef was dominated continuously by *Acropora*, without intervening transitions to other community configurations (Stemann and Johnson 1995).

More paleontological evidence is needed to test the generality of these patterns. There is, for example, evidence of late Holocene declines of *A. palmata* populations in the Bahamas, Barbados, and Florida (Lighty 1981; Lewis 1984; Shinn et al. 1989). We agree with Jackson and Budd (1996) that ecological change in fossil reef communities is best understood from the relative abundance of species and not just from presence-absence data.

On the biological side, we know precious little about diseases on coral reefs. Do pollution and other forms of disturbance render corals more susceptible to disease (Edmunds 1991; Harvell et al. 1999)? We also must learn more about the sexual systems, fertilization, larval transport and behavior, settlement, recruitment, and post-settlement mortality of the coral species that have been displaced and those that have begun to colonize in their wake (see Bak and Meesters 1999; Mumby 1999). Reproductive processes interact in complex ways to influence the distribution of adult colonies and the ecology of reef communities. Self-fertilization, rather than brooding, could be the key to success after disturbance, and the predominance of brooding in Caribbean corals could be a nonadaptive consequence of species sorting after all.

By analogy with current models of natural selection at multiple taxonomic levels, Wilson (1997) argued that selection at the level of the community is at least conceivable and should not be rejected out of hand. One model of coral reef ecology holds that the persistence of reef communities turns on the positive interactions between corals and herbivorous fishes: Corals provide the structure necessary to shelter fish, and fish control the algae that would otherwise impede coral growth (e.g., Hay and Goertemiller 1983; Hay and Taylor 1985; Szmant 1997). Coral species may not be mutually dependent, but perhaps compositionally variable assemblages of coral species are dependent on compositionally variable assemblages of herbivorous fish species, and vice versa. Does this postulated "diffuse mutualism" between corals and fish represent community integration? Was it responsible for faunal stasis of Caribbean reefs during the Pleistocene–Holocene? Only by long-term monitoring can we attempt to answer these questions. We must follow the ecological dynamics of reef communities on the multidecadal scale appropriate to the turnover times of the fish and corals, while recognizing that some hypotheses will be untestable in our lifetimes.

There remains a potential link between the individualistic behavior of coral species in ecological communities and integrated dynamics of reef metacommunities, but it is highly speculative. The current decline of the three primary framework builders in the Caribbean could promote bioerosion and slow future reef accretion (cf. Glynn and Colgan 1992). As a result, reefs could drown in the face of rising sea level (Smith and Buddemeier 1992; Graus and Macintyre 1998). In such a case, the trajectory of the future community would be influenced by species in the current community. Biological interaction, interspecific interdependence, and metacommunity integration could thus occur on a geological timescale, despite independent species fluctuations within communities on an ecological scale.

Of course, the recent coral-to-macroalgal transition in the Caribbean has not yet been tied to extinction or speciation. The *Acropora* spp. appear to be surviving, and they may yet recover to their former large population sizes. What the transition portends for the diversity of corals in the Caribbean is a question for the future.

Ecology and paleobiology reciprocally illuminate each other, and they can be used in tandem to attain a broad understanding of biological systems (Gould 1981; Peterson 1984; Hubbard 1988; Jackson 1988; Bennett 1997). The most fruitful approach to the paleoecology of coral reefs is to combine paleontology with ecological observations. And just as surely, reef ecology cannot be understood without considering the historical context provided by paleoecology (e.g., Colgan 1990; Aronson and Precht 1997). Fossil reefs are time-averaged, highlighting long-term patterns; living reef communities supply information on the mechanisms underlying those long-term trends. Our impression is that paleobiologists pay close attention to ecology, but ecologists largely ignore the fossil record. If the important questions are to be answered, the only option is to combine the two disciplines into an evolutionary paleo-ecology of coral reefs.

ACKNOWLEDGMENTS

This chapter is dedicated to the biologists and geologists who worked at the Discovery Bay Marine Laboratory in the 1970s. They set the standard for coral reef research by the novelty of their discoveries and by the quality of their science. R.B.A. wishes to thank Sharon Ohlhorst for providing a college freshman with his first trip to Discovery Bay in 1976. W.F.P. is equally indebted to David J. Thomas for encouraging and advising his first coral reef research project at Discovery Bay in 1978.

Many colleagues helped us develop our ideas over the years. Individuals who had specific input to this chapter include Dan Brazeau, Andy Bruckner, John Bythell, Lyndon DeVantier, Terry Done, Bill Fitt, Danny Gleason, Mark Hay, Ken Heck, Denny Hubbard, Jeremy Jackson, Lisa Kellogg, Don Levitan, Diego Lirman, Laurence McCook, Ian Macintyre, Thad Murdoch, John Ogden, Esther Peters, Caroline Rogers, Robbie Smith, Bob Steneck, Chris Thomann, Emre Turak, John Valentine, and Jeremy Woodley. We are particularly grateful to Warren Allmon, Ann Budd, Evan Edinger, Pete Edmunds, and the late Jack Sepkoski for their critiques of the manuscript. We also thank the students of Northeastern University's East/West Program for asking questions that forced us to consolidate our thoughts about Caribbean reefs. Peter Edmunds kindly shared unpublished data from Jamaica. Our recent studies of coral reefs in Jamaica, St. Croix, the Florida Keys, and Belize have been funded by NOAA's National Undersea Research Program, the National Geographic Society, the National Science Foundation, and the Smithsonian Institution. This is Contribution No. 295 of the Dauphin Island Sea Lab and Contribution No. 596 of the Discovery Bay Marine Laboratory, University of the West Indies.

REFERENCES

Adey, W. H. and R. B. Burke. 1977. Holocene bioherms of Lesser Antilles: Geologic control of development. In S. H. Frost, M. P. Weiss, and J. B. Saunders, eds., *Reefs and Related Carbonates: Ecology and Sedimentology*, pp. 67–82. Studies in Geology No. 4. Tulsa OK: American Association of Petroleum Geologists.

Adey, W. H., P. J. Adey, R. Burke, and L. Kaufman. 1977. The Holocene reef systems of eastern Martinique, French West Indies. *Atoll Research Bulletin* 218:1–40.

Allen, T. F. H. and T. W. Hoekstra. 1992. *Toward a Unified Ecology*. New York: Columbia University Press.

Andres, N. G. and J. D. Witman. 1995. Trends in community structure on a Jamaican reef. *Marine Ecology Progress Series* 118:305–310.

Antonius, A. 1981. The "band" diseases in coral reefs. *Proceedings of the Fourth International Coral Reef Symposium*, Manila, Philippines, v. 2, pp. 6–14.

Antonius, A. 1995. Incidence and prevalence of coral diseases on coral reefs: What progress in research? *Coral Reefs* 14:224.

Antonius, A. and B. Riegl. 1997. A possible link between coral diseases and a corallivorous snail (*Drupella cornus*) outbreak in the Red Sea. *Atoll Research Bulletin* 447:1–9.

Aronson, R. B. 1994. Scale-independent biological interactions in the marine environment. *Oceanography and Marine Biology: An Annual Review* 32:435–460.

Aronson, R. B. and W. F. Precht. 1995. Landscape patterns of reef coral diversity: A test of the intermediate disturbance hypothesis. *Journal of Experimental Marine Biology and Ecology* 192:1–14.

Aronson, R. B. and W. F. Precht. 1997. Stasis, biological disturbance, and community structure of a Holocene coral reef. *Paleobiology* 23:326–346.

Aronson, R. B. and W. F. Precht. 2000. Herbivory and algal dynamics on the coral reef at Discovery Bay, Jamaica. *Limnology and Oceanography* 45:251–255.

Aronson, R. B., J. P. Ebersole, and K. P. Sebens. 1994. Hurricane Hugo's impact on Salt River Submarine Canyon, St. Croix, U.S. Virgin Islands. In R. N. Ginsburg, compiler, *Proceedings of the Colloquium on Global Aspects of Coral Reefs: Health, Hazards and History, 1993*, pp. 189–195. Miami: Rosenstiel School of Marine and Atmospheric Science, University of Miami.

Aronson, R. B., P. J. Edmunds, W. F. Precht, D. W. Swanson, and D. R. Levitan. 1994. Large-scale, long-term monitoring of Caribbean coral reefs: Simple, quick, inexpensive techniques. *Atoll Research Bulletin* 421:1–19.

Aronson, R. B., W. F. Precht, and I. G. Macintyre. 1998. Extrinsic control of species replacement on a Holocene reef in Belize: The role of coral disease. *Coral Reefs* 17:223–230.

Aronson, R. B., W. F. Precht, I. G. Macintyre, and T. J. T. Murdoch. 2000. Coral bleach-out in Belize. *Nature* 405:36.

Bak, R. P. M. 1983. Aspects of community organization in Caribbean stony corals. In J. C. Ogden, and E. H. Gladfelter, eds., *Coral Reefs, Seagrass Beds and Mangroves: Their Interaction in the Coastal Zones of the Caribbean*, pp. 51–68. UNESCO Reports in Marine Science 23. Montivideo: UNESCO Regional Office for Science and Technology for Latin America and the Caribbean.

Bak, R. P. M. and S. R. Criens. 1981. Survival after fragmentation of colonies of *Madracis mirabilis, Acropora palmata* and *A. cervicornis* (Scleractinia) and the subsequent impact of coral disease. *Proceedings of the Fourth International Coral Reef Symposium*, Manila, Philippines, v. 2, pp. 221–228.

Bak, R. P. M. and M. S. Engel. 1979. Distribution, abundance and survival of juvenile hermatypic corals (Scleractinia) and the importance of life history strategies in the parent coral community. *Marine Biology* 54:341–352.

Bak, R. P. M. and B. E. Luckhurst. 1980. Constancy and change in coral reef habitats along depth gradients at Curaçao. *Oecologia* 47:145–155.

Bak, R. P. M. and E. H. Meesters. 1999. Population structure as a response of coral communities to global change. *American Zoologist* 39:56–65.

Ball, M. M., E. A. Shinn, and R. W. Stockman. 1967. The geologic effects of Hurricane Donna in south Florida. *Journal of Geology* 75:583–597.

Barnes, D. J. and B. E. Chalker. 1990. Calcification and photosynthesis in reef-building corals and algae. In Z. Dubinsky, ed., *Ecosystems of the World 25: Coral Reefs*, pp. 109–131. Amsterdam: Elsevier.

Bell, P. R. F. and T. Tomascik. 1994. The demise of the fringing coral reefs of Barbados and of regions in the Great Barrier Reef (GBR) lagoon: Impacts of eutrophication. In R. N. Ginsburg, compiler, *Proceedings of the Colloquium on Global Aspects of Coral Reefs: Health, Hazards and History, 1993*, pp. 319–325. Miami: Rosenstiel School of Marine and Atmospheric Science, University of Miami.

Bengtsson, J., H. Jones, and H. Setälä. 1997. The value of biodiversity. *Trends in Ecology and Evolution* 12:334–336.

Bennett, K. D. 1997. *Evolution and Ecology: The Pace of Life*. Cambridge: Cambridge University Press.

Bennington, J. B. and R. K. Bambach. 1996. Statistical testing for paleocommunity recurrence: Are similar fossil assemblages ever the same? *Palaeogeography, Palaeoclimatology, Palaeoecology* 127:107–133.

Birkeland, C. 1977. The importance of rate of biomass accumulation in early successional stages of benthic communities to the survival of coral recruits. *Proceedings of the Third International Coral Reef Symposium*, Miami, v. 1, pp. 15–21.

Birkeland, C. 1996. Why some species are especially influential on coral-reef communities and others are not. *Galaxea* 13:77–84.

Blair, S. M., T. L. McIntosh, and B. J. Mostkoff. 1994. Impacts of Hurricane Andrew on the offshore reef systems of central and northern Dade County, Florida. *Bulletin of Marine Science* 54:961–973.

Blanchon, P. 1997. Architectural variation in submerged shelf-edge reefs: The hurricane control hypothesis. *Proceedings of the Eighth International Coral Reef Symposium*, Panama, v. 1, pp. 547–554.

Blanchon, P., B. Jones, and W. Kalbfleisch. 1997. Anatomy of a fringing reef around Grand Cayman: storm rubble, not coral framework. *Journal of Sedimentary Research* 67:1–16.

Boss, S. K. and W. D. Liddell. 1987. Back-reef and fore-reef analogs in the Pleistocene of north Jamaica: Implications for facies recognition and sediment flux in fossil reefs. *Palaios* 2:219–228.

Boucot, A. J. 1983. Does evolution take place in an ecological vacuum? II. "'The time has come' the Walrus said . . ." *Journal of Paleontology* 57:1–30.

Boucot, A. J. 1990. Evolution of communities. In D. E. G. Briggs and P. R. Crowther, eds. *Palaeobiology: A Synthesis*, pp. 391–394. Oxford, U.K.: Blackwell.

Brazeau, D. A., D. F. Gleason, and M. E. Morgan. 1998. Self-fertilization in brooding hermaphroditic Caribbean corals: Evidence from molecular studies. *Journal of Experimental Marine Biology and Ecology* 231:225–238.

Brett, C. E. 1998. Sequence stratigraphy, paleoecology, and evolution: Biotic clues and responses to sea-level fluctuations. *Palaios* 13:241–262.

Brett, C. E. and G. C. Baird. 1995. Coordinated stasis and evolutionary ecology of Silurian to Middle Devonian faunas in the Appalachian Basin. In D. Erwin and R. L. Anstey, eds., *New Approaches to Speciation in the Fossil Record,* pp. 285–315. New York: Columbia University Press.

Brett, C. E., L. C. Ivany, and K. M. Schopf. 1996. Coordinated stasis: An overview. *Palaeogeography, Palaeoclimatology, Palaeoecology* 127:1–20.

Brown, B. E. 1987. Worldwide death of corals: Natural cyclical events or man-made pollution. *Marine Pollution Bulletin* 18:9–13.

Brown, B. E. 1997. Adaptations of reef corals to physical environmental stress. *Advances in Marine Biology* 31:221–299.

Bruckner, A. W. and R. J. Bruckner. 1997. Outbreak of coral disease in Puerto Rico. *Coral Reefs* 16:260.

Bruckner, A. W. and R. J. Bruckner. 1998. Rapid Wasting Syndrome or coral predation by stoplight parrotfish? *Reef Encounter* 23:18–22.

Bruckner, R. J., A. W. Bruckner, and E. H. Williams Jr. 1997. Life history strategies of *Coralliophila abbreviata* Lamarck (Gastropoda: Coralliophilidae) on the southwest coast of Puerto Rico. *Proceedings of the Eighth International Coral Reef Symposium,* Panama, v. 1, pp. 627–632.

Bruno, J. F. and P. J. Edmunds. 1997. Clonal variation for phenotypic plasticity in the coral *Madracis mirabilis. Ecology* 78:2177–2190.

Budd, A. F. and K. G. Johnson. 1997. Coral reef community dynamics over 8 million years of evolutionary time: Stasis and turnover. *Proceedings of the Eighth International Coral Reef Symposium,* Panama, v. 1, pp. 423–428.

Budd, A. F. and K. G. Johnson. 1999. Origination preceding extinction during late Cenozoic turnover of Caribbean reefs. *Paleobiology* 25:188–200.

Budd, A. F., T. A. Stemann, and R. H. Stewart. 1992. Eocene Caribbean reef corals: A unique fauna from the Gatuncillo Formation of Panama. *Journal of Paleontology* 66:570–594.

Budd, A. F., K. G. Johnson, and J. B. C. Jackson. 1994. Patterns of replacement in late Cenozoic Caribbean coral communities (abstract). *Geological Society of America Abstracts With Programs* 26:A454.

Budd, A. F., K. G. Johnson, and T. A. Stemann. 1994. Plio-Pleistocene extinctions and the origin of the modern Caribbean reef-coral fauna. In R. N. Ginsburg, compiler, *Proceedings of the Colloquium on Global Aspects of Coral Reefs: Health, Hazards and History, 1993,* pp. 7–13. Miami: Rosenstiel School of Marine and Atmospheric Science, University of Miami.

Budd, A. F., T. A. Stemann, and K. G. Johnson. 1994. Stratigraphic distributions of genera and species of Neogene to Recent Caribbean reef corals. *Journal of Paleontology* 68:951–977.

Budd, A. F., K. G. Johnson, and T. A. Stemann. 1996. Plio-Pleistocene turnover and extinctions in the Caribbean reef coral fauna. In J. B. C. Jackson, A. F. Budd, and A. G. Coates, eds., *Evolution and Environment in Tropical America*, pp. 168–204. Chicago: University of Chicago Press.

Budd, A. F., R. A. Petersen, and D. F. McNeill. 1998. Stepwise faunal change during evolutionary turnover: An example from the Neogene of Curaçao, Netherlands Antilles. *Palaios* 13:170–188.

Burke, R. B. 1982. Reconnaissance study of the geomorphology and benthic communities of the outer barrier reef platform, Belize. In K. Rützler and I. G. Macintyre, eds., *The Atlantic Barrier Reef Ecosystem at Carrie Bow Cay, Belize: I. Structure and Communities*, pp. 509–526. Smithsonian Contributions to the Marine Sciences 12. Washington DC: Smithsonian Institution Press.

Burke, R. B. 1994. How have Holocene sea level rise and antecedent topography influenced Belize barrier reef development? In R. N. Ginsburg, compiler, *Proceedings of the Colloquium on Global Aspects of Coral Reefs: Health, Hazards and History, 1993*, pp. 14–20. Miami: Rosenstiel School of Marine and Atmospheric Science, University of Miami.

Burke, C. D., T. M. McHenry, W. D. Bischoff, and S. J. Mazzullo. 1998. Coral diversity and mode of growth of lateral expansion patch reefs at Mexico Rocks, northern Belize shelf, Central America. *Carbonates and Evaporites* 13:32–42.

Burns, T. P. 1985. Hard-coral distribution and cold-water disturbances in south Florida. *Coral Reefs* 4:117–124.

Buzas, M. A. and S. J. Culver. 1994. Species pool and dynamics of marine paleocommunities. *Science* 264:1439–1441.

Buzas, M. A. and S. J. Culver. 1998. Assembly, disassembly, and balance in marine paleocommunities. *Palaios* 13:263–275.

Bythell, J. and C. Sheppard. 1993. Mass mortality of Caribbean shallow corals. *Marine Pollution Bulletin* 26:296–297.

Bythell, J. C., M. Bythell, and E. H. Gladfelter. 1993. Initial results of a long-term coral reef monitoring program: Impact of Hurricane Hugo at Buck Island Reef National Monument, St. Croix, U.S. Virgin Islands. *Journal of Experimental Marine Biology and Ecology* 172:171–183.

Bythell, J. C., E. H. Gladfelter, W. B. Gladfelter, K. E. French, and Z. Hillis. 1989. Buck Island Reef National Monument: Changes in modern reef community structure since 1976. In D. K. Hubbard, ed., *Terrestrial and Marine Geology of St. Croix, U.S. Virgin Islands*, pp. 145–153. *West Indies Laboratory Special Publication* 8. St. Croix: West Indies Laboratory, Fairleigh Dickinson University.

Carlon, D. B. and R. R. Olson. 1993. Larval dispersal distance as an explanation for adult spatial pattern in two Caribbean reef corals. *Journal of Experimental Marine Biology and Ecology* 173:247–263.

Carpenter, R. C. 1981. Grazing by *Diadema antillarum* (Philippi) and its effects on the benthic algal community. *Journal of Marine Research* 39:749–765.

Carpenter, R. C. 1986. Partitioning herbivory and its effects on coral reef algal communities. *Ecological Monographs* 56:345–363.

Carpenter, R. C. 1988. Mass mortality of a Caribbean sea urchin: Immediate effects on community metabolism and other herbivores. *Proceedings of the National Academy of Sciences of the United States of America* 85:511–514.

Carpenter, R. C. 1990a. Mass mortality of *Diadema antillarum:* I. Long-term effects on sea urchin population-dynamics and coral reef algal communities. *Marine Biology* 104:67–77.

Carpenter, R. C. 1990b. Mass mortality of *Diadema antillarum:* II. Effects on population densities and grazing intensity of parrotfishes and surgeonfishes. *Marine Biology* 104:79–86.

Carpenter, R. C. 1997. Invertebrate predators and grazers. In C. Birkeland, ed., *Life and Death of Coral Reefs*, pp. 198–229. New York: Chapman and Hall.

Chiappone, M. and K. M. Sullivan. 1996. Distribution, abundance and species composition of juvenile scleractinian corals in the Florida Reef Tract. *Bulletin of Marine Science* 58:555–569.

Choat, J. H. 1991. The biology of herbivorous fishes on coral reefs. In P. F. Sale, ed., *The Ecology of Fishes on Coral Reefs*, pp. 120–155. San Diego: Academic Press.

Clements, F. E. 1916. Plant succession, an analysis of the development of vegetation. *Carnegie Institution of Washington Publication* 242:1–512.

Clements, F. E. 1936. Nature and structure of the climax. *Journal of Ecology* 24:252–284.

Colgan, M. W. 1987. Coral reef recovery on Guam (Micronesia) after catastrophic predation by *Acanthaster planci*. *Ecology* 68:1592–1605.

Colgan, M. W. 1990. El Niño and the history of eastern Pacific reef building. In P. W. Glynn, ed., *Global Ecological Consequences of the 1982–83 El Niño-Southern Oscillation*, pp. 183–232. Amsterdam: Elsevier.

Connell, J. H. 1978. Diversity in tropical rain forests and coral reefs. *Science* 199:1302–1310.

Connell, J. H. 1997. Disturbance and recovery of coral assemblages. *Coral Reefs* 16:S101–S113.

Connell, J. H., T. P. Hughes, and C. C. Wallace. 1997. A 30-year study of coral abundance, recruitment, and disturbance at several scales in space and time. *Ecological Monographs* 67:461–488.

Cortés, J. 1994. A reef under siltation stress: A decade of degradation. In R. N. Ginsburg, compiler, *Proceedings of the Colloquium on Global Aspects of Coral Reefs: Health, Hazards and History, 1993*, pp. 240–246. Miami: Rosenstiel School of Marine and Atmospheric Science, University of Miami.

Cortés, J. 1997. Biology and geology of eastern Pacific coral reefs. *Coral Reefs* 16:S39–S46.

Curran, H. A., D. P. Smith, L. C. Meigs, A. E. Pufall, and M. L. Greer. 1994. The health and short-term change of two coral patch reefs, Fernandez Bay, San Salvador Island, Bahamas. In R. N. Ginsburg, compiler, *Proceedings of the Colloquium on Global Aspects of Coral Reefs: Health, Hazards and History, 1993*, pp. 147–153. Miami: Rosenstiel School of Marine and Atmospheric Science, University of Miami.

Davis, G. E. 1982. A century of natural change in coral distribution at the Dry Tortugas: A comparison of reef maps from 1881 and 1976. *Bulletin of Marine Science* 32:608–623.

Davis, M. B. 1986. Climatic instability, time lags, and community disequilibrium. In J. Diamond and T. J. Case, eds. *Community Ecology*, pp. 269–284. New York: Harper and Row.

Dayton, P. K., M. J. Tegner, P. B. Edwards, and K. L. Riser. 1998. Sliding baselines, ghosts, and reduced expectations in kelp forest communities. *Ecological Applications* 8:309–322.

D'Elia, C. F., R. W. Buddemeier, and S. V. Smith. 1991. *Workshop on Coral Bleaching, Coral Reef Ecosystems and Global Change: Report of Proceedings*. College Park: Maryland Sea Grant College, University of Maryland.

de Ruyter van Steveninck, E. D., L. L. van Mulekom, and A. M. Breeman. 1988. Growth inhibition of *Lobophora variegata* (Lamoroux) Womersley by scleractinian corals. *Journal of Experimental Marine Biology and Ecology* 115:169–178.

DiMichele, W. A. 1994. Ecological patterns in time and space. *Paleobiology* 20:89–92.

DiMichele, W. A., H. W. Pfefferkorn, and T. L. Phillips. 1996. Persistence of Late Carboniferous tropical vegetation during glacially driven climatic and sea-level fluctuations. *Palaeogeography, Palaeoclimatology, Palaeoecology* 125:105–128.

Dodge, R. E., R. C. Aller, and J. Thomson. 1974. Coral growth related to resuspension of bottom sediments. *Nature* 247:574–577.

Done, T. J. 1992a. Phase shifts in coral reef communities and their ecological significance. *Hydrobiologia* 247:121–132.

Done, T. J. 1992b. Constancy and change in some Great Barrier Reef coral communities: 1980–1990. *American Zoologist* 32:655–662.

Done, T. J., J. C. Ogden, W. J. Wiebe, and B. R. Rosen. 1996. Biodiversity and ecosystem function of coral reefs. In H. A. Mooney, J. H. Cushman, E. Medina, O. E. Sala, and E.-D. Schulze, eds. *Functional Roles of Biodiversity: A Global Perspective*, pp. 393–429. Chichester, U.K.: Wiley.

Dubinsky, Z. and N. Stambler. 1996. Marine pollution and coral reefs. *Global Change Biology* 2:511–526.

Dustan, P. 1975. Growth and form in the reef-building coral *Montastrea annularis*. *Marine Biology* 33:101–107.

Eakin, C. M. 1989. Microhabitat selection in juveniles of the damselfishes *Pomacentrus planifrons* and *P. partitus* (abstract). *Proceedings of the Association of Marine Laboratories of the Caribbean* 22:35–36.

Ebersole, J. P. 1985. Niche separation of two damselfish species by aggression and differential microhabitat utilization. *Ecology* 66:14–20.

Edinger, E. N. and M. J. Risk. 1994. Oligocene-Miocene extinction and geographic restriction of Caribbean corals: Roles of turbidity, temperature, and nutrients. *Palaios* 9:576–598.

Edinger, E. N. and M. J. Risk. 1995. Preferential survivorship of brooding corals in a regional extinction. *Paleobiology* 21:200–219.

Edmunds, P. J. 1991. Extent and effect of Black Band Disease on a Caribbean reef. *Coral Reefs* 10:161–165.

Edmunds, P. J. and J. D. Witman. 1991. Effect of Hurricane Hugo on the primary framework of a reef along the south shore of St. John, U.S. Virgin Islands. *Marine Ecology Progress Series* 78:201–204.

Edmunds, P. J. and J. F. Bruno. 1996. The importance of sampling scale in ecology: Kilometer-wide variation in coral reef communities. *Marine Ecology Progress Series* 143:165–171.

Edmunds, P. J., R. B. Aronson, D. W. Swanson, D. R. Levitan, and W. F. Precht. 1998. Photographic versus visual census techniques for the quantification of juvenile corals. *Bulletin of Marine Science* 62:937–946.

Endean, R., A. M. Cameron, and L. M. DeVantier. 1988. *Acanthaster* predation on massive corals: The myth of rapid recovery of devastated reefs. *Proceedings of the Sixth International Coral Reef Symposium*, Townsville, Australia, v. 2, pp. 143–148.

Fabricius, K. E. 1997. Soft coral abundance on the central Great Barrier Reef: Effects of *Acanthaster planci*, space availability, and aspects of the physical environment. *Coral Reefs* 16:159–167.

Fadlallah, Y. H. 1983. Sexual reproduction, development and larval biology in scleractinian corals: A review. *Coral Reefs* 2:129–150.

Fairbanks, R. G. 1989. A 17,000-year glacio-eustatic sea level record: Influence of glacial melting rates on the Younger Dryas event and deep-ocean circulation. *Nature* 342:637–642.

Fenner, D. P. 1991. Effects of Hurricane Gilbert on coral reefs, fishes and sponges at Cozumel, Mexico. *Bulletin of Marine Science* 48:719–730.

Fitt, W. K., H. J. Spero, J. Halas, M. W. White, and J. W. Porter. 1993. Recovery of the coral *Montastrea annularis* in the Florida Keys after the 1987 Caribbean "bleaching event." *Coral Reefs* 12:57–64.

Foster, S. A. 1987. The relative impact of grazing by Caribbean coral reef fishes and *Diadema*: Effects of habitat surge. *Journal of Experimental Marine Biology and Ecology* 105:1–20.

Frost, S. H. 1977a. Miocene to Holocene evolution of Caribbean Province reef-building corals. *Proceedings of the Third International Coral Reef Symposium*, Miami, v. 2, pp. 353–359.

Frost, S. H. 1977b. Ecologic controls of Caribbean and Mediterranean Oligocene reef coral communities. *Proceedings of the Third International Coral Reef Symposium*, Miami, v. 2, pp. 367–373.

Frost, S. H., and R. L. Langenheim, Jr. 1974. *Cenozoic Reef Biofacies: Tertiary Larger Foraminifera and Scleractinian Corals from Chiapas, Mexico.* De Kalb: Northern Illinois University Press.

Garzón-Ferreira, J. and M. Kielman. 1994. Extensive mortality of corals in the Colombian Caribbean during the last two decades. In R. N. Ginsburg, compiler, *Proceedings of the Colloquium on Global Aspects of Coral Reefs: Health, Hazards and History, 1993,* pp. 247–253. Miami: Rosenstiel School of Marine and Atmospheric Science, University of Miami.

Geister, J. 1977. The influence of wave exposure on the ecological zonation of Caribbean coral reefs. *Proceedings of the Third International Coral Reef Symposium,* Miami, v. 1, pp. 1:23–29.

Geister, J. 1980. Calm-water reefs and rough-water reefs of the Caribbean Pleistocene. *Acta Palaeontologia Polonica* 25:541–556.

Geister, J. 1983. Holozäne westindische Korallenriffe: Geomorphologie, Ökologie und Fazies. *Facies* 9:173–284.

Gilmore, M. D. and B. R. Hall. 1976. Life history, growth habits, and constructional roles of *Acropora cervicornis* in the patch reef environment. *Journal of Sedimentary Petrology* 40:519–522.

Ginsburg, R. N., compiler. 1994. *Proceedings of the Colloquium on Global Aspects of Coral Reefs: Health, Hazards and History, 1993.* Miami: Rosenstiel School of Marine and Atmospheric Science, University of Miami.

Ginsburg, R. N. and E. A. Shinn. 1994. Preferential distribution of reefs in the Florida Reef Tract: The past is the key to the present. In R. N. Ginsburg, compiler, *Proceedings of the Colloquium on Global Aspects of Coral Reefs: Health, Hazards and History, 1993,* pp. 21–26. Miami: Rosenstiel School of Marine and Atmospheric Science, University of Miami.

Gladfelter, W. B. 1982. White band disease in *Acropora palmata:* Implications for the structure and growth of shallow reefs. *Bulletin of Marine Science* 32:639–643.

Gleason, H. A. 1926. The individualistic concept of plant association. *Bulletin of the Torrey Botanical Club* 53:7–26.

Gleason, M. G. 1996. Coral recruitment in Moorea, French Polynesia: The importance of patch type and temporal variation. *Journal of Experimental Marine Biology and Ecology* 207:79–101.

Gleason, D. F. and D. A. Brazeau. 1997. Self-fertilization in brooding hermaphroditic Caribbean corals (abstract). *American Zoologist* 37:15A.

Glynn, P. W. 1990a. Feeding ecology of selected coral-reef macroconsumers: Patterns and effects on coral community structure. In Z. Dubinsky, ed., *Ecosystems of the World 25: Coral Reefs,* pp. 365–400. Amsterdam: Elsevier.

Glynn, P. W. 1990b. Coral mortality and disturbances to coral reefs in the tropical Eastern Pacific. In P. W. Glynn, ed., *Global Ecological Consequences of the 1982–83 El Niño-Southern Oscillation,* pp. 55–126. Amsterdam: Elsevier.

Glynn, P. W. 1993. Coral reef bleaching: Ecological perspectives. *Coral Reefs* 12:1–17.

Glynn, P. W. 1996. Coral reef bleaching: Facts, hypotheses and implications. *Global Change Biology* 2:495–510.

Glynn, P. W. and M. W. Colgan. 1992. Sporadic disturbances in fluctuating coral reef environments: El Niño and coral reef development in the Eastern Pacific. *American Zoologist* 32:707–718.

Glynn, P. W., L. R. Almodóvar, and J. G. González. 1964. Effects of Hurricane Edith on marine life in La Parguera, Puerto Rico. *Caribbean Journal of Science* 4:335–345.

Goreau, T. F. 1959. The ecology of Jamaican coral reefs: I. Species composition and zonation. *Ecology* 40:67–90.

Goreau, T. F. 1969. Post Pleistocene urban renewal in coral reefs. *Micronesica* 5:323–326.

Goreau, T. F. and N. I. Goreau. 1973. The ecology of Jamaican coral reefs: II. Geomorphology, zonation and sedimentary phases. *Bulletin of Marine Science* 23:399–464.

Goreau, T. F. and J. W. Wells. 1967. The shallow water Scleractinia of Jamaica: Revised list of species and their vertical distribution range. *Bulletin of Marine Science* 17:442–453.

Gould, S. J. 1981. Palaeontology plus ecology as palaeobiology. In R. M. May, ed., *Theoretical Ecology: Principles and Applications*, 2nd ed., pp. 295–317. Sunderland MA: Sinauer Associates.

Grassle, J. F. 1973. Variety in coral reef communities. In O. A. Jones and R. Endean, eds., *Biology and Geology of Coral Reefs, Volume II (Biology 1)*, pp. 247–270. New York: Academic Press.

Graus, R. R. and I. G. Macintyre. 1982. Variation in growth form of the reef coral *Montastrea annularis* (Ellis and Solander): A quantitative evaluation of growth response to light distribution using computer simulation. In K. Rützler and I. G. Macintyre, eds., *The Atlantic Barrier Reef Ecosystem at Carrie Bow Cay, Belize, I. Structure and Communities*, pp. 441–464. Smithsonian Contributions to the Marine Sciences 12. Washington DC: Smithsonian Institution Press.

Graus, R. R. and I. G. Macintyre. 1989. The zonation of Caribbean coral reefs as controlled by wave and light energy input, bathymetric setting and reef morphology: Computer simulation experiments. *Coral Reefs* 8:9–18.

Graus, R. R. and I. G. Macintyre. 1998. Global warming and the future of Caribbean coral reefs. *Carbonates and Evaporites* 13:43–47.

Greenstein, B. J., H. A. Curran, and J. M. Pandolfi. 1998. Shifting ecological baselines and the demise of *Acropora cervicornis* in the Western Atlantic and Caribbean Province: A Pleistocene perspective. *Coral Reefs* 17:249–261.

Grigg, R. W. and S. J. Dollar. 1990. Natural and anthropogenic disturbance on coral reefs. In Z. Dubinsky, ed., *Ecosystems of the World 25: Coral Reefs*, pp. 439–452. Amsterdam: Elsevier.

Hagman, D. K., S. R. Gittings, and P. D. Vize. 1998. Fertilization in broadcast-spawning corals of the Flower Garden Banks National Marine Sanctuary. *Gulf of Mexico Science* 16:180–187.

Hallock, P. 1988. The role of nutrient availability in bioerosion: Consequences to carbonate buildups. *Palaeogeography, Palaeoclimatology, Palaeoecology,* 63:275–291.

Hallock, P. and W. Schlager. 1986. Nutrient excess and the demise of coral reefs and carbonate platforms. *Palaios* 1:389–398.

Harrison, P. L., and C. C. Wallace. 1990. Reproduction, dispersal and recruitment of scleractinian corals. In Z. Dubinsky, ed., *Ecosystems of the World 25: Coral Reefs,* pp. 133–207. Amsterdam: Elsevier.

Harvell, C. D. and T. H. Suchanek. 1987. Partial predation on tropical gorgonians by *Cyphoma gibbosum* (Gastropoda). *Marine Ecology Progress Series* 38:37–44.

Harvell, C. D., K. Kim, J. M. Burkholder, R. R. Colwell, P. R. Epstein, D. J. Grimes, E. E. Hofmann, E. K. Lipp, A. D. M. E. Osterhaus, R. M. Overstreet, G. W. Smith, and G. R. Vastra. 1999. Emerging marine diseases—climate links and anthropogenic factors. *Science* 285:1505–1510.

Hatcher, B. G. 1984. A maritime accident provides evidence for alternate stable states in benthic communities on coral reefs. *Coral Reefs* 3:199–204.

Hatcher, B. G., R. E. Johannes, and A. I. Robertson. 1989. Review of research relevant to the conservation of shallow tropical marine ecosystems. *Oceanography and Marine Biology: An Annual Review* 27:337–414.

Hay, M. E. 1981. Herbivory, algal distribution, and the maintenance of between-habitat diversity on a tropical fringing reef. *American Naturalist* 118:520–540.

Hay, M. E. 1984. Patterns of fish and urchin grazing on Caribbean coral reefs: Are previous results typical? *Ecology* 65:446–454.

Hay, M. E. 1985. Spatial patterns of herbivore impact and their importance in maintaining algal species richness. *Proceedings of the Fifth International Coral Reef Congress,* Tahiti, v. 4, pp. 29–34.

Hay, M. E. 1991. Fish-seaweed interactions on coral reefs: Effects of herbivorous fishes and adaptations of their prey. In P. F. Sale, ed., *The Ecology of Fishes on Coral Reefs,* pp. 96–119. San Diego: Academic Press.

Hay, M. E. 1997. The ecology and evolution of seaweed-herbivore interactions on coral reefs. *Coral Reefs* 16:S67-S76.

Hay, M. E. and T. Goertemiller. 1983. Between-habitat differences in herbivore impact on Caribbean coral reefs. In M. L. Reaka, ed., *The Ecology of Deep and Shallow Coral Reefs,* pp. 97–102. Rockville: NOAA Symposia Series for Undersea Research, Volume 3.

Hay, M. E., Q. E. Kappel, and W. Fenical. 1994. Synergisms in plant defenses against herbivores: Interactions of chemistry, calcification, and plant quality. *Ecology* 75:1714–1726.

Hay, M. E. and P. R. Taylor. 1985. Competition between herbivorous fishes and urchins on Caribbean reefs. *Oecologia* 65:591–598.

Highsmith, R. C. 1980. Geographic patterns of coral bioerosion: A productivity hypothesis. *Journal of Experimental Marine Biology and Ecology* 46:177–196.

Highsmith, R. C. 1982. Reproduction by fragmentation in corals. *Marine Ecology Progress Series* 7:207–226.

Highsmith, R. C., R. L. Lueptow, and S. C. Schonberg. 1983. Growth and bioerosion of three massive corals on the Belize barrier reef. *Marine Ecology Progress Series* 13:261–271.

Hixon, M. A. 1997. Effects of reef fishes on corals and algae. In C. Birkeland, ed., *Life and Death of Coral Reefs*, pp. 230–248. New York: Chapman and Hall.

Hoffman, A. 1979. Community paleoecology as an epiphenomenal science. *Paleobiology* 5:357–379.

Holterhoff, P. F. 1996. Crinoid biofacies in Upper Carboniferous cyclothems, midcontinent North America: Faunal tracking and the role of regional processes in biofacies recurrence. *Palaeogeography, Palaeoclimatology, Palaeoecology* 127:47–82.

Hubbard, D. K. 1988. Controls of modern and fossil reef development: Common ground for biological and geological research. *Proceedings of the Sixth International Coral Reef Symposium*, Townsville, Australia, v. 1, pp. 243–252.

Hubbard, D. K., K. M. Parsons, J. C. Bythell, and N. D. Walker. 1991. The effects of Hurricane Hugo on the reefs and associated environments of St. Croix, U.S. Virgin Islands. *Journal of Coastal Research, Special Issue* 8:33–48.

Hubbard, D. K., E. H. Gladfelter, and J. C. Bythell. 1994. Comparison of biological and geological perspectives of coral-reef community structure at Buck Island, U.S. Virgin Islands. In R. N. Ginsburg, compiler, *Proceedings of the Colloquium on Global Aspects of Coral Reefs: Health, Hazards and History, 1993*, pp. 201–207. Miami: Rosenstiel School of Marine and Atmospheric Science, University of Miami.

Hubbell, S. P. 1997. Niche assembly, dispersal limitation, and the maintenance of diversity in tropical tree communities and coral reefs. *Proceedings of the Eighth International Coral Reef Symposium*, Panama, v. 1, pp. 387–396.

Hudson, J. H. 1981a. Growth rates in *Montastraea annularis:* A record of environmental change in Key Largo Coral Reef Marine Sanctuary, Florida. *Bulletin of Marine Science* 31:444–459.

Hudson, J. H. 1981b. Response of *Montastrea annularis* to environmental change in the Florida Keys. *Proceedings of the Fourth International Coral Reef Symposium*, Manila, Philippines, v. 2, pp. 233–240.

Hughes, T. P. 1985. Life histories and population dynamics of early successional corals. *Proceedings of the Fifth International Coral Reef Congress*, Tahiti, v. 4, pp. 101–106.

Hughes, T. P. 1989. Community structure and diversity of coral reefs: The role of history. *Ecology* 70:275–279.

Hughes, T. P. 1992. Monitoring of coral reefs: A bandwagon. *Reef Encounter* 11:9–12.

Hughes, T. P. 1994. Catastrophes, phase shifts and large-scale degradation of a Caribbean coral reef. *Science* 265:1547–1551.

Hughes, T. P. and J. H. Connell. 1999. Multiple stressors on coral reefs: A long-term perspective. *Limnology and Oceanography* 44:932–940.

Hughes, T. P., D. C. Reed, and M.-J. Boyle. 1987. Herbivory on coral reefs: Community structure following mass mortalities of sea urchins. *Journal of Experimental Marine Biology and Ecology* 113:39–59.

Hughes, T. P., D. Ayre, and J. H. Connell. 1992. The evolutionary ecology of corals. *Trends in Ecology and Evolution* 7:292–295.

Hunte, W. and M. Wittenberg. 1992. Effects of eutrophication and sedimentation on juvenile corals: II. Settlement. *Marine Biology* 114:625–631.

Hunter, C. L. and C. W. Evans. 1995. Coral reefs in Kaneohe Bay, Hawaii: Two centuries of Western influence and two decades of data. *Bulletin of Marine Science* 57:501–515.

Hunter, I. G. and B. Jones. 1996. Coral associations of the Pleistocene Ironshore Formation, Grand Cayman. *Coral Reefs* 15:249–267.

Huston, M. A. 1985a. Patterns of species diversity on coral reefs. *Annual Review of Ecology and Systematics* 16:149–177.

Huston, M. A. 1985b. Patterns of species diversity in relation to depth at Discovery Bay, Jamaica. *Bulletin of Marine Science* 37:928–935.

Hutchings, P. A. 1986. Biological destruction of coral reefs. *Coral Reefs* 4:239–252.

Jaap, W. C. 1979. Observations on zooxanthellae expulsion at Middle Sambo Reef, Florida Keys. *Bulletin of Marine Science* 29:414–422.

Jaap, W. C., J. C. Halas, and R. G. Muller. 1988. Community dynamics of stony corals (Milleporina and Scleractinia) at Key Largo National Marine Sanctuary, Florida, during 1981–1986. *Proceedings of the Sixth International Coral Reef Symposium*, Townsville, Australia, v. 2, pp. 237–243.

Jablonski, D. and J. J. Sepkoski, Jr. 1996. Paleobiology, community ecology, and scales of ecological pattern. *Ecology* 77:1367–1378.

Jackson, J. B. C. 1988. Does ecology matter? (book review). *Paleobiology* 14:307–312.

Jackson, J. B. C. 1991. Adaptation and diversity of reef corals. *BioScience* 41:475–482.

Jackson, J. B. C. 1992. Pleistocene perspectives of coral reef community structure. *American Zoologist* 32:719–731.

Jackson, J. B. C. 1994a. Community unity? *Science* 264:1412–1413.

Jackson, J. B. C. 1994b. Constancy and change of life in the sea. *Philosophical Transactions of the Royal Society of London, Series B* 344:55–60.

Jackson, J. B. C. 1997. Reefs since Columbus. *Coral Reefs* 16:S23-S32.

Jackson, J. B. C. and A. F. Budd. 1996. Evolution and environment: Introduction and overview. In J. B. C. Jackson, A. F. Budd, and A. G. Coates, eds., *Evolution and Environment in Tropical America*, pp. 1–20. Chicago: University of Chicago Press.

Jackson, J. B. C., A. F. Budd, and J. M. Pandolfi. 1996. The shifting balance of natural communities? In D. Jablonski, D. H. Erwin and J. H. Lipps, eds. *Evolutionary Paleobiology*, pp. 89–122. Chicago: University of Chicago Press.

James, N. P. and I. G. Macintyre. 1985. Carbonate depositional environments: Modern and ancient: I. Reefs: zonation, depositional facies, and diagenesis. *Colorado School of Mines Quarterly* 80(3):1–70.

Johnson, K. G., A. F. Budd, and T. A. Stemann. 1995. Extinction selectivity and ecology of Neogene Caribbean reef corals. *Paleobiology* 21:52–73.

Jokiel, P. L., Z. Dubinsky, and N. Stambler. 1994. Results of the 1991 United States–Israel workshop, "Nutrient Limitation in the Symbiotic Association Between Zooxanthellae and Reef-Building Corals." *Pacific Science* 48:215–218.

Jordán-Dahlgren, E. 1992. Recolonization patterns of *Acropora palmata* in a marginal environment. *Bulletin of Marine Science* 51:104–117.

Karlson, R. H. and L. E. Hurd. 1993. Disturbance, coral reef communities, and changing ecological paradigms. *Coral Reefs* 12:117–125.

Karlson, R. H. and H. V. Cornell. 1998. Scale-dependent variation in local vs. regional effects on coral species richness. *Ecological Monographs* 68:259–274.

Kaufman, L. 1977. The three spot damselfish: Effects on benthic biota of Caribbean coral reefs. *Proceedings of the Third International Coral Reef Symposium*, Miami, v. 1, pp. 559–564.

Kaufman, L. 1981. There was biological disturbance on Pleistocene coral reefs. *Paleobiology* 7:527–532.

Kaufman, L. S. 1983. Effects of Hurricane Allen on reef fish assemblages near Discovery Bay, Jamaica. *Coral Reefs* 2:43–47.

Kaufman, L. S. and K. F. Liem. 1982. Fishes of the Suborder Labroidei (Pisces: Perciformes): Phylogeny, ecology, and evolutionary significance. *Breviora* 472:1–19.

Kauffman, E. G. and J. A. Fagerstrom. 1993. The Phanerozoic evolution of reef diversity. In R. E. Ricklefs and D. Schluter, eds., *Species Diversity in Ecological Communities: Historical and Geographical Perspectives*, pp. 315–329. Chicago: University of Chicago Press.

Kinzie, R. A. III. 1973. The zonation of West Indian gorgonians. *Bulletin of Marine Science* 23:93–155.

Kinzie, R. A. III. 1997. Reproduction, symbiosis and the fossil record: Do geologists care about sex? (abstract). *American Zoologist* 37:99A.

Kjerfve, B. J. and S. P. Dinnel. 1983. Hindcast hurricane characteristics on the Belize Barrier Reef. *Coral Reefs* 1:203–207.

Kjerfve, B. J., K. E. Magill, J. W. Porter, and J. D. Woodley. 1986. Hindcasting of hurricane characteristics and observed storm damage on a fringing reef, Jamaica, West Indies. *Journal of Marine Research* 44:119–148.

Knowlton, N. 1992. Thresholds and multiple stable states in coral reef community dynamics. *American Zoologist* 32:674–682.

Knowlton, N., J. C. Lang, M. C. Rooney, and P. Clifford. 1981. Evidence for delayed mortality in hurricane-damaged Jamaican staghorn corals. *Nature* 294:251–252.

Knowlton, N., J. C. Lang, and B. D. Keller. 1990. Case study of natural population collapse: Post-hurricane predation on Jamaican staghorn corals. *Smithsonian Contributions to the Marine Sciences* 31:1–25.

Knowlton, N., E. Weil, L. A. Weigt, and H. M. Guzmán. 1992. Sibling species in *Montastraea annularis*, coral bleaching, and the coral climate record. *Science* 255:330–333.

Knowlton, N., J. L. Maté, H. M. Guzmàn, R. Rowan, and J. Jara. 1997. Direct evidence for reproductive isolation among the three species of the *Montastraea annularis* complex in Central America (Panamà and Honduras). *Marine Biology* 127:705–711.

Kobluk, D. R. and M. A. Lysenko. 1992. Storm features on a southern Caribbean fringing reef. *Palaios* 7:213–221.

Kojis, B. L. and N. J. Quinn. 1994. Biological limits to Caribbean reef recovery: A comparison with western South Pacific reefs. In R. N. Ginsburg, compiler, *Proceedings of the Colloquium on Global Aspects of Coral Reefs: Health, Hazards and History, 1993*, pp. 353–359. Miami: Rosenstiel School of Marine and Atmospheric Science, University of Miami.

Kuta, K. and L. L. Richardson. 1996. Abundance and distribution of black band disease on coral reefs in the northern Florida Keys. *Coral Reefs* 15:219–223.

Lang, J. C. 1973. Interspecific aggression by scleractinian corals: II. Why the race is not only to the swift. *Bulletin of Marine Science* 23:260–279.

Lang, J. C. and E. A. Chornesky. 1990. Competition between scleractinian corals: A review of mechanisms and effects. In Z. Dubinsky, ed., *Ecosystems of the World 25: Coral Reefs*, pp. 209–252. Amsterdam: Elsevier.

Lang, J. C., H. R. Lasker, E. H. Gladfelter, P. Hallock, W. C. Jaap, F. J. Losada, and R. G. Muller. 1992. Spatial and temporal variability during periods of "recovery" after mass bleaching on western Atlantic coral reefs. *American Zoologist* 32:696–706.

Lapointe, B. E. 1997. Nutrient thresholds for bottom-up control of macroalgal blooms on coral reefs in Jamaica and southeast Florida. *Limnology and Oceanography* 42:1119–1131.

Lasker, H. R. and M. A. Coffroth. 1988. Temporal and spatial variability among grazers: Variability in the distribution of the gastropod *Cyphoma gibbosum* on octocorals. *Marine Ecology Progress Series* 43:285–295.

Laydoo, R. 1984. Inference of a "White Band" epidemic in the elkhorn coral, *Acropora palmata*, populations in Tobago, W. I. (abstract). *Proceedings of the Association of Island Marine Laboratories of the Caribbean* 18:12.

Lessios, H. A. 1988. Mass mortality of *Diadema antillarum* in the Caribbean: What have we learned? *Annual Review of Ecology and Systematics* 19:371–393.

Lessios, H. A., D. R. Robertson, and J. D. Cubit. 1984. Spread of *Diadema* mass mortality through the Caribbean. *Science* 226:335–337.

Levitan, D. R. 1988. Algal-urchin biomass responses following mass mortality of *Diadema antillarum* Philippi at Saint John, U.S. Virgin Islands. *Journal of Experimental Marine Biology and Ecology* 119:167–178.

Levitan, D. R. 1992. Community structure in times past: Influence of human fishing pressure on algal-urchin interactions. *Ecology* 73:1597–1605.

Levitan, D. R. and C. Petersen. 1995. Sperm limitation in the sea. *Trends in Ecology and Evolution* 10:228–231.

Lewis, J. B. 1984. The *Acropora* inheritance: A reinterpretation of the development of fringing reefs in Barbados, West Indies. *Coral Reefs* 3:117–122.

Lewis, S. M. 1986. The role of herbivorous fishes in the organization of a Caribbean reef community. *Ecological Monographs* 56:183–200.

Lewis, S. M. and P. C. Wainwright. 1985. Herbivore abundance and grazing intensity on a Caribbean coral reef. *Journal of Experimental Marine Biology and Ecology* 87:215–228.

Liddell, W. D. and S. L. Ohlhorst. 1986. Changes in benthic community composition following the mass mortality of *Diadema antillarum. Journal of Experimental Marine Biology and Ecology* 95:271–278.

Liddell, W. D. and S. L. Ohlhorst. 1987. Patterns of reef community structure, north Jamaica. *Bulletin of Marine Science* 40:311–329.

Liddell, W. D. and S. L. Ohlhorst. 1993. Ten years of disturbance and change on a Jamaican fringing reef. *Proceedings of the Seventh International Coral Reef Symposium,* Guam, v. 1, pp. 144–150.

Liddell, W. D., S. L. Ohlhorst, and A. G. Coates. 1984. *Modern and Ancient Carbonate Environments of Jamaica: Sedimenta* X. Miami: Rosenstiel School of Marine and Atmospheric Science, University of Miami.

Lighty, R. G. 1981. Fleshy-algal domination of a modern Bahamian barrier reef: Example of an alternate climax reef community (abstract). *Proceedings of the Fourth International Coral Reef Symposium,* Manila, Philippines, v. 1, p. 722.

Lighty, R. G., I. G. Macintyre, and R. Stuckenrath. 1982. *Acropora palmata* reef framework: A reliable indication of sea level in the western Atlantic for the past 10,000 years. *Coral Reefs* 1:125–130.

Lirman, D. and P. Fong. 1996. Sequential storms cause zone-specific damage on a reef in the northern Florida reef tract: Evidence from Hurricane Andrew and the 1993 Storm of the Century. *Florida Scientist* 59:50–64.

Lirman, D. and P. Fong. 1997a. Patterns of damage to the branching coral *Acropora palmata* following Hurricane Andrew: Damage and survivorship of hurricane-generated asexual recruits. *Journal of Coastal Research* 13:67–72.

Lirman, D. and P. Fong. 1997b. Susceptibility of coral communities to storm intensity, duration, and frequency. *Proceedings of the Eighth International Coral Reef Symposium,* Panama, v. 1, pp. 561–566.

Littler, M. M. and D. S. Littler. 1985. Factors controlling relative dominance of primary producers on biotic reefs. *Proceedings of the Fifth International Coral Reef Congress,* Tahiti, v. 4, pp. 35–39.

Littler, M. M., P. R. Taylor, D. S. Littler, R. H. Sims, and J. N. Norris. 1987. Dominant macrophyte standing stocks, productivity and community structure on a Belizean barrier reef. *Atoll Research Bulletin* 302:1–24.

Littler, M. M., D. S. Littler, and E. A. Titlyanov. 1991. Comparisons of N- and P-limited productivity between high granitic islands *versus* low carbonate atolls in the Seychelles Archipelago: A test of the relative-dominance paradigm. *Coral Reefs* 10:199–209.

Lobel, P. S. 1980. Herbivory by damselfishes and their role in coral reef community ecology. *Bulletin of Marine Science* 30:273–289.

McClanahan, T. R. 1995. A coral reef ecosystem-fisheries model: Impacts of fishing intensity and catch selection on reef structure and processes. *Ecological Modelling* 80:1–19.

McClanahan, T. R. 1997. Primary succession of coral-reef algae: Differing patterns on fished versus unfished reefs. *Journal of Experimental Marine Biology and Ecology* 218:77–102.

McClanahan, T. R. and N. A. Muthiga. 1988. Changes in Kenyan coral reef community structure and function due to exploitation. *Hydrobiologia* 166:269–276.

McClanahan, T. R. and N. A. Muthiga. 1989. Patterns of predation on a sea urchin, *Echinometra mathaei* (de Blainville), on Kenyan coral reefs. *Journal of Experimental Marine Biology and Ecology* 126:77–94.

McClanahan, T. R. and N. A. Muthiga. 1998. An ecological shift among patch reefs of Glovers Reef Atoll, Belize over 25 years. *Environmental Conservation* 25:122–130.

McClanahan, T. R., V. Hendrick, M. J. Rodrigues, and N. V. C. Polunin. 1999. Varying responses of herbivorous and invertebrate-feeding fishes to macroalgal reduction on a coral reef. *Coral Reefs* 18:195–203.

McCook, L. 1996. Effects of herbivores and water quality on *Sargassum* distribution on the central Great Barrier Reef: Cross-shelf transplants. *Marine Ecology Progress Series* 139:179–192.

McCook, L. J. 1999. Macroalgae, nutrients and phase shifts on coral reefs: Scientific issues and management consequences for the Great Barrier Reef. *Coral Reefs* 18:357–367.

McCook, L. J., I. R. Price, and D. W. Klumpp. 1997. Macroalgae on the GBR: Causes or consequences, indicators or models of reef degradation? *Proceedings of the Eighth International Coral Reef Symposium*, Panama, v. 2, pp. 1851–1856.

McIntosh, R. P. 1995. H. A. Gleason's 'individualistic concept' and theory of animal communities: A continuing controversy. *Biological Reviews of the Cambridge Philosophical Society* 70:317–357.

McNeill, D. F., A. F. Budd, and P. F. Borne. 1997. Earlier (late Pliocene) first appearance of the Caribbean reef-building coral *Acropora palmata*: Stratigraphic and evolutionary implications. *Geology* 25:891–894.

Macintyre, I. G. 1988. Modern coral reefs of Western Atlantic: New geological perspective. *American Association of Petroleum Geologists Bulletin* 72:1360–1369.

Macintyre, I. G. 1997. Reevaluating the role of crustose coralline algae in the construction of coral reefs. *Proceedings of the Eighth International Coral Reef Symposium*, Panama, v. 1, pp. 725–730.

Macintyre, I. G. and R. B. Aronson. 1997. Field guidebook to the reefs of Belize. *Proceedings of the Eighth International Coral Reef Symposium*, Panama, v. 1, pp. 203–222.

Macintyre, I. G., R. B. Burke, and R. Stuckenrath. 1977. Thickest recorded Holocene reef section, Isla Pérez core hole, Alacran Reef, Mexico. *Geology* 5:749–754.

Mah, A. J. and C. W. Stearn. 1986. The effect of Hurricane Allen on the Bellairs fringing reef, Barbados. *Coral Reefs* 4:169–176.

Mann, P., F. W. Taylor, K. Burke, and R. Kulstad. 1984. Subaerially exposed Holocene coral reef, Enriquillo Valley, Dominican Republic. *Geological Society of America Bulletin* 95:1084–1092.

Marubini, F. and P. S. Davies. 1996. Nitrate increases zooxanthellae population density and reduces skeletogenesis in corals. *Marine Biology* 127:319–328.

Meesters, E. and R. P. M. Bak. 1993. Effects of coral bleaching on tissue regeneration potential and colony survival. *Marine Ecology Progress Series* 96:189–198.

Mesolella, K. J. 1967. Zonation of uplifted Pleistocene coral reefs on Barbados, West Indies. *Science* 156:638–640.

Meyer, J. L., E. T. Shultz, and G. S. Helfman. 1983. Fish schools: An asset to corals. *Science* 220:1047–1049.

Miller, A. I. 1997. Coordinated stasis or coincident relative stability? *Paleobiology* 23:155–164.

Miller, M. W. 1998. Coral/seaweed competition and the control of reef community structure within and between latitudes. *Oceanography and Marine Biology: An Annual Review* 36:65–96.

Miller, M. W. and M. E. Hay. 1998. Effects of fish predation and seaweed competition on the survival and growth of corals. *Oecologia* 113:231–238.

Miller, M. W., M. E. Hay, S. L. Miller, D. Malone, S. Sotka, and A. M. Szmant. 1999. Effects of nutrients versus herbivores on reef algae: A new method for manipulating nutrients on coral reefs. *Limnology and Oceanography* 44:1847–1861.

Miller, W. III. 1996. Ecology of coordinated stasis. *Palaeogeography, Palaeoclimatology, Palaeoecology* 127:177–190.

Morris, P. J., L. C. Ivany, K. M. Schopf, and C. E. Brett. 1995. The challenge of paleoecological stasis: Reassessing sources of evolutionary stability. *Proceedings of the National Academy of Sciences of the United States of America* 92:11269–11273.

Morrison, D. 1988. Comparing fish and urchin grazing in shallow and deeper coral reef algal communities. *Ecology* 69:1367–1382.

Mumby, P. J. 1999. Can Caribbean coral populations be modelled at metapopulation scales? *Marine Ecology Progress Series* 180:2175–288.

Munro, J. L. 1983. Caribbean coral reef fishery resources. *ICLARM Studies and Reviews (Manila)* 7:1–276.

Muthiga, N. A. and T. R. McClanahan. 1987. Population changes of a sea urchin (*Echinometra mathaei*) on an exploited fringing reef. *African Journal of Ecology* 25:1–8.

Neese, D. G. and R. K. Goldhammer. 1981. Constructional framework in buttress zone: Role of *Acropora cervicornis* and *Agaricia* (abstract). *American Association of Petroleum Geologists Bulletin* 65:1015.

Neigel, J. E. and J. C. Avise. 1983. Clonal diversity and population structure in a reef-building coral, *Acropora cervicornis*: Self-recognition analysis and demographic interpretation. *Evolution* 37:437–453.

Neumann, A. C. and I. Macintyre. 1985. Reef response to sea level rise: Keep-up, catch-up or give-up. *Proceedings of the Fifth International Coral Reef Congress*, Tahiti, v. 3, pp. 105–110.

Neumann, C. J., B. R. Jarvinen, A. C. Pike, and J. D. Elms. 1987. *Tropical Cyclones of the North Atlantic Ocean, 1871–1986, Third Revision.* Asheville NC: NOAA National Climatic Data Center.

Newell, N. D. 1971. An outline history of tropical organic reefs. *American Museum Novitates* 2465:1–37.

Ogden, J. C. and P. S. Lobel. 1978. The role of herbivorous fishes and urchins in coral reef communities. *Environmental Biology of Fishes* 3:49–63.

Ogden, J. C. and N. B. Ogden. 1994. The coral reefs of the San Blas Islands: Revisited after 20 years. In R. N. Ginsburg, compiler, *Proceedings of the Colloquium on Global Aspects of Coral Reefs: Health, Hazards and History, 1993*, pp. 267–272. Miami: Rosenstiel School of Marine and Atmospheric Science, University of Miami.

Ogden, J. C. and R. Wicklund, eds. 1988. Mass Bleaching of Reef Corals in the Caribbean: A Research Strategy. NOAA National Undersea Research Program Research Report 88–2, pp. 1–51.

Ogden, J. C., D. P. Abbott, and I. Abbott, eds. 1973. *Studies on the Activity and Food of the Echinoid* Diadema antillarum *Philippi on a West Indian Patch Reef. West Indies Laboratory Special Publication* 2. St. Croix: West Indies Laboratory, Fairleigh Dickinson University.

Ohlhorst, S. L. 1984. Spatial competition on a Jamaican coral reef. In P. W. Glynn, P. K. Swart, and A. M. Szmant-Froelich, eds., *Advances in Reef Science*, pp. 91–92. Miami: Rosenstiel School of Marine and Atmospheric Science, University of Miami.

Pandolfi, J. M. 1996. Limited membership in Pleistocene reef coral assemblages from the Huon Peninsula, Papua New Guinea: Constancy during global change. *Paleobiology* 22:152–176.

Pandolfi, J. M. and J. B. C. Jackson. 1997. The maintenance of diversity on coral reefs: Examples from the fossil record. *Proceedings of the Eighth International Coral Reef Symposium, Panama*, v. 1, pp. 397–404.

Patzkowsky, M. E. and S. M. Holland. 1997. Patterns of turnover in Middle and Upper Ordovician brachiopods of the eastern United States: A test of coordinated stasis. *Paleobiology* 23:420–443.

Paulay, G. 1990. Effects of late Cenozoic sea-level fluctuations on the bivalve faunas of tropical oceanic islands. *Paleobiology* 16:415–434.

Paulay, G. 1997. Diversity and distribution of reef organisms. In C. Birkeland, ed., *Life and Death of Coral Reefs*, pp. 298–353. New York: Chapman and Hall.

Pearson, R. G. 1981. Recovery and recolonization of coral reefs. *Marine Ecology Progress Series* 4:105–122.

Perkins, R. D. and P. Enos. 1968. Hurricane Betsy in the Florida-Bahamas area—geologic effects and comparison with Hurricane Donna. *Journal of Geology* 76:710–717.

Peters, E. C. 1993. Diseases of other invertebrate phyla: Porifera, Cnidaria, Ctenophora, Annelida, Echinodermata. In J. A. Couch and J. W. Fournie, eds., *Pathobiology of Marine and Estuarine Organisms*, pp. 393–449. Boca Raton FL: CRC Press.

Peters, E. C. 1997. Diseases of coral-reef organisms. In C. Birkeland, ed., *Life and Death of Coral Reefs*, pp. 114–139. New York: Chapman and Hall.

Peters, E. C., N. J. Gassman, J. C. Firman, R. H. Richmond, and E. A. Power. 1997. Ecotoxicology of tropical marine ecosystems. *Environmental Toxicology and Chemistry* 16:12–40.

Peters, E. C. and H. B. McCarty. 1996. Carbonate crisis? *Geotimes* 41(4):20–23.

Peterson, C. H. 1984. The pervasive biological explanation (book review). *Paleobiology* 9:429–436.

Porter, J. W. 1974. Community structure of coral reefs on opposite sides of the Isthmus of Panama. *Science* 186:543–545.

Porter, J. W. and O. W. Meier. 1992. Quantification of loss and change in Floridian reef coral populations. *American Zoologist* 32:625–640.

Porter, J. W., J. D. Woodley, G. J. Smith, J. E. Neigel, J. F. Battey, and D. G. Dallmeyer. 1981. Population trends among Jamaican reef corals. *Nature* 294:249–250.

Porter, J. W., J. F. Battey, and J. G. Smith. 1982. Perturbation and change in coral reef communities. *Proceedings of the National Academy of Sciences of the United States of America* 79:1678–1681.

Porter, J. W., W. K. Fitt, H. J. Spero, C. S. Rogers, and M. W. White. 1989. Bleaching in reef corals: Physiological and stable isotopic responses. *Proceedings of the National Academy of Sciences of the United States of America* 86:9342–9346.

Porter, J. F., O. W. Meier, L. Chiang, and T. Richardson. 1993. Quantification of coral reef change (part 2): The establishment and computer analysis of permanent photostations in the Florida "SEAKEYS" survey (abstract). *Proceedings of the Seventh International Coral Reef Symposium,* Guam, v. 1, pp. 168.

Potts, D.C. 1977. Suppression of coral populations by filamentous algae within damselfish territories. *Journal of Experimental Marine Biology and Ecology* 28:207–216.

Potts, D.C. 1984. Generation times and the Quaternary evolution of reef-building corals. *Paleobiology* 10:48–58.

Potts, D. C. and R. L. Garthwaite. 1991. Evolution of reef-building corals during periods of rapid global change. In E. C. Dudley, ed., *The Unity of Evolutionary Biology: Proceedings of the Fourth International Congress of Systematic and Evolutionary Biology, Volume I,* pp. 170–178. Portland OR: Dioscorides Press.

Precht, W. F. 1990. Geologic and ecologic perspectives of catastrophic storms and disturbance on coral reefs: Lessons from Discovery Bay, Jamaica (abstract). *Geological Society of America Annual Meeting, Abstracts with Programs* 22:A331.

Precht, W. F. 1993. Holocene coral patch reef ecology and sedimentary architecture, northern Belize, Central America. *Palaios* 8:499–503.

Precht, W. F. and W. H. Hoyt. 1991. Reef facies distribution patterns, Pleistocene (125 Ka) Falmouth Formation, Rio Bueno, Jamaica, W. I. (abstract). *American Association of Petroleum Geologists Bulletin* 75:656–657.

Precht, W. F., R. B. Aronson, P. J. Edmunds, and D. R. Levitan. 1993. Hurricane Andrew's effect on the Florida reef tract (abstract). *American Association of Petroleum Geologists Bulletin* 77:1473.

Reinthal, P. N., B. Kensley, and S. M. Lewis. 1984. Dietary shifts in the queen trigger-fish, *Balistes vetula*, in the absence of its primary food item, *Diadema antillarum*. *Pubblicazioni della Stazione Zoologica di Napoli I: Marine Ecology* 5:191–195.

Richardson, L. L. 1998. Coral diseases: What is really known? *Trends in Ecology and Evolution* 13:438–443.

Richmond, R. H. 1987. Energetics, competency, and long-distance dispersal of plan-ula larvae of the coral *Pocillopora damicornis*. *Marine Biology* 93:527–533.

Richmond, R. H. 1988. Competency and dispersal potential of planula larvae of a spawning versus a brooding coral. *Proceedings of the Sixth International Coral Reef Symposium*, Townsville, Australia, v. 2, pp. 827–831.

Richmond, R. H. 1997. Reproduction and recruitment in corals: Critical links in the persistence of reefs. In C. Birkeland, ed., *Life and Death of Coral Reefs*, pp. 175–197. New York: Chapman and Hall.

Richmond, R. H. and C. L. Hunter. 1990. Reproduction and recruitment of corals: Comparisons among the Caribbean, the Tropical Pacific, and the Red Sea. *Marine Ecology Progress Series* 60:185–203.

Roberts, C. M. 1995. Effects of fishing on the ecosystem structure of coral reefs. *Conservation Biology* 9:988–995.

Roberts, C. M. 1997. Connectivity and management of Caribbean coral reefs. *Science* 278:1454–1457.

Roberts, H. H., L. J. Rouse Jr., N. D. Walker, and J. H. Hudson. 1982. Cold-water stress in Florida Bay and northern Bahamas: A product of winter cold-air out-breaks. *Journal of Sedimentary Petrology* 52:145–155.

Robertson, D. R. 1991. Increases in surgeonfish populations after mass mortality of the sea urchin *Diadema antillarum* in Panama indicate food limitation. *Marine Biology* 111:437–444.

Rogers, C. S. 1985. Degradation of Caribbean and western Atlantic coral reefs and decline of associated fisheries. *Proceedings of the Fifth International Coral Reef Congress*, Tahiti, v. 6, pp. 491–496.

Rogers, C. S. 1990. Responses of coral reefs and reef organisms to sedimentation. *Marine Ecology Progress Series* 62:185–202.

Rogers, C. S. 1993a. Hurricanes and coral reefs: The intermediate disturbance hypothesis revisited. *Coral Reefs* 12:127–137.

Rogers, C. S. 1993b. A matter of scale: Damage from Hurricane Hugo (1989) to U.S. Virgin Islands reefs at the colony, community, and whole reef level. *Proceedings of the Seventh International Coral Reef Symposium*, Guam, v. 1, pp. 127–133.

Rogers, C. S. 1994. Hurricanes and anchors: Preliminary results from the National Park Service Regional Reef Assessment Program. In R. N. Ginsburg, compiler, *Proceedings of the Colloquium on Global Aspects of Coral Reefs: Health, Hazards and History, 1993*, pp. 214–219. Miami: Rosenstiel School of Marine and Atmospheric Science, University of Miami.

Rogers, C. S. 2000. Is *Acropora palmata* (elkhorn coral) making a comeback in the Virgin Islands? *Reef Encounter* 27:15–17.

Rogers, C. S., T. H. Suchanek, and F. Pecora. 1982. Effects of Hurricanes David and Frederic (1979) on shallow *Acropora palmata* reef communities: St. Croix, U.S. Virgin Islands. *Bulletin of Marine Science* 32:532–548.

Rogers, C. S., H. C. Fitz III, M. Gilnack, J. Beets, and J. Hardin. 1984. Scleractinian coral recruitment patterns at Salt River Submarine Canyon, St. Croix, U.S. Virgin Islands. *Coral Reefs* 3:69–76.

Rogers, C. S., L. N. McLain, and C. R. Tobias. 1991. Effects of Hurricane Hugo (1989) on a coral reef in St. John, USVI. *Marine Ecology Progress Series* 78:189–199.

Rogers, C. S., V. Garrison, and R. Grober-Dunsmore. 1997. A fishy story about hurricanes and herbivory: Seven years of research on a reef in St. John, U.S. Virgin Islands. *Proceedings of the Eighth International Coral Reef Symposium*, Panama, v. 1, pp. 555–560.

Rosesmyth, M. C. 1984. Growth and survival of sexually produced *Acropora* recruits: A post-hurricane study at Discovery Bay. In P. W. Glynn, P. K. Swart, and A. M. Szmant-Froelich, eds., *Advances in Reef Science*, pp. 105–106. Miami: Rosenstiel School of Marine and Atmospheric Science, University of Miami.

Rougerie, F., J. A. Fagerstrom, and C. Andrie. 1992. Geothermal endo-upwelling: A solution to the reef nutrient paradox? *Continental Shelf Research* 12:785–798.

Roughgarden, J. 1989. The structure and assembly of communities. In J. Roughgarden, R. M. May, and S. A. Levin, eds., *Perspectives in Ecological Theory*, pp. 203–226. Princeton NJ: Princeton University Press.

Roy, K., D. Jablonski, and J. W. Valentine. 1995. Thermally anomalous assemblages revisited: Patterns in the extraprovincial latitudinal range shifts of Pleistocene marine mollusks. *Geology* 23:1071–1074.

Rützler, K. and I. G. Macintyre. 1982. The habitat distribution and community structure of the barrier reef complex at Carrie Bow Cay, Belize. In K. Rützler and I. G. Macintyre, eds., *The Atlantic Barrier Reef Ecosystem at Carrie Bow Cay, Belize, I. Structure and Communities*, pp. 9–45. Smithsonian Contributions to the Marine Sciences 12. Washington DC: Smithsonian Institution Press.

Rützler, K., D. L. Santavy, and A. Antonius. 1983. The Black Band Disease of Atlantic reef corals: III. Distribution, ecology, and development. *Pubblicazioni della Stazione Zoologica di Napoli I: Marine Ecology* 4:329–358.

Rylaarsdam, K. W. 1983. Life histories and abundance patterns of colonial corals on Jamaican reefs. *Marine Ecology Progress Series* 13:249–260.

Sakai, K. 1997. Gametogenesis, spawning, and planula brooding by the reef coral *Goniastrea aspersa* (Scleractinia) in Okinawa, Japan. *Marine Ecology Progress Series* 151:67–72.

Sammarco, P. W. 1980. *Diadema* and its relationship to coral spat mortality: Grazing, competition, and biological disturbance. *Journal of Experimental Marine Biology and Ecology* 45:245–272.

Sammarco, P. W. 1982. Echinoid grazing as a structuring force in coral communities: Whole reef manipulations. *Journal of Experimental Marine Biology and Ecology* 61:31–55.

Sammarco, P. W. 1985. The Great Barrier Reef vs. the Caribbean: Comparisons of grazers, coral recruitment patterns and reef recovery. *Proceedings of the Fifth International Coral Reef Congress,* Tahiti, v. 4, pp. 391–397.

Sammarco, P. W. and J. C. Andrews. 1989. The Helix experiment: Differential localized dispersal and recruitment patterns in Great Barrier Reef corals. *Limnology and Oceanography* 34:896–912.

Santavy, D. L. and E. C. Peters. 1997. Microbial pests: Coral diseases in the western Atlantic. *Proceedings of the Eighth International Coral Reef Symposium,* Panama, v. 1, pp. 607–612.

Shackleton, N. J. 1985. Oceanic carbon isotope constraints on oxygen and carbon dioxide in the Cenozoic atmosphere. *American Geophysical Union Monograph* 32:412–417.

Sheppard, C. 1993. Coral reef environmental science: Dichotomies, not the Cassandras, are false. *Reef Encounter* 14:12–13.

Sheppard, C. R. C., K. Mathieson, J. C. Bythell, P. Murphy, C. Blair Myers, and B. Blake. 1995. Habitat mapping in the Caribbean for management and conservation: Use and assessment of aerial photography. *Aquatic Conservation: Marine and Freshwater Ecosystems* 5:277–298.

Shick, J. M., M. P. Lesser, and P. L. Jokiel. 1996. Effects of ultraviolet radiation on corals and other coral reef organisms. *Global Change Biology* 2:527–545.

Shinn, E. A. 1966. Coral growth rate, an environmental indicator. *Journal of Paleontology* 40:233–240.

Shinn, E. A. 1976. Coral reef recovery in Florida and the Persian Gulf. *Environmental Geology* 1:241–254.

Shinn, E. A. 1989. What is really killing the corals. *Sea Frontiers* 35:72–81.

Shinn, E. A., J. H. Hudson, D. M. Robbin, and B. Lidz. 1981. Spurs and grooves revisited: Construction versus erosion, Looe Key Reef, Florida. *Proceedings of the Fourth International Coral Reef Symposium,* Manila, Philippines, v. 1, pp. 475–483.

Shinn, E. A., B. H. Lidz, R. B. Halley, J. H. Hudson, and J. L. Kindinger. 1989. *Reefs of Florida and the Dry Tortugas.* Field Trip Guidebook T176, 28th International Geological Congress. Washington DC: American Geophysical Union.

Shulman, M. J. and D. R. Robertson. 1996. Changes in the coral reefs of San Blas, Caribbean Panama: 1983 to 1990. *Coral Reefs* 15:231–236.

Smith, D. B. 1991. The reproduction and recruitment of *Porites panamensis* verrill at Uva Island, Pacific Panama. M.S. thesis, University of Miami, Coral Gables FL.

Smith, S. R. 1988. Recovery of a disturbed reef in Bermuda: Influence of reef structure and herbivorous grazers on algal and sessile invertebrate recruitment. *Proceedings of the Sixth International Coral Reef Symposium,* Townsville, Australia, v. 2, pp. 267–272.

Smith, S. R. 1992. Patterns of coral recruitment and post-settlement mortality on Bermuda's reefs: Comparison to Caribbean and Pacific reefs. *American Zoologist* 32:663–673.

Smith, S. R. 1997. Patterns of coral settlement, recruitment and juvenile mortality with depth at Conch Reef, Florida. *Proceedings of the Eighth International Coral Reef Symposium,* Panama, v. 2, pp. 1197–1202.

Smith, S. V. and R. W. Buddemeier. 1992. Global change and coral reef ecosystems. *Annual Review of Ecology and Systematics* 23:89–118.

Stanley, S. M. 1986. Anatomy of a regional mass extinction: Plio-Pleistocene decimation of the western Atlantic bivalve fauna. *Palaios* 1:17–36.

Stemann, T. A. and K. G. Johnson. 1992. Coral assemblages, biofacies, and ecological zones in the mid-Holocene reef deposits of the Enriquillo Valley, Dominican Republic. *Lethaia* 25:231–241.

Stemann, T. A. and K. G. Johnson. 1995. Ecologic stability and spatial continuity in a Holocene reef, Lago Enriquillo, Dominican Republic (abstract). *Geological Society of America Annual Meeting, Abstracts with Programs* 27:A166.

Steneck, R. S. 1988. Herbivory on coral reefs: A synthesis. *Proceedings of the Sixth International Coral Reef Symposium,* Townsville, Australia, v. 1, pp. 37–49.

Steneck, R. S. 1994. Is herbivore loss more damaging to reefs than hurricanes? Case studies from two Caribbean reef systems (1978–1988). In R. N. Ginsburg, compiler, *Proceedings of the Colloquium on Global Aspects of Coral Reefs: Health, Hazards and History, 1993,* pp. 220–226. Miami: Rosenstiel School of Marine and Atmospheric Science, University of Miami.

Steneck, R. S. 1997. Crustose corallines, other algal functional groups, herbivores and sediments: Complex interactions along reef productivity gradients. *Proceedings of the Eighth International Coral Reef Symposium,* Panama, v. 1, pp. 695–700.

Stoddart, D. R. 1963. Effects of Hurricane Hattie on the British Honduras reefs and cays, October 30–31, 1961. *Atoll Research Bulletin* 95:1–142.

Stoddart, D. R. 1969. Post-hurricane changes on the British Honduras reefs and cays: Re-survey of 1965. *Atoll Research Bulletin* 131:1–25.

Stoddart, D. R. 1974. Post-hurricane changes on the British Honduras reefs: Re-survey of 1972. *Proceedings of the Second International Coral Reef Symposium,* Brisbane, Australia, v. 2, pp. 473–483.

Szmant, A. M. 1986. Reproductive ecology of Caribbean reef corals. *Coral Reefs* 5:43–53.

Szmant, A. M. 1997. Nutrient effects on coral reefs: A hypothesis on the importance of topographic and trophic complexity to reef nutrient dynamics. *Proceedings of the Eighth International Coral Reef Symposium,* Panama, v. 2, pp. 1527–1532.

Szmant, A. M. and A. Forrester. 1996. Water column and sediment nitrogen and phosphorus distribution patterns in the Florida Keys, USA. *Coral Reefs* 15:21–41.

Szmant, A. M., E. Weil, M. W. Miller, and D. E. Colón. 1997. Hybridization within the species complex of the scleractinian coral *Montastraea annularis. Marine Biology* 129:561–572.

Szmant-Froelich, A., M. Ruetter, and L. Riggs. 1985. Sexual reproduction of *Favia fragum* (Esper): Lunar patterns of gametogenesis, embryogenesis and planulation in Puerto Rico. *Bulletin of Marine Science* 37:880–892.

Tang, C. M. and D. J. Bottjer. 1996. Long-term faunal stasis without evolutionary coordination: Jurassic benthic marine paleocommunities, Western Interior, United States. *Geology* 24:815–818.

Tanner, J. E. 1995. Competition between scleractinian corals and macroalgae: An experimental investigation of coral growth, survival and reproduction. *Journal of Experimental Marine Biology and Ecology* 190:151–168.

Taylor, F. W., P. Mann, S. Valastro Jr., and K. Burke. 1985. Stratigraphy and radiocarbon chronology of a subaerially exposed Holocene coral reef, Dominican Republic. *Journal of Geology* 93:311–332.

Tolimieri, N. 1998. Contrasting effects of microhabitat use on large-scale adult abundance in two families of Caribbean reef fishes. *Marine Ecology Progress Series* 167:227–239.

Tomascik, T. 1991. Settlement patterns of Caribbean scleractinian corals on artificial substrata along a eutrophication gradient, Barbados, West Indies. *Marine Ecology Progress Series* 77:261–269.

Tomascik, T. and F. Sander. 1987. Effects of eutrophication on reef-building corals: II. Structure of scleractinian coral communities on fringing reefs, Barbados, West Indies. *Marine Biology* 94:53–75.

Tremel, E., M. Colgan, and M. Keevican. 1997. Hurricane disturbance and coral reef development: A geographic information system (GIS) analysis of 501 years of hurricane data from the Lesser Antilles. *Proceedings of the Eighth International Coral Reef Symposium*, Panama, v. 1, pp. 541–546.

Tunnicliffe, V. 1981. Breakage and propagation of the stony coral *Acropora cervicornis*. *Proceedings of the National Academy of Sciences of the United States of America* 78:2427–2431.

Tunnicliffe, V. 1983. Caribbean staghorn coral populations: Pre-Hurricane Allen conditions in Discovery Bay, Jamaica. *Bulletin of Marine Science* 33:132–151.

Umar, M. J., L. J. McCook, and I. R. Price. 1998. Effects of sediment deposition on the seaweed *Sargassum* on a fringing coral reef. *Coral Reefs* 17:169–177.

Underwood, A. J., and E. J. Denley. 1984. Paradigms, explanations, and generalizations in models for the structure of intertidal communities on rocky shores. In D. R. Strong Jr., D. Simberloff, L. G. Abele and A. B. Thistle, eds., *Ecological Communities: Conceptual Issues and the Evidence*, pp. 151–180. Princeton NJ: Princeton University Press.

Valentine, J. W. and D. Jablonski. 1993. Fossil communities: Compositional variation at many time scales. In R. E. Ricklefs and D. Schluter, eds., *Species Diversity in Ecological Communities: Historical and Geographical Perspectives*, pp. 341–349. Chicago: University of Chicago Press.

van Duyl, F. C. 1982. The distribution of *Acropora palmata* and *Acropora cervicornis* along the coasts of Curaçao and Bonaire, Netherlands Antilles (abstract). In *Inter-*

disciplinary Studies in Coral Reef Research, p. 27. Leiden: International Society for Reef Studies.

van Duyl, F. C. 1985. *Atlas of the Living Reefs of Curaçao and Bonaire (Netherlands Antilles.* Utrecht: Natuurwetenschappelijke Studiekring voor Suriname en de Nederlandse Antillen [Foundation for Scientific Research in Surinam and the Netherlands Antilles].

Veron, J. E. N. 1985. Aspects of the biogeography of hermatypic corals. *Proceedings of the Fifth International Coral Reef Congress,* Tahiti, v. 4, pp. 83–88.

Veron, J. E. N. 1995. *Corals in Space and Time: The Biogeography and Evolution of the Scleractinia.* Ithaca NY: Cornell University Press.

Vicente, V. P. 1994. Structural changes and vulnerability of a coral reef (Cayo Enrique) in La Parguera, Puerto Rico. In R. N. Ginsburg, compiler, *Proceedings of the Colloquium on Global Aspects of Coral Reefs: Health, Hazards and History, 1993,* pp. 227–232. Miami: Rosenstiel School of Marine and Atmospheric Science, University of Miami.

Vrba, E. S. 1985. Environment and evolution: Alternative causes of the temporal distribution of evolutionary events. *South African Journal of Science* 81:229–236.

Walter, G. H. and H. E. H. Paterson. 1994. The implications of palaeontological evidence for theories of ecological communities and species richness. *Australian Journal of Ecology* 19:241–250.

Ward, S. 1992. Evidence for broadcast spawning as well as brooding in the scleractinian coral *Pocillopora damicornis.* Marine Biology 112:641–646.

Weiss, M. P. and D. A. Goddard. 1977. Man's impact on coastal reefs: An example from Venezuela. In S. H. Frost, M. P. Weiss, and J. B. Saunders, eds., *Reefs and Related Carbonates: Ecology and Sedimentology,* pp. 111–124. Studies in Geology No. 4. Tulsa OK: American Association of Petroleum Geologists.

Wells, S. M., ed. 1988. *Coral Reefs of the World. Volume 1: Atlantic and Eastern Pacific.* Nairobi, Gland, and Cambridge: United Nations Environment Programme/International Union for Conservation of Nature and Natural Resources.

Westrop, S. R. 1996. Temporal persistence and stability of Cambrian biofacies: Sunwaptan (Upper Cambrian) trilobite faunas of North America. *Palaeogeography, Palaeoclimatology, Palaeoecology* 127:33–46.

Wilkinson, C. R. 1993. Coral reefs of the world are facing widespread devastation: Can we prevent this through sustainable management practices? *Proceedings of the Seventh International Coral Reef Symposium,* Guam, v. 1, pp. 11–21.

Williams, A. H. 1978. Ecology of threespot damselfish: Social organization, age structure, and population stability. *Journal of Experimental Marine Biology and Ecology* 34:197–213.

Williams, E. H. Jr. and L. Bunkley-Williams. 1990. The world-wide coral reef bleaching cycle and related sources of coral mortality. *Atoll Research Bulletin* 335:1–71.

Wilson, D. S. 1997. Biological communities as functionally organized units. *Ecology* 78:2018–2024.

Wittenberg, M. and W. Hunte. 1992. Effects of eutrophication and sedimentation on juvenile corals: I. Abundance, mortality and community structure. *Marine Biology* 112:131–138.

Wood, R. 1993. Nutrients, predation and the history of reef-building. *Palaios* 8:526–543.

Woodley, J. D. 1989. The effects of Hurricane Gilbert on coral reefs at Discovery Bay. In P. R. Bacon, ed., *Assessment of the Economic Impacts of Hurricane Gilbert on Coastal and Marine Resources in Jamaica*, pp. 71–73. Nairobi: UNEP Regional Seas Reports and Studies 110, United Nations Environment Programme.

Woodley, J. D. 1992. The incidence of hurricanes on the north coast of Jamaica since 1870: Are the classic reef descriptions atypical? *Hydrobiologia* 247:133–138.

Woodley, J. D., E. A. Chornesky, P. A. Clifford, J. B. C. Jackson, L. S. Kaufman, N. Knowlton, J. C. Lang, M. P. Pearson, J. W. Porter, M. C. Rooney, K. W. Rylaarsdam, V. J. Tunnicliffe, C. M. Wahle, J. L. Wulff, A. S. G. Curtis, M. D. Dallmeyer, B. P. Jupp, M. A. R. Koehl, J. Neigel, and E. M. Sides. 1981. Hurricane Allen's impact on Jamaican coral reefs. *Science* 214:749–755.

Woodley, J. D., K. De Meyer, P. Bush, G. Ebanks-Petrie, J. Garzón-Ferreira, E. Klein, L. P. J. J. Pors, and C. M. Wilson. 1997. Status of coral reefs in the south central Caribbean. *Proceedings of the Eighth International Coral Reef Symposium*, Panama, v. 1, pp. 357–362.

Zea, S., J. Geister, J. Garzón-Ferreira, and J. M. Díaz. 1998. Biotic changes in the reef of San Andrés Island (southwestern Caribbean Sea, Colombia) occurring over nearly three decades. *Atoll Research Bulletin* 456:1–30.

10

Rates and Processes of Terrestrial Nutrient Cycling in the Paleozoic: The World Before Beetles, Termites, and Flies

Anne Raymond, Paul Cutlip, and Merrill Sweet

RATES OF ORGANIC decomposition and weathering directly affect the primary productivity of terrestrial ecosystems because they control nutrient availability (Beerbower 1985; Jordan 1985; Perry 1994). Of these two factors, organic decomposition may have greater influence on terrestrial primary productivity through geologic time. Although rates of physical and chemical weathering probably changed with the evolution of land plants, since the appearance of trees in the Late Devonian, the evolution of new plant groups and morphologies may not have affected weathering rates, especially in moist lowland habitats (Robinson 1991; see Knoll and James 1987 for another view). In modern forest ecosystems, most nutrients come from organic decomposition, which results from interactions between bacteria, fungi, and invertebrate detritivores (Lavelle et al. 1993, 1995; Perry 1994). The importance of organic decomposition in determining the productivity of modern terrestrial ecosystems suggests that changes in the rates and processes of organic decomposition through geologic time could have had a major influence on the primary productivity of terrestrial ecosystems.

A wealth of evidence suggests that the rate of terrestrial decomposition has increased since the Paleozoic. Robinson (1990) linked both low levels of atmospheric CO_2, predicted by models of the carbon cycle through geologic time (Berner 1990, 1991), and the vast amount of Late Carboniferous coal to slow rates of terrestrial decomposition. The occurrence of Early Carboniferous coals at anomalous, arid paleolatitudes may reflect both the weak climatic zonation of the Early Carboniferous and slow rates of terrestrial decomposition (Raymond 1997). Some Late Carboniferous permineralized peats contain delicate tissues and organs such as phloem, pollen tubes, and cellular megagametophytes (Rothwell 1972; Stubblefield and Rothwell 1981; Taylor and Taylor 1993) that seldom occur in modern peats; their presence in ancient peats may signal slower rates of terrestrial decomposition in the late Paleozoic than at present. If rates of terrestrial decomposition were slower during the Paleozoic, then ancient terrestrial ecosystems would have had lower levels of primary productivity than modern terrestrial ecosystems from similar climatic zones.

Possible causes for apparent differences in the rate of terrestrial decomposition during the Late Carboniferous include evolutionary changes in land plants and decomposer groups (basidiomycete fungi and invertebrate detritivores). At the onset of the Late Carboniferous, arborescent lycopsids dominated many lowland ecosystems. Gymnosperms (seed-ferns, coniferophytes, and cycadophytes) began to dominate lowland habitats at the end of the Late Carboniferous and remained dominant into the Mesozoic. In the Late Cretaceous and Tertiary, angiosperms replaced conifers as the dominant plants in these ecosystems.

The evolutionary ascendancy of angiosperms in the Late Mesozoic–Early Tertiary may have increased rates of terrestrial decomposition. Conifer wood contains more lignin and different hemicellulose compounds than angiosperm wood (Eriksson, Blanchette, and Ander 1990; Robinson 1990; Shearer, Moore, and Demchuk 1995) and few decomposer groups can attack lignin (Rayner and Boddy 1988). Shearer, Moore, and Demchuk (1995) found that most of the fossil wood in Tertiary lignites derives from gymnosperm conifers, even in paleomires dominated by angiosperm pollen, suggesting that angiosperm wood decomposed more quickly than conifer wood in these habitats. However, there is less evidence for changes in the rate of terrestrial decomposition tied to replacement of lycopsids and other pteridophytes by gymnosperms beginning in the Carboniferous and culminating in the Mesozoic.

Robinson (1990) and others (Taylor 1993; Taylor and Osborn 1992) suggested that the rarity or absence of lignolytic fungi contributed to slow rates of terrestrial decomposition in the Late Carboniferous. Nevertheless, Late Carboniferous wood shows abundant evidence of decomposition by lignolytic fungi (Raymond, Heise, and Cutlip, in review). In addition, 18S ribosomal RNA gene sequence data for 37 fungal species calibrated using fossil evidence suggests the presence of lignolytic basidiomycetes in the Late Carboniferous (Berbee and Taylor 1993). Thus, the evolution of new invertebrate detritivores and plants may have had a greater influence on the rate of terrestrial decomposition than the evolution of new fungi.

Evolutionary changes in invertebrate detritivores could have increased the rate of terrestrial decomposition after the Late Carboniferous (Raymond and Heise 1994; Labandeira, Phillips, and Norton 1997). Detritivores enhance rates of organic decomposition by fragmenting large particles, which increases the surface area available for fungal and microbial attack (Swift, Heal, and Anderson 1979; Perry 1994). They also influence decomposition processes, often enhancing bacterial decomposition and inhibiting fungal decay (Perry 1994). The probable arthropod detritivores of the Paleozoic include oribatid mites, diplopods, arthropleurids, Collembola, as well as cockroaches and other primitive insects (Rolfe 1985; Shear and Kukalova-Peck 1990; Labandeira 1998). Many important modern detritivore and wood-boring groups appeared after the Late Carboniferous, including beetles, flies, termites, wood wasps, horntails, and ants (Labandeira 1994).

In this contribution, we focus on the evolution of terrestrial detritivory and its influence on decomposition and productivity within tropical paleomires. Tropical peat provides an excellent opportunity to investigate decomposition and terrestrial nutrient cycling through time. Today, these organic sediments accumulate within a narrow range of wetland habitats, in which there is no, or only a feeble, dry season (Morley 1981; Anderson 1983; Thompson and Hamilton 1983; Ziegler et al. 1987). Late Carboniferous equatorial coals frequently contain concretions of permineralized peat, also known as coal balls (Phillips 1980). Permineralized wood from these peats lacks regular growth banding, suggesting that they also accumulated in ever-wet, ever-warm habitats. Based on peat characteristics such as matrix frequency, root percentage, and leaf-mat thickness, we argue that most of the ancient peat in our study sample has experienced less and slower decomposition than modern tropical peats. In the Williamson No. 3 peat deposit, the frequency of detritivore fecal pellets correlates positively with the amount of decomposition, suggesting that ancient detritivores did influence terrestrial decomposition rates and processes. Based on the distribution of arthropod coprolites and fecal pellets in this ancient peat, we evaluate the relative rate of nutrient recycling in paleomire communities dominated by *Medullosa* and *Cordaites* and present evidence suggesting that arthropod-mediated detritivory in Late Carboniferous tropical peats differed from that of the present day.

Comparing Nutrient Cycling in Ancient and Modern Peat

The Late Carboniferous peats used in this study come from four coal mines, three in Iowa (the Williamson No. 3 Coal Mine in Lucas Co., the Urbandale Mine in Polk Co., and the Shuler Mine in Dallas Co.) and one in Illinois (the Sahara Coal Mine in Carrier Mills). All four coal deposits lie near the Late Carboniferous paleoequator and accumulated in tropical paleomires (Phillips 1980). *Cordaites,* a coniferophyte tree or shrub, *Medullosa* seed ferns, and *Calamites* trees predominate in the three Iowa deposits, which may have accumulated in paleomires that contained both freshwater and brackish environments (Raymond and Phillips 1983). The Iowa material used in this study came from the Harvard Botanical Museum. Lesnikowska and Willard (1997) placed the Shuler deposit in the late Bolsovian–early Westphalian D (late Atokan–Desmoinesian). The similarity of these three Iowa floras (Shuler, Urbandale, and Williamson No. 3), as well as the proximity of the Shuler and Urbandale Mines (Andrews 1945; Andrews and Kernen 1946), suggest a similar age for the Williamson No. 3 and Urbandale deposits. Lycopsid trees predominate in the Herrin (No. 6) Coal from Illinois, which probably accumulated in a freshwater paleomire (Phillips and DiMichele 1981). The Illinois material used in this study came from the University of Illinois Paleoherbarium; Phillips (1980) placed the Herrin (No. 6) Coal of Illinois in the Desmoisnesian.

The study sample consists of cellulose acetate peels made from the cut surfaces of concretions or coal balls. Because modern peat data come from cores that are constrained in width, the number of cm measured perpendicular to bedding may provide a better comparison of the size of ancient and modern peat samples than the number of cm^2. When possible, we used previously uncut concretions for this study. Because of the small number of uncut Williamson No. 3 concretions in the Harvard Botanical Museum, we supplemented the uncut study sample of 42 concretions with 67 previously cut concretions from the curated Museum collection. Concretions in the curated collection generally contain more leaves, reproductive material, and wood than previously uncut material. Table 10.1 lists the number of concretions used from each mine, their total height perpendicular to bedding, their combined surface area, and the types of data gathered from each deposit.

The relationship between peat characteristics (e.g., particle-size distributions and root percentages) and mire habitats is poorly understood, primarily due to the small number of comprehensive studies of modern peat. Accordingly, we compare these Late Carboniferous permineralized peats to modern peats from a variety of salinities, depositional settings, and climate zones

TABLE 10.1. Ancient Peat Samples

| Mine | Total No. of Concretions | Combined Height (cm) | Total Surface Area (cm²) | Data Available For Each Deposit: | | | | | |
				Matrix Freq.	Root %	Leaf Mat	Coprolite Survey	Pelletization %
Williamson No. 3	109[a]	419.5[b]	3,896[b]	+	+	+	+	+
Urbandale	103	625.7	5725	na	+	+	na	na
Shuler	119	701.0	4498	na	na	+	na	na
Sahara, Herrin (No. 6) Coal	53	482.0	4299	na	+	na	na	na

[a] Of these, 41 were previously uncut, and constitute the least biased sample of Williamson No. 3 peat available.
[b] These values are the combined height and total surface of the previously uncut sample.

(table 10.2). The closest modern analogues for *Cordaites–Medullosa* peats in terms of plant growth habit and climate may be the tropical freshwater domed peats from Malaysia and Indonesia (Esterle 1989). Both these and Late Carboniferous peats formed in tropical forests (Esterle 1989; Raymond and Phillips 1983). Nevertheless, environmental differences could complicate comparisons of ancient and modern tropical peats. The peats of Indonesia and Malaysia accumulated above the water table in freshwater domed mires (Esterle 1989). The relatively high mineral content of Late Carboniferous coals from Iowa (average 13.6%: Olin, Kinne, and Hale 1929) suggests that *Cordaites–Medullosa* peats accumulated in planar rather than domed paleomires and some *Cordaites* peat may have formed in brackish or marine swamps (Raymond 1988).

In the absence of root percentage and matrix frequency data for tropical peats from planar mires, comparisons of Late Carboniferous peats to subtropical mangrove and freshwater planar peats from the Florida Everglades (Cohen 1968; Cohen and Spackman 1977; Raymond 1987) may have some relevance. Both subtropical mangrove and Late Carboniferous peats formed in forested swamps (Cohen and Spackman 1977; Raymond 1987, 1988). Except that the Florida Everglades are subtropical, this mire appears to fit the depositional setting and salinity gradient proposed for Late Carboniferous *Cordaites–Medullosa* peats (Raymond 1988). Nevertheless, differences in plant growth habit and peat thickness complicate comparisons of Everglades and Late Carboniferous *Cordaites–Medullosa* peats. Although freshwater forest and shrub peats do occur in the Florida Everglades, herbaceous marsh plants formed most of the freshwater peat from the Everglades (Cohen 1968; Spackman et al. 1976). Because marsh peats have significantly higher average root percentages than forest peats (Raymond 1987), we exclude marsh peats from our comparison of the distribution of root percentages in ancient and modern peats. Peat thickness and mineral matter may pose a more serious problem. Peats from the Florida Everglades are too thin (91–381 cm) to form an economic coal given our understanding of peat-to-coal compaction ratios (Cecil et al. 1982). If relative peat thickness provided an indirect measure of decomposition rates, we would expect ancient *Cordaites–Medullosa* peats to yield lower matrix frequencies and root percentages than Everglades peats.

Cecil et al. (1993) suggested that thick Indonesian peats from Siak Kanan and Bengkalis Island, Sumatra, started as planar (topogenous) peats and gradually became domed. In these deposits, both planar and domed peats had low percentages of mineral matter, generally considered characteristic of domed peats. If tropical peats formed in both planar and domed settings also share similar matrix frequencies and root percentages, the lack of an exact modern

TABLE 10.2. Average Matrix Frequencies and Root Percentages of Modern Peat Deposits

Climate Zone, Salinity, Vegetation Type	Location	Av. Matrix Frequency (Range)	Av. Root Percentage (Range)	Sample Size, Source
Warm temperate freshwater swamp-marsh peat	Billy's Lake Okefenokee Swamp Georgia, USA	0.50	57%	203 cm peat Spackman et al., 1976 Figure 16, Table 3
Warm temperate freshwater swamp-marsh peat	Bird Wing Run Okefenokee Swamp Georgia, USA	0.54	75%	335 cm peat Spackman et al., 1976 Figure 16, Table 3
Warm temperate freshwater swamp-marsh peat	Suwanee Canal Okefenokee Swamp Georgia, USA	0.53	55%	137 cm peat Spackman et al., 1976 Figure 31, Table 3
Subtropical mangrove peat	Little Shark River Core Everglades Natl. Park Florida, USA	0.49 (0.19–0.75)	92% (52%–99%)	381 cm peat 20 observations Table 2, Cohen 1968
Subtropical mangrove peat	Lane River Core Everglades Natl. Park Florida, USA	0.52 (0.33–0.72)	96% (87%–100%)	91 cm peat 10 observations Table 2, Cohen 1968
Subtropical freshwater swamp-marsh peat	Loxahatchee B-4 Core Everglades Natl. Park Florida, USA	0.57 (0.29–0.79)	85% (55%–99%)	274 cm peat 29 observations Table 2, Cohen 1968
Tropical freshwater (Dipterocarp) swamp peat	Batang Hari, Jambi Province, Sumatra Indonesia	0.36 (1.14–0.65)	49% (22%–73%)	1650 cm peat 30 observations Esterle 1989
Tropical freshwater (Dipterocarp) swamp peat	Baram River Sarawak, Malaysia	0.30 (0.03–0.57)	55% (29%–81%)	4700 cm peat 43 observations Esterle 1989

analogue for *Cordaites–Medullosa* peats may not affect our interpretation of the results of this study.

Measures of the Amount and Rate of Peat Decomposition

Matrix Frequency

As peat decomposes, the particle size decreases (Boelter 1969; Levesque and Mathur 1979). Cohen (1968) and Cohen and Spackman (1977) used the frequency of matrix, defined as organic peat components with all dimensions less than 100 mm, to compare peats based on particle size. High matrix frequencies indicate relatively decomposed peats; low matrix frequencies indicate relatively undecomposed peats or tidally flushed mangrove peats (Cohen and Spackman 1977).

Both Cohen (1968) and Esterle (1989) used the point count method to determine the matrix frequency of modern peat. Cohen (1968) compiled the matrix frequency of a variety of swamp and marsh peats from the Florida Everglades. Because mangrove peats had lower average matrix frequencies than freshwater peats from the same climate zone (Cohen and Spackman 1977; Raymond 1987), we keep subtropical mangrove and freshwater peats separate in our comparison of matrix frequency values in ancient and modern peat. Because freshwater herbaceous and forested peats have similar matrix frequency values, we use all Cohen's (1968) subtropical freshwater peat samples in our comparison of ancient and modern peats.

We determine the matrix frequency of ancient peat from the Williamson No. 3 deposit by laying a transparent cm^2 grid sheet over a cellulose acetate peel from each concretion in the random sample and measuring the peat constituent touching the bottom left corner of each grid square. Some Williamson No. 3 concretions have been secondarily pyritized, obliterating small peat constituents and organic matrix particles. Accordingly, we exclude nine pyritic concretions, having a combined height of 94 cm, from the matrix frequency study. Pyritized concretions contain both *Cordaites* and *Medullosa* peats, suggesting that their exclusion does not bias the results in favor of either brackish or freshwater environments.

We present our results as a series of histograms showing the percentage of modern tropical freshwater domed peat, modern subtropical forest peat, and ancient *Cordaites–Medullosa* peat from the Williamson No. 3 deposit in each category of matrix frequency. Because concretions of permineralized peat vary in size, we use the height of each concretion measured perpendicular to bedding divided by the sum of heights of concretions in the random sample

(274 cm) to weight ancient matrix frequencies. Weighting ancient matrix frequency data by height minimizes the effect of small concretions, which may yield extreme values because they contain a single large organ. We use the Kolmogorov–Smirnov test (Campbell 1974) to indicate whether the distribution of these parameters differs significantly in ancient and modern peats; this test shows less sensitivity to highly skewed distributions than the χ^2 test.

Root Percentage

Spackman et al. (1976) identified the ratio of root debris to shoot debris (leaves, stems, bark, and aerial reproductive organs) in peat, herein referred to as root percentage, as one of the best indicators of relative decomposition state. Because shoots and roots become incorporated into peat in different ways, the root percentage of peat also indicates the relative rate of peat decomposition. Much of the decomposition that takes place during peat formation occurs near the surface, in the acrotelm, defined as the peat that lies above the permanent water table (Clymo 1983, 1984; Cohen, Spackman, and Raymond 1987). By analogy with shelly marine deposits (Davies, Powell, and Stanton 1989), the acrotelm corresponds to the taphonomically active zone of the peat. Shoot debris decomposes faster than root debris in part because all shoot debris passes through the acrotelm. Although roots in the acrotelm decay rapidly after death, roots that bypass the acrotelm and grow into the underlying catotelm, defined as the permanently water-logged zone of a peat deposit, decay more slowly than all shoot debris (Moore 1987). Because plants continually add both roots and shoots to peat, and shoots decompose faster than roots, the root percentage of peat indicates the relative rate of decomposition. Within a given habitat, small root percentages indicate relatively slow rates of decomposition; large root percentages indicate relatively rapid rates of decomposition.

Both environment and plant architecture may influence the root percentage of peat, which complicates between-habitat comparisons based on root percentage. Mangrove peats may have high root percentages due to tidal flushing of aerial debris (Cohen and Spackman 1977). Late Carboniferous plants derived structural support from a wider variety of tissues and organs than do modern plants, which could influence the root percentage of Late Carboniferous peat. Nevertheless, *Cordaites* trees, modern conifer trees, and modern dicotyledonous angiosperm trees derive structural support from woody trunks and probably had similar initial ratios of roots to aerial organs. Some *Medullosa* seed ferns and *Psaronius* tree ferns had an umbrella-like architecture similar to modern palms, cycad trees, and tree ferns (Taylor and Taylor

1993) and may have initial subaerial root percentages similar to these extant groups. Because arborescent lycopsids, which derived structural support from bark rather than wood, have no modern analogue, we can not predict the initial subaerial root percentage of this group.

Cohen (1968) compiled the root percentage of a variety of swamp and marsh peats from the Florida Everglades. Because mangrove peats have higher average root percentages than freshwater forested peats from the same climate zone (Cohen and Spackman 1977; Raymond 1987), we keep subtropical mangrove and freshwater peats separate in our comparison of root percentages in ancient and modern peat. Because freshwater herbaceous peats have higher root percentages than freshwater forested peats (Cohen 1968; Cohen and Spackman 1977), we use only those freshwater Everglades peats that contain organs of trees and shrubs in our comparison of root percentages in ancient and modern peat: *Myrica-Persea-Salix* peat and *Mariscus*-fern-*Myrica* peat.

Esterle (1989) used nine categories to describe the framework constituents (constituents with one dimension greater than 100 mm) of tropical freshwater domed peats, including three root categories, three categories for leaves, wood, and bark respectively, a seed and fruit category, and two categories for unidentified plant tissue. In this compilation, we consider wood and all unidentified plant tissues as aerial debris. Because some of these tissues may derive from roots, we probably underestimate the root percentage of modern tropical freshwater peats.

We compile the root percentage of Late Carboniferous permineralized peats from the Williamson No. 3, Urbandale, and Herrin (No. 6) Coal deposits using the grid method of Phillips, Kuntz, and Mickish (1977) to determine the relative abundance of root debris and shoot debris in one cellulose acetate peel from each concretion in the random sample. Because most Late Carboniferous swamp plants have distinctive root morphologies, we can identify most fossil organs as either root or shoot. However, we exclude decorticated *Cordaites* wood having no leaf or branch traces, unidentifiable fragments of charcoal, and unidentifiable plant tissues from our compilation of ancient root-shoot ratios because we can not determine their origin. Accordingly, we exclude five concretions with a combined height of 44 cm from the Williamson No. 3 sample and eight concretions with a combined height of 42.9 cm from the Urbandale sample. The excluded concretions contain only *Cordaites* wood.

As with matrix frequency, we present our results as a series of histograms showing the percentage of modern tropical freshwater domed peat, modern subtropical peat, and ancient peat in each root percentage category, and we weight ancient root percentage data based on concretion height. We use the

Kolmogorov–Smirnov test (Campbell 1974) to indicate whether the distribution of these parameters differs significantly in ancient and modern peats.

Leaf Mat Thickness

In modern peats, the rate of decomposition and the rate of litter fall, which correlate in general with rates of above ground net primary productivity (Perry 1994), control the thickness of surficial leaf mats (Heal, Latter, and Howson 1978). Assuming that the above ground net primary productivity of ancient plants did not exceed that of modern plants, leaf mat thickness provides an additional indication of the relative rate of decomposition in Late Carboniferous and modern peats.

Thick leaf mats can accumulate in water-filled depressions on the surface of peat because the acidic, anoxic water of freshwater wetlands inhibits all decomposers, especially fungi and detritivores (Cohen and Spackman 1977; Gastaldo and Staub 1996). We use the following criteria to identify surficial, subaerially exposed leaf mats in permineralized peats: (1) concretions composed primarily of imbricated leaves and roots (figure 10.1A). (2) Groups of fecal pellets produced by terrestrial detritivores in the peat matrix (figure 10.1B). Few terrestrial detritivores can tolerate standing water having low pH and low dissolved O_2, such as commonly occurs in peat-accumulating wetlands (Kühnelt 1955; Speight and Blackith 1983; Rader 1994; Kok and Van der Velde 1994). Although pelletized debris and coprolites might fall into a surface pond, groups of fecal pellets in the matrix indicate the presence of terrestrial detritivores (probably oribatid mites) in the peat. (3) Tree rootlets growing between leaves and into aerial organs (figure 10.1C). In peats, nearly all nutrient-gathering rootlets grow above the permanent water table (Crawford 1983). The smallest *Cordaites* rootlets range from 0.4 to 1.0 mm in diameter (Cridland 1964; figure 10.1A, 10.1C), which is comparable to the size of nutrient-gathering rootlets in Recent tropical and subtropical swamp peats (Esterle 1989; Cohen 1968). (4) Pelletized debris, detritivore coprolites, and leaves with the epidermis and cuticle shrunken around the resistant sclerenchyma bands (figures 10.1A and 10.2A–C). Cohen and Spackman (1977) identified subaerially exposed mangrove leaves based on their shrunken epidermis and cuticle. Although these peat components could form subaerially and fall into water-filled depressions, their common occurrence in Late Carboniferous leaf mats argues for subaerial exposure of these mats prior to permineralization, in some cases long enough for small woody rootlets to grow through the leaf mat (figure 10.1A).

Few data exist on the thickness of surficial leaf mats in modern tropical freshwater peats; nevertheless, Gastaldo (personal communication, 1994)

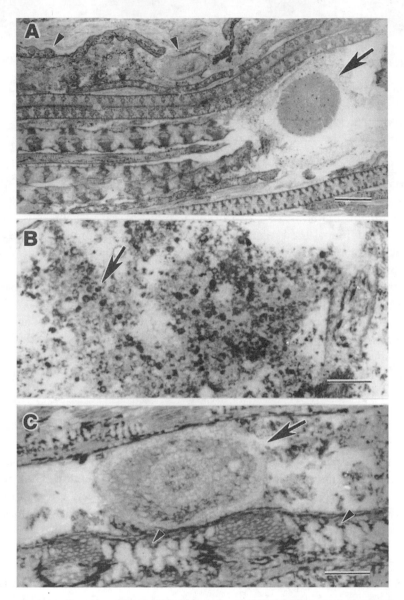

FIGURE 10.1. Evidence for subaerial exposure of leaf mats from the Williamson No. 3 deposit. (A) *Cordaites* leaf mat showing leaves with the epidermis and cuticle shrunken around the veins and sclerenchymatous strands (upper, small arrow pointing left), intruded by a *Cordaites* root (large arrow with shaft) and rootlets (upper, small arrow pointing right), Specimen W3–97–1, scale = 1 mm. (B) Fecal pellets in the matrix of a mixed *Cordaites-Medullosa* leaf mat. Specimen W3–2pc, scale = 200 mm. (C) *Cordaites* rootlet (left-pointing arrow with shaft) growing between two *Cordaites* leaves in a *Cordaites* leaf mat. The small triangular arrows point to mesophyll strands of the lower *Cordaites* leaf, Specimen W3–97–1, scale = 200 mm.

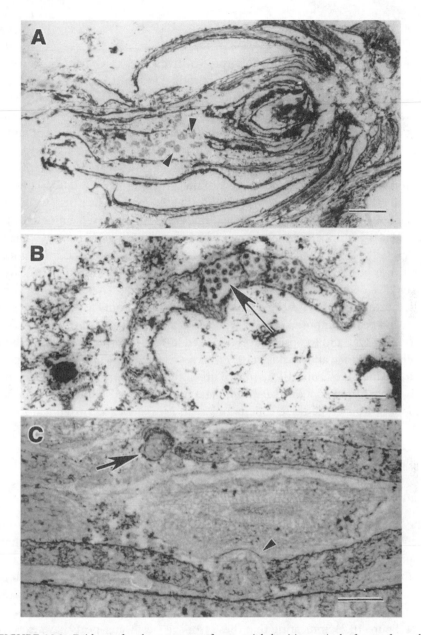

FIGURE 10.2. Evidence for the presence of terrestrial detritivores in leaf mats from the Williamson No. 3 deposit. (A) Spore or pollen-filled fecal pellets associated with a *Cordaites* pollen cone in a *Cordaites* leaf mat. Terrestrial microarthropods produced these fecal pellets, probably after the cone fell from the tree, Specimen W3–4pc, scale = 1 mm. (B) Microarthropod fecal pellets in a *Medullosa* (*Alethopteris sp.*) pinnule in a mixed *Cordaites-Medullosa* leaf mat, Specimen W3–2pc, scale = 500 mm. (C) *Cordaites* leaf mat with a finely macerated detritivore coprolite (right-pointing arrow with shaft) and a *Cordaites* root with secondary wood giving rise to a rootlet (small arrow pointing left), Specimen W3–97–1, scale 500 mm.

observed that these were generally three to five leaves thick. We supplement this information with measures of leaf mat thickness in 15 cores of mangrove peat from the Florida Everglades collected by Raymond (1987). Ten of these cores have surficial leaf mats; none contain buried leaf mats. Interestingly, the single core collected from a shallow pond in the l987 study does not contain a surficial leaf mat.

We present data on the frequency and thickness of leaf mats from three *Cordaites–Medullosa* peats. Forty-nine of the 331 *Cordaites–Medullosa* concretions used in this study contain leaf mats that we can measure perpendicular to bedding. Forty-three of these 49 leaf mats are well preserved enough for us to count the number of leaves perpendicular to bedding. We consider peats composed primarily of imbricated *Cordaites* leaves and roots as *Cordaites* leaf mats; using the grid method of Phillips, Kuntz, and Mickish (1977), leaves cover 25% or more of the peel in these concretions. We consider peats composed primarily of imbricated *Medullosa* leaves and roots as *Medullosa* leaf mats and peats composed primarily of imbricated roots and *Medullosa, Psaronius, Cordaites,* and lycopsid leaves as mixed leaf mats. Using the grid method of Phillips, Kuntz, and Mickish (1977), leaves covered 10% or more of the peel in a *Medullosa* or mixed leaf mat. Three concretions from the Shuler and Urbandale Mines contained *Psaronius* leaf mats or mixed *Psaronius-Cordaites* leaf mats; we include these concretions with *Medullosa* and mixed leaf mats.

Evidence for Less and Slower Decomposition in Late Carboniferous Tropical Peat

Matrix Frequency

The Kolmogorov–Smirnov test reveals that peat samples (concretions) from the Williamson No. 3 deposit have significantly lower matrix frequencies than tropical freshwater domed peats (figure 10.3, tables 10.2, 10.3, and 10.4), suggesting that this Late Carboniferous peat experienced less decomposition. Modern tropical freshwater domed peat has significantly lower matrix frequencies than both mangrove and freshwater subtropical peat (tables 10.2 and 10.4, figure 10.3), indicating that these tropical freshwater peats have experienced less decomposition than the subtropical peats in our data set. Subtropical mangrove peats do have significantly lower matrix frequencies than subtropical freshwater planar peats (table 10.4, figure 10.3; Cohen 1968; Cohen and Spackman 1977), perhaps due to tidal flushing. Comparisons of matrix frequency in ancient and modern peat using only subtropical freshwater peat derived from tree and shrub communities yield the same results.

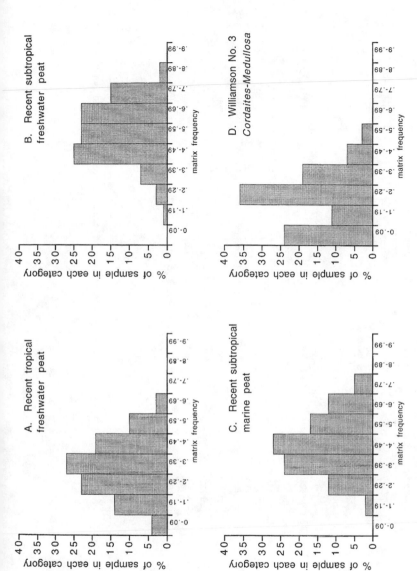

FIGURE 10.3. Distribution of matrix frequency values in modern and ancient peats. (A) Tropical freshwater domed peats from Indonesia and Malaysia studied by Esterle (1989), $n = 73$. (B) Subtropical freshwater peats from the Florida Everglades studied by Cohen (1968), $n = 87$. (C) Subtropical mangrove peats from the Florida Everglades studied by Cohen (1968), $n = 41$. (D) *Cordaites-Medullosa* peat from the Williamson No. 3 Coal Mine, $n = 33$. Ancient *Cordaites-Medullosa* peats have significantly less matrix than modern peats.

TABLE 10.3. Peat Types of the Williamson No. 3 Deposit

Peat Type (Concretions Studied)	% of Unbiased Sample (Concretions)	Matrix Frequency (Range)	Root Percentage (Range)	% Concretion Surface (cm2) with Fecal Pellets (Range)
Cordaites leaf mat 5 concretions	04% 3 concretions	0.10 0–0.20	35% 24–57%	05% 0–20%
Cordaites wood 10 concretions	15% 8 concretions	0.05 0–0.25	41%[a] 0–100%	04% 0–18%
Cordaites root-*Calamites* 16 concretions	34% 15 concretions	0.27 0.08–0.48	65% 33–100%	25% 6–40%
Psaronius 3 concretions	04% 2 concretions	0.27 0.24–0.30	42% 15–70%	34% 26–46%
Medullosa stem 11 concretions	29% 11 concretions	0.27 0.21–0.45	34% 0–64%	36% 19–72%
Medullosa root 1 concretion	02% 1 concretion	0.33	87%	51%
Medullosa leaf mat 3 concretions	11% 1 concretion	0.29 0.24–0.33	51% 33–65%	34% 26–60%
all *Cordaites* peat	53%	0.18	58%	16%
all *Medullosa* peat	42%	0.28	41%	36%
all Williamson #3 peat		0.22	49%	25%

[a] This value is based on five of the ten concretions of *Cordaites* wood peat. We could not determine whether the five excluded concretions contained root or stem wood.

TABLE 10.4. Kolmogorov-Smirnov Test Results Comparing the Distribution of Matrix Frequencies in Ancient and Modern Peats

	Ancient Peat:	*Modern Peats:*		
	Cordaites-Medullosa	*Tropical Domed*	*Subtropical Planar*	
	Williamson No. 3	*Freshwater*	*Freshwater*	*Mangrove*
Williamson No. 3 (33 observations)	—	—	—	—
Tropical domed freshwater (73 observations)	significantly different $p = 0.01$	—	—	—
Subtropical planar freshwater (87 observations)	significantly different[a] $p = 0.001$	significantly different[a] $p = 0.001$	—	—
Subtropical planar mangrove (41 observations)	significantly different $p = 0.001$	significantly different $p = 0.01$	significantly different[a] $p = 0.01$	—

[a] Comparisons using only the subtropical freshwater planar peats that contain debris from trees or shrubs yield the same results as comparisons using all samples of freshwater planar peat.

The low matrix frequency values of permineralized peats may reflect preservational bias. Because peat porosity decreases as decomposition proceeds (Clymo 1983; Boelter 1969), permineralization probably favors nearly pristine peat with high porosity and low matrix frequency over highly decomposed peat with low porosity and high matrix frequency. The peats that escaped permineralization to become coal probably had matrix frequencies and decomposition states similar to modern peats. Furthermore, decreases in peat porosity linked to increases in decomposition may contribute to the abundance of permineralized peat in the late Paleozoic and its scarcity in the Mesozoic and Cenozoic. Although silicified peats occur sporadically from the Devonian onward, calcium carbonate permineralizations of peat (coal balls) appear confined to the Late Carboniferous and earliest Permian (Phillips 1980).

Root Percentage

Although permineralized peats may represent a biased "least-decomposed" sample of all Late Carboniferous peats, root percentage data suggest that the rate of peat decomposition was slower in the Late Carboniferous than at present. Results of the Kolmogorov–Smirnov test reveal that the three Late Carboniferous peats (*Cordaites–Medullosa* peat from the Williamson No. 3 and

Urbandale Mines and lycopsid peat from the Sahara Mine) have significantly more samples with low root percentages than all modern peats used in this study ($p = 0.01$ to 0.001) (tables 10.2, 10.3, and 10.5; figure 10.4). All of the ancient peats have similar root percentage distributions, as do both modern freshwater peats (tables 10.2, 10.3, and 10.5; figure 10.4). Modern freshwater peats (tropical domed and subtropical planar) have significantly lower root percentage values compared to mangrove peat ($p = 0.001$ and $p = 0.01$, respectively) (tables 10.2 and 10.5).

The average shoot-root ratios of Late Carboniferous permineralized peat deposits from North America and Europe suggest that many ancient peats experienced lower rates of terrestrial decomposition and nutrient cycling than modern peats. Average shoot-root ratios of Late Carboniferous peats range in value from 0.42 to 8.6, corresponding to root percentages of 70 to 11%. ($n = 49$: Phillips, Peppers, and DiMichele 1985; Pryor 1996; this study). The average root percentage of modern peat deposits ranges from 96 to 49% (table 10.2).

Leaf Mat Thickness

In conjunction with the data on root percentages, the thickness of Late Carboniferous leaf mats provides additional evidence for slower rates of terrestrial decomposition and nutrient cycling in the Late Carboniferous than at present. Leaf mats compose from 11 to 16% of the random samples of *Cordaites-Medullosa* peat from the three ancient deposits (table 10.6). Most of these leaf mats contain leaves with shrunken epidermis and cuticle, indicating that they experienced subaerial exposure prior to permineralization. We could not assess the condition of leaf epidermis in extremely pyritic leaf-root concretions; however, leaves with shrunken epidermis occurred in 86% or more of *Cordaites* leaf mats and 60% or more of *Medullosa*, *Psaronius* and mixed leaf mats, consisting of *Medullosa*, *Psaronius*, or lycopsid leaves in addition to *Cordaites* leaves (table 10.6).

Ancient leaf mats are usually much thicker than modern leaf mats (figure 10.5). Leaf mats from the three *Cordaites-Medullosa* deposits have an average thickness of 25 leaves (range 5 to 78 leaves, $n = 43$; figure 10.5). The leaf mats observed in modern mangrove peats have an average thickness of four leaves (range = 1–8 leaves, $n = 6$; figure 10.5). As previously discussed, the surficial leaf mats observed in freshwater tropical peats were seldom more than three to five leaves thick (Gastaldo, personal communication, 1994). Considering tropical ecosystems as a whole, Lavelle et al. (1993) suggested short half-lives for tropical leaf litter, on the order of a few weeks, although they note that this general statement covers a wide range of values. Because the leaf mats of modern

TABLE 10.5. Kolmogorov-Smirnov Test Results Comparing the Distribution of Root Percentages in Ancient and Modern Peats

| | Ancient Peats: | | | Modern Peats: | | |
| | Cordaites-Medullosa | | Lycopsid | Tropical Domed | Subtropical Planar | |
Av. Root % of Sample	Williamson 49%	Urbandale 41%	Herrin 40%	Freshwater 49–55%	Freshwater 76%	Mangrove 91%
Williamson No. 3 (37 observations)	—	—	—	—	—	—
Urbandale (95 observations)	NS	—	—	—	—	—
Sahara (Herrin Coal) (53 observations)	NS	NS	—	—	—	—
Tropical domed freshwater (73 observations)	significantly different $p = 0.001$	significantly different $p = 0.001$	significantly different $p = 0.001$	—	—	—
Subtropical planar freshwater (11 observations)	significantly different $p = 0.01$	significantly different $p = 0.001$	significantly different $p = 0.001$	NS	—	—
Subtropical planar mangrove (41 observations)	significantly different $p = 0.001$	significantly different $p = 0.001$	significantly different $p = 0.001$	significantly different $p = 0.001$	significantly different $p = 0.01$	—

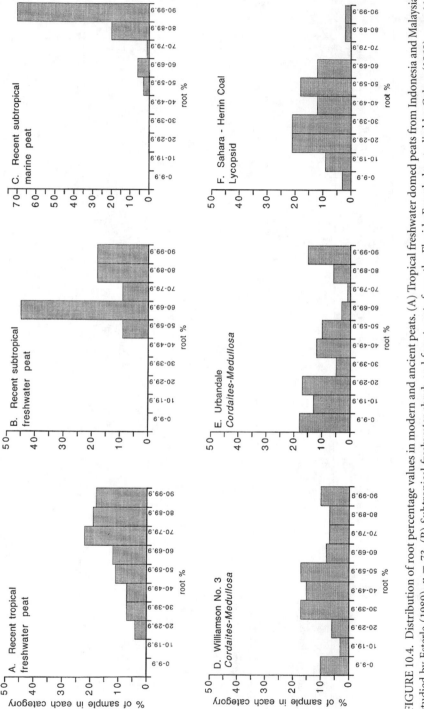

FIGURE 10.4. Distribution of root percentage values in modern and ancient peats. (A) Tropical freshwater domed peats from Indonesia and Malaysia studied by Esterle (1989), n = 73. (B) Subtropical freshwater shrub and forest peats from the Florida Everglades studied by Cohen (1968), n = 11. (C) Subtropical mangrove peats from the Florida Everglades studied by Cohen (1968), n = 41. (D) Cordaites-Medullosa peat from the Williamson No. 3 Coal Mine, n = 37. (E) Cordaites-Medullosa peat from the Urbandale Coal Mine studied by Raymond (1988), n = 95. (F) Lycopsid peat from the Herrin (No. 6) Coal, Sahara Mine studied by Raymond (1988), N = 53. All three ancient peats have significantly lower root percentages than modern peats.

TABLE 10.6. Leaf Mats in the Williamson No. 3, Shuler and Urbandale Deposits

Deposit	Total % of All Leaf-Mat Peat[a] (Concretions)	Cordaites Leaf Mats		Medullosa, Psaronius or Mixed Leaf Mats[b]	
		% of Total (Concretions)	% Having Leaves w/ Collapsed Epidermis (Concretions)	% of Total (Concretions)	% Having Leaves w/Collapsed Epidermis (Concretions)
Williamson No. 3	15%[c] (4)[c]	04%[c] (3)[c]	87% (13/15)	11%[c] (1)[c]	100% (3/3)
Shuler	11% (12)	06% (7)	86% (6/7)	05% (5)	60% (3/5)
Urbandale	16% (13)	10% (7)	100% (7/7)	06% (6)	83% (5/6)

[a] Percentages based on height perpendicular to bedding plane.

[b] Seven of the nine mixed leaf mats consisted of Medullosa (Alethopteris), Cordaites, Psaronius, or lycopsid leaves; two consisted of Psaronius and Cordaites leaves. Four concretions contained primarily Medullosa (Alethopteris) leaves; one contained primarily Psaronius leaves.

[c] Percentage in the sample of previously uncut concretions. In addition to the four leaf mats found in the previously uncut sample, we included an additional 14 leaf mats from the curated museum collection of Williamson No. 3 coal balls, for a total of 18 Williamson No. 3 leaf-mats.

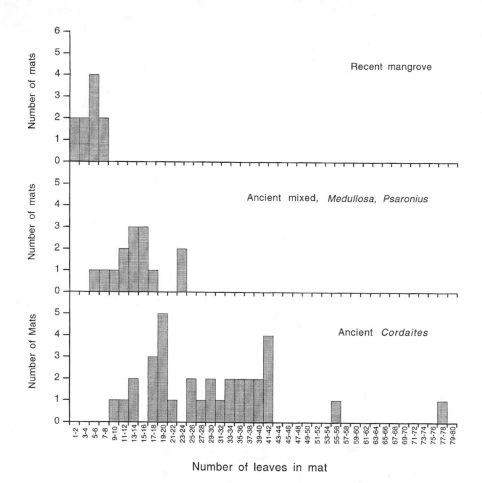

FIGURE 10.5. Leaf mat thickness in modern and ancient peats. Data on the thickness of Recent mangrove leaf mats from 15 cores collected in Everglades National Park by Raymond (1987). Data on the thickness of ancient *Cordaites* and mixed *Cordaites-Medullosa-Psaronius* leaf mats from the Williamson No. 3, Urbandale, and Shuler Mines in Iowa.

mires are so thin, even instantaneous permineralization of a modern peat would seldom result in a thick leaf-root peat, because leaves decay before thick mats intruded by nutrient-gathering rootlets can accumulate.

Among ancient leaf mats, *Cordaites* mats are usually thicker than *Medullosa, Psaronius,* or mixed leaf mats (figure 10.5). *Cordaites* leaf mats had an average thickness of 30 leaves (range 8 to 78 leaves, $n = 29$); *Medullosa, Psaronius* and mixed leaf mats had an average thickness of 16 leaves (range 5 to 24 leaves, $n = 13$). The thickness of *Cordaites* leaf mats relative to *Medullosa* leaf mats in these three deposits suggests that the rate of decomposition and nutrient cycling was slower in *Cordaites*-dominated communities than in *Medullosa*-dominated or mixed *Medullosa-Psaronius*-lycopsid communities. Although leaf productivity could conceivably control the relative thickness of *Cordaites, Medullosa,* and *Psaronius* leaf mats, Swift, Heal, and Anderson (1979) concluded that variations in the rate of decomposition tied to climate and environment are the primary controls on the accumulation of organic matter in modern terrestrial ecosystems.

Causes of Slow Terrestrial Decomposition in the Paleozoic

Late Carboniferous permineralized peats appear to have less matrix, fewer roots, and thicker surficial leaf mats than modern peat. Preferential permineralization may account for the low matrix frequencies observed for the Williamson No. 3 peat: Peat matrix increases during decomposition, and permineralization probably favors porous peat with little matrix. Nonetheless, preferential permineralization does not explain the low root percentages and thick surficial leaf mats of some Late Carboniferous peats. Slow rates of terrestrial decomposition in the Late Carboniferous would explain both observations and would be consistent with the low matrix frequency values observed for Williamson No. 3 peat. Possible causes include changes in stem architecture, differences in the depositional environment of ancient and modern peats, different fungi, and different detritivores.

Differences in stem architecture probably cannot account for slower rates of decomposition in ancient peats relative to modern peats. Based on stem architecture, we would expect relatively rapid decomposition of ancient seed ferns and tree ferns, which had little wood and derived structural support respectively from parenchymatous cortex and aerial roots, and relatively slow decomposition of *Cordaites,* modern conifers and dicotyledonous angiosperm trees, which derive structural support from woody trunks. Indeed, root percentage values for dominant plant groups in the Herrin (No. 6) Coal at Carrier Mills, IL (Sahara Mine) suggest that bark-supported and root-supported trees (respectively, lycopsids and *Psaronius* tree-ferns) decomposed faster than wood-supported trees, such as *Cordaites.* In this deposit, *Psaronius* tree ferns had a shoot-root ratio of 0.69, corresponding to a root percentage of 59%; *Cordaites* had a shoot-root ratio of 1.94, corresponding to a root percentage of 34% (Raymond 1988), suggesting that *Cordaites* debris decomposed more slowly than *Psaronius* debris in this freshwater paleomire. Lycopsid trees from the Sahara Mine had a root percentage of 45% (shoot-root ratio = 1.22: Raymond 1988), again suggesting that *Cordaites* debris decomposed more slowly than lycopsid debris in this deposit. This analysis assumes that arborescent lycopsids, *Psaronius* tree ferns, and *Cordaites* had similar initial shoot-to-root ratios. Because roots function to support trees and to absorb water and nutrients, the structural constraints of arborescence may require that trees growing on similar substrates with similar moisture availability have equivalent shoot-to-root ratios.

Variations in swamp habitat do not account for the relative lack of decomposition in ancient peats relative to modern peats or for the slow rate of decomposition in the Late Carboniferous peats we studied. Like modern peats, Late Carboniferous permineralized peats probably accumulated in a wide range of habitats, including tropical domed freshwater swamp forests; tropical flooded (topogenous) swamp forests; tropical mangrove forests; and temperate swamp forests (Raymond 1988; DiMichele and Phillips 1994). Few Late Carboniferous permineralized peats derive from herbaceous communities, with the possible exception of *Chaloneria*-dominated peats (DiMichele, Mahaffy, and Phillips 1979). Because of this, we confine our modern comparison sample to forest and shrub peats, especially for peat characteristics like root percentage, which appears to differ systematically between marsh and forest peats (Raymond 1987).

Robinson (1990) suggested that Late Carboniferous basidiomycetes with the ability to decompose lignin were either rare or absent, leading to slow rates of terrestrial decomposition and massive peat accumulation. Nevertheless, most *Cordaites* wood from the Williamson No. 3 deposit shows evidence of both simultaneous and selective delignification (Raymond, Heise and Cutlip, in review), indicating that Late Carboniferous basidiomycetes could decompose lignin. Rayner and Boddy (1988) distinguished selective delignification, which destroys the compound middle lamellae between tracheids creating fibrous, stringy rotted wood, from simultaneous delignification (white pocket rot), which destroys the entire tracheid wall leading to a honeycomb pattern of decay. Evidence of selective delignification in an ancient ecosystem is particularly significant because selective delignification accounts for more lignin decomposition in modern ecosystems than simultaneous delignification (Rayner and Boddy 1988).

A relatively inefficient detritivore community rather than inefficient fungi may account for slow rates of terrestrial decomposition in the Late Carboniferous. Whereas fungi and bacteria perform the chemical transformations involved in decomposition and nutrient cycling, invertebrate detritivores, which influence the environment on far larger temporal and geographic scales, mediate fungal and bacterial decomposition (Mills and Sinha 1971; Moore 1988; Lavelle et al. 1995). Edwards, Reichle, and Crossley (1970) found that litter decomposes at a rate proportional to the number of invertebrates in the litter and underlying soil. Kurcheva (1960) suggested that arthropods doubled the rate of leaf-litter recycling.

Detritivores increase the rate of decomposition by fragmenting plant debris and increasing the surface area available for bacterial and fungal attack (Witkamp and Crossley 1966; Swift, Heal, and Anderson 1979; Perry 1994). Groups that tunnel through wood can have the same effect: Carpenter ants create particles as they tunnel in wood for shelter. Detritivores also influence rates and processes of decomposition through their feces, which provide a favorable environment for bacteria (Perry 1994); for up to one month after deposition, detritivore feces contained higher bacterial counts and different relative abundances of bacterial groups than the surrounding soil in a peat-accumulating wetland (Kozlovskaya and Zagural'skaya 1976; Kozlovskaya and Belous 1976). Although the plant material within detritivore feces appears chemically similar to the parent material (Webb 1976), feces differ from soil and plant detritus in having a higher pH, a greater capacity to retain water, and smaller fragments (Crossley 1976). Detritivores and bacteria maintain a mutualistic relationship, which is particularly important for nutrient recycling in warm, humid tropical environments (Lavelle et al. 1995), such as those proposed for most Late Carboniferous permineralized peats. Most of the standing microbial biomass within the soil lies essentially dormant (Jenkinson and Ladd 1981) until ingested by detritivores (Lavelle et al. 1995). Detritivores that move through litter enhance fungal and bacterial decomposition by mixing nutrients (Perry 1994).

The ability of modern detritivores to influence rates and processes of plant decomposition in modern ecosystems suggests that evolutionary changes in the detritivore community could have had a major effect on rates of terrestrial decomposition and levels of primary productivity. Ancient detritivores (mites, diplopods, Collembola, cockroaches, and ancient insects) may have been less efficient than modern detritivores (Labandeira et al. 1994; Raymond and Heise 1994). Five modern groups that play an important role in the decomposition and comminution of plant litter and wood—termites (Isoptera), flies (Diptera), beetles (Coleoptera) ants (Hymenoptera-Formicidae), and wood wasps and horntails (Hymenoptera-Siricoidea)—appeared after the Late Carboniferous (Carpenter 1976; Labandeira 1994; Hasiotis and Demko 1996). In the absence of modern detritivores and large wood-tunneling insects, all decomposition processes, including fungal and bacterial decomposition, may have proceeded more slowly during the Late Carboniferous than at present. The size of tunnels in late Paleozoic and modern wood provides a graphic illustration of this point. Late Carboniferous wood contains small tunnels, 45–450 mm in diameter (Cichan and Taylor 1982; Labandeira, Phillips, and

Norton 1997); tunnels excavated by modern wood-boring insects are much larger [e.g., 2 mm for tipulid craneflies (Diptera) and 5 mm for caddisflies (Tricoptera); Maser and Sedell 1994].

In the second part of this contribution, we present evidence, in the form of amount of fecal pellet accumulations, that detritivores influenced the amount of decomposition in Late Carboniferous peat from the Williamson No. 3 deposit. We evaluate the size, diversity, and distribution of coprolites in Williamson No. 3 peat to determine relative rates of nutrient cycling in different paleomire communities; the size of arthropod detritivores (and herbivores); and the diversity of detritivore feeding strategies. We look at the taphonomy of *Cordaites* leaves to evaluate the role of microarthropod detritivores (oribatid mites and Collembola) in leaf decomposition.

Detritivore–Detritus Interactions in Late Carboniferous Peat

Fecal Pellet Accumulations

Cohen and Spackman (1977) concluded that detritivores increase the matrix constituents of modern peat at the expense of framework constituents. If ancient detritivores also acted to create matrix from framework, the frequency of matrix and detritivore activity should correlate in ancient peat. Although we can not measure the activity of ancient detritivores directly, Late Carboniferous peats contain abundant traces of detritivore activity in the form of coprolites and accumulations of fecal pellets. Fecal pellets, which range from 20 to 100 mm in minor axis length, occur in nearly all of the organs found in Late Carboniferous permineralized peat, including roots, leaves, stems, and within tunnels in wood (Kubiena 1955; Baxendale 1979; Cichan and Taylor 1982; Scott and Taylor 1983; Labandeira, Phillips, and Norton 1997; figure 10.2A,B) and provide an indirect measure of the amount of detritivory in Late Carboniferous mires.

For the Williamson No. 3 deposit, we use percent pelletization (the percentage of cm² grid cells on each peel that contain fecal pellet accumulations) to measure the amount of detritivory experienced by that sample of permineralized peat. We use regression analysis and Pearson's product moment correlation to establish the nature and strength of the relationship between the amount of detritivore activity (percent pelletization) and matrix frequency in the Williamson No. 3 deposit. Because we could not determine the matrix and pelletization frequency of pyritic concretions, the regression analysis sample consists of 33 concretions from the previously uncut sample supplemented by seven previously cut concretions.

Coprolites

The diversity and size of coprolite types indicates the diversity of detritivore feeding strategies and the size of arthropod detritivores; coprolites can also indicate the presence of herbivory (Baxendale 1979; Scott and Taylor 1983; Rolfe 1985; and Labandeira 1998). We classify the arthropod coprolites present in the Williamson No. 3 deposit based on their shape, texture, and color, and note the distribution of coprolites in a representative sample of some common Williamson No. 3 peat types: *Cordaites* leaf-root peat (171 cm^2); *Cordaites* root peat (303 cm^2); *Medullosa* leaf-root peat (161 cm^2); *Medullosa* stem peat (315 cm^2); and *Medullosa* root peat (33 cm^2). *Cordaites* wood contains almost no coprolites, and those found in the peat surrounding pieces of *Cordaites* wood are similar in size and type to the coprolites of equivalent peat types, primarily *Cordaites* root and *Cordaites* leaf-root peat. Most coprolites appear brown in color and contain fragments of seed coats and vegetative cell walls; golden or orange-yellow coprolites contain cuticle, pollen, and spores (Scott and Taylor 1983; Baxendale 1979).

Leaf Taphonomy

The taphonomy of *Cordaites* leaves in the Williamson No. 3 peat may help to elucidate factors that contribute to the formation and preservation of thick leaf mats in Late Carboniferous *Cordaites–Medullosa* peats. Accordingly, we look at the condition of 112 *Cordaites* leaves along transects parallel to bedding, in seven concretions, using cellulose-acetate peels. Four of these concretions contain *Cordaites* leaf mats ranging from 9 to 34 leaves thick (average, 17 leaves); two contain mixed leaf mats, each 13 leaves thick. The seventh concretion contains a leaf-rich *Cordaites* root peat. For each *Cordaites* leaf, we note the condition of the epidermis and mesophyll and the presence of fecal pellets. Although both *Medullosa* and *Psaronius* leaves occur in the two mixed leaf mats, we include only *Cordaites* leaves in this analysis. *Medullosa* (*Alethopteris* and *Neuropteris*), *Psaronius*, and *Cordaites* leaves all have different anatomy and probably decomposed differently as well.

Arthropod-Mediated Nutrient Cycling in Late Carboniferous Mires

Fecal Pellets

Within the Williamson No. 3 deposit, the percentage of pelletization varies among peat types (table 10.3). *Cordaites* wood and leaf-root peats have low pelletization percentages (average: 4% and 5%, respectively; range: 0–20%).

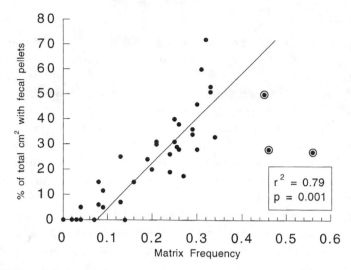

FIGURE 10.6. Relationship between matrix frequency and the frequency of fecal pellets in permineralized peat from the Williamson No. 3 Mine. As the matrix frequency of concretions increases, so does the pelletization percentage. The regression line and r^2 value reflect all of the data shown on the plot. Because pelletization becomes harder to see as matrix frequency increases, the pelletization percentage of matrix-rich samples (matrix frequency >0.4) may underestimate the true frequency of fecal pellets. Omission of the three matrix-rich samples would increase the r^2 value to 0.88.

Other peat types have higher average pelletization percentages (25–51%) and individual values that range from 6 to 72%. Overall, *Cordaites* peat has a lower pelletization percentage than *Medullosa* peat (table 10.3), suggesting that *Cordaites* peat experienced less detritivory and slower rates of decomposition than *Medullosa* peat.

Within this deposit, the frequencies of matrix and fecal pellet accumulations display a significant positive correlation (Pearson's product-moment correlation coefficient [r^2] = 0.79, n = 41, sig. = 0.001; figure 10.6), suggesting that detritivores contributed to matrix formation in permineralized peat from the Williamson No. 3 Mine. As peats decompose, pelletization becomes harder to see; elimination of the highly decomposed matrix-rich samples (matrix frequency >0.40) increases the correlation coefficient [r^2] to 0.88 (n = 39; sig. = 0.001).

Coprolites

All Williamson No. 3 peat types contain coprolites, ranging from 0.10 to 6.7 mm in minor axis length, probably derived from arthropod detritivores and herbivores. Most coprolite types occur in both *Cordaites* and *Medullosa* peats; although some occur more commonly in one peat type than another. For example, oxidized coprolites occur more commonly in *Cordaites* peat (57% of the leaf-root peat sample, 72% of the root-peat sample, 66% of the overall sample) than in *Medullosa* peat (27% of the leaf-root peat sample, 51% of the *Medullosa* stem peat sample, 39% of the overall sample; table 10.7). Small golden coprolites (0.18–0.80 mm minor axis length, figure 10.7A) occur primarily in *Cordaites* leaf-root peat; large golden coprolites (1.3–5.6 mm minor axis length, figure 10.7B) occur only in *Medullosa* stem peat (table 10.7). This distribution pattern holds for the entire sample of Williamson No. 3 peat: A survey of all the available concretions from this deposit revealed six additional large golden coprolites, all in *Medullosa* peat. None occur in *Cordaites* peat.

In addition to size, the following characteristics distinguish small and large golden coprolites: Small golden coprolites are solid, shiny yellow in color, and appear to contain only spores and pollen (figure 10.7A). These invariably co-occur with sporangia containing spores or pollen organs containing pollen. None contain fecal pellets produced by small coprophagous detritivores. Large golden coprolites have a reddish-orange tinge, probably due to the presence of cuticle as well as spores and pollen (figure 10.7B). They commonly contain void spaces and all contain accumulations of fecal pellets produced by small coprophagous detritivores (figure 10.7C).

Within each coprolite type, the populations found in *Medullosa* peats have higher average minor axis lengths than those from *Cordaites* peats. With the exception of oxidized coprolites, the largest coprolites within each category occur in *Medullosa* peat (table 10.7). *Medullosa* peat has higher frequencies of pelletized debris and a higher frequency of coprophagy than *Cordaites* peat (tables 10.3 and 10.7), especially considering that no coprophagy occurs in *Cordaites* wood peat, which constitutes 15% of the entire Williamson No. 3 sample and 28% of all *Cordaites* peat.

Leaf Taphonomy

Leaf taphonomy data confirm that most leaf-root peats from the Williamson No. 3 deposit contain dried leaves and probably represent surficial leaf mats. Of 112 *Cordaites* leaves encountered in the seven transects, 13% have epidermis collapsed around the vascular traces and sclerenchymatous strands (figure 10.1A), a condition indicative of subaerial dessication (Cohen and Spackman 1977).

Pelletized leaves provide direct evidence of endophagous leaf detritivory, in which microarthropod detritivores consume the mesophyll and vascular tissue of leaves, leaving the epidermis and cuticle largely intact (Jacot 1939). In

TABLE 10.7. Distribution of Coprolite Types and Coprophagy in the Williamson No. 3 Deposit

Sample	Cordaites Leaf-Root 171 cm	Cordaites Root 303 cm	Medullosa Leaf-Root 161 cm	Medullosa Stem 315 cm	Medullosa Root 33 cm[a]
Coprolite type:	Number (Percentage of Sample) Average Minor Axis Length in mm (Range)				
Brown					
Oxidized	62 (57%) 0.4 (0.1–2.2)	130 (72%) 0.2 (0.1–2.0)	14 (27%) 0.6 (0.3–2.0)	25 (51%) 0.35 (0.1–0.9)	9 (82%) 0.14 (0.1–0.2)
Finely macerated	26 (24%) 0.6 (0.2–1.4)	36 (20%) 0.65 (0.1–1.6)	27 (53%) 1.1 (0.4–2.7)	12 (24%) 1.1 (0.5–3.5)	2 (18%) 0.95 (0.3–1.6)
Coarsely macerated[b]	4 (4%) 2.8 (0.6–3.9)	2 (1%) 0.75 (0.7–0.8)	8 (16%) 1.9 (1.0–5.0)	2 (4%) 4.0 (1.2–5.5)	— —
Debris string	2 (2%) 0.25 (0.2–0.3)	— —	— —	2 (4%) 4.6 (3.9–5.3)	— —
Dispersed matrix	— —	11 (6%) 0.65 (0.4–0.9)	2 (4%) 1.35 (1.2–1.5)	6 (12%) 0.5 (0.1–0.9)	— —
Golden					
Small	14 (13%) 0.4 (0.2–0.8)	2 (1%) 0.3 (0.2–0.4)	— —	— —	— —
Large	— —	— —	— —	2 (4%) 3.2 (1.3–5.6)[c]	— —
Total Coprolites	108 (.63/cm)	181 (1.4/cm)	51 (.31/cm)	49 (.15/cm)	11 (.33/cm)
Coprophagy	9 (8%)	36 (20%)	6 (12%)	16 (34%)	1 (9%)

[a] The entire Medullosa root peat sample consisted of 33 cm².

[b] These may be cross-sections of debris strings.

[c] Average size and range based all six large golden coprolites found in the Williamson No. 3 sample.

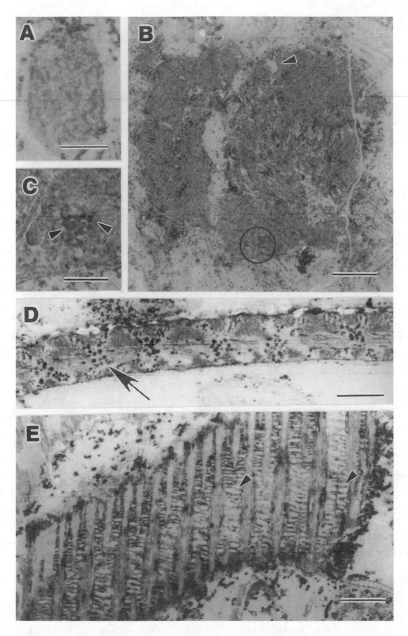

FIGURE 10.7. Golden, or spore and pollen-filled, coprolites and *Cordaites* leaf taphonomy in the Williamson No. 3 deposit. (A) Small golden, or spore and pollen-filled, coprolites from a *Cordaites* leaf mat, Specimen W3–21pc, scale = 200 mm. (B) Large golden, or spore and pollen-filled, coprolite from a *Medullosa* stem peat, Specimen W3–19d, scale = 1 mm. The arrow indicates the location of a void space; the circle indicates the location of the coprophagy shown in figure 10.7C. (C) Fecal pellet accumulation indicating coprophagy in a large golden, or spore and pollen-filled, coprolite from a *Medullosa* stem peat, Specimen W3–19d, scale = 200 mm. (D) Microarthropod fecal pellets in a *Cordaites* leaf from a *Cordaites* root peat, Specimen W3–25pc, scale = 200 mm. (E) *Cordaites* leaf in oblique cross-section showing strands of tissue derived from the mesophyll plates (small triangular arrows), Specimen W3–15pc, scale = 200 mm.

the root-peat transect, 35% of *Cordaites* leaves (9 of 26) contain pellets (figures 10.1F and 10.7D); no pelletized *Cordaites* leaves occur in the leaf-mat transects, suggesting that endophagous microarthropod detritivores feeding on *Cordaites* leaves did not live in the surficial litter layers of this deposit. The high percentage of leaves retaining mesophyll tissue in the leaf-mat transects supports this hypothesis. Most leaves from the six leaf-mat transects (91%, 78 of 86 leaves) retain some vestige of mesophyll, either in the form of cells or anastamozing strands of tissue derived from the mesophyll plates, or both (table 10.8, figures 10.1C and 10.7E). In comparison, only 46% of leaves from the root-peat transect (12 of 26) retain mesophyll tissue.

Fecal pellets associated with the surfaces or broken edges of leaves provide evidence of leaf skeletonization, in which microarthropods consume the cuticle, epidermis, and mesophyll of leaves, leaving a network of veins or "skeleton" (Eisenbeis and Wichard 1987). No leaves from the transect study show direct evidence of leaf skeletonizers in the form of associated fecal pellets. Indeed most leaves and leaf fragments retain their epidermis (78%, 87 of 112 leaves), although many are shredded (37%, 41 of 112) or have small cracks and splits in their epidermis (12%, 14 of 112 leaves: table 10.8, figure 10.8A,B). Leaves having eroded or missing epidermis may provide indirect evidence of leaf skeletonization. Some areas of epidermal erosion on *Cordaites* leaves look similar to those produced by Collembola feeding on the surfaces of fallen conifer needles (compare figure 10.8C to Zachariae 1963). However, other processes of decomposition in addition to leaf skeletonization could result in eroded or missing epidermis. Eroded and missing epidermis occurs more commonly among root-peat leaves (42%, 11 of 26 leaves) than leaf mat leaves (16%, 14 of 86 leaves) and six of the eight pelletized leaves have areas of eroded epidermis.

Implications for Community Paleoecology

The size of coprolites, the frequency of pelletized plant debris, and the frequency of pelletized coprolites (resulting from coprophagy) indicate the relative nutrient levels and rates of nutrient cycling in different peat-forming communities. Large detritivores require higher nutrient levels and higher rates of nutrient cycling to sustain viable populations than do small detritivores (Perry 1994). *Medullosa* peat has larger coprolites, more coprophagy, and higher frequencies of pelletized debris than *Cordaites* peat, indicating that *Medullosa* communities had more nutrients and faster rates of nutrient cycling. *Medullosa* and mixed *Medullosa-Cordaites* leaf mats are thinner than

TABLE 10.8. Condition of *Cordaites* Leaves in the Williamson No. 3 Deposit

Mesophyll Condition	Cells with Distinct Walls	Cells with Indistinct Walls	Cells and Tissue Strands	Tissue Strands	Absent	Total — Leaf Mat (# out of 86) / Root Peat (# out of 26)	All
Epidermis Condition							
Leaf entire[a], epidermis intact							
Leaf mat[b]	2	8	7	14	1	37% (32)	29% (32)
Root peat[b]	—	—	—	—	—	—	
Leaf entire, epidermis cracked							
Leaf mat	1	4	—	6	—	13% (11)	12% (14)
Root peat	—	—	—	1	2	12% (03)	
Leaf shredded[c]							
Leaf mat	4	12	7	3	3	34% (29)	37% (41)
Root peat	—	—	2	7	3	46% (12)	
Epidermis eroded							
Leaf mat	1	1	2	5	2	13% (11)	15% (17)
Root peat	—	—	—	2	4	23% (06)	
Epidermis missing on one side							
Leaf mat	—	—	—	—	1	01% (01)	03% (03)
Root peat	—	—	—	—	2	08% (02)	
No epidermis							
Leaf mat	—	—	1	—	1	02% (02)	04% (05)
Root peat	—	—	—	—	3	12% (03)	
Leaf mat total % (# out of 86)	09% (08)	29% (25)	20% (17)	32% (28)	09% (08)		
Root peat total % (# out of 26)	—	—	08% (02)	38% (10)	54% (14)		
Total % (# out of 112)	07% (08)	22% (25)	17% (19)	34% (38)	20% (22)		

[a] Includes leaves with no cracks in the epidermis that continued off the edge of the concretion.

[b] The leaf mat sample consists of six concretions; the root peat sample consists of one concretion.

[c] All shredded leaves have cracked or split epidermis.

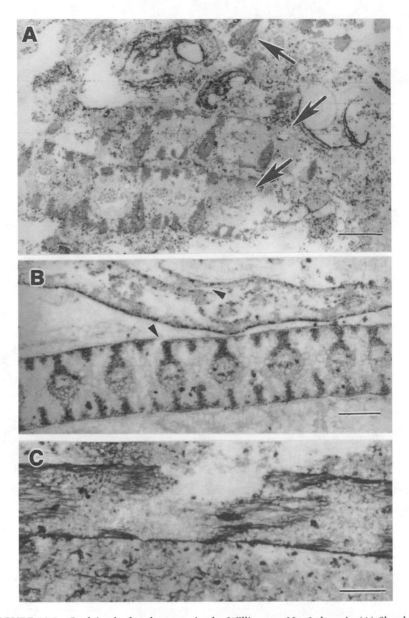

FIGURE 10.8. *Cordaites* leaf taphonomy in the Williamson No. 3 deposit. (A) Shredded *Cordaites* leaf in a *Cordaites* leaf mat. This leaf is shredded parallel to the veins (left-pointing arrows) and portions have disintegrated into scattered sclerenchymatous strands, (upper, right-pointing arrow), Specimen W3–21pc, scale = 500 mm. (B) *Cordaites* leaves with broken epidermis in a *Cordaites* leaf mat. The upper leaf has overlapping epidermis caused by shifting during decomposition; the lower leaf has cracked epidermis, Specimen W3–97–1, scale = 250 mm. (C) Eroded epidermis in a *Cordaites* leaf from a *Cordaites* leaf mat, Specimen W3–2ug, scale = 200 mm.

Cordaites leaf mats, again suggesting that *Medullosa* communities experienced faster rates of nutrient recycling.

Based on coprolite size, the nutrient levels of *Medullosa* and *Cordaites* communities from the Williamson No. 3 coal appear higher than those of lycopsid communities from the Lewis Creek Coal in Eastern Kentucky. The largest coprolites from the lycopsid-dominated Lewis Creek Coal measured 2 mm or more in diameter (Scott and Taylor 1983) compared to 3.9 mm and 6.7 mm, respectively, for coprolites from Williamson No. 3 *Cordaites* and *Medullosa* peats.

The correlation between rates of nutrient cycling and primary productivity (Jordan 1985) may indicate that ancient terrestrial ecosystems were less productive than modern terrestrial ecosystems from the same climate zone. Decomposition rates and ecosystem productivity likely rose after the Late Carboniferous with the evolution of new detritivore groups. The appearance of angiosperms may have contributed to increased rates of nutrient cycling in the Cretaceous and Tertiary because angiosperms decompose more quickly than conifers (Knoll and James 1987; Shearer, Moore, and Demchuk 1995). Shearer, Moore, and Demchuk (1995) linked increases in the thickness of coal seams and vitrain bands during the Tertiary to enhanced terrestrial productivity due to the rapid recycling of angiosperm detritus. Bambach (1993) argued that the supply of terrestrial nutrients to the oceans during the Late Cretaceous increased due in part to the rise of angiosperms (although see Vermeij 1995 for a contrasting view). Evolutionary changes in the detritivore community may have mediated increases in the rate of terrestrial nutrient cycling that have been linked to the radiation of angiosperms.

Detritivory and Herbivory in the Late Carboniferous

The record of detritivore–detritus, and possibly herbivore–plant, interactions in Late Carboniferous peat indicate that some modern detritivore niches did not exist in the Late Carboniferous and that some ancient niches may have disappeared. Lavelle et al. (1995) identified three digestive systems among invertebrate detritivores: external rumen (reingesting feces to take advantage of nutrients released during microbial metabolism); facultative mutualism (providing favorable conditions for microbial activity within the gut); and symbiosis (maintaining specific and permanent gut symbionts that enable the host to digest cellulose and possibly lignin). Some external ruminants maintain mutualistic associations with fungi to aid in the digestion of cellulose. Siricid wood wasps inoculate wood with fungi during oviposition, and the larvae utilize ingested fungal enzymes to metabolize cellulose, as do macrotermites and fungus-growing ants (Martin 1984). Examples of terrestrial detritivore niches that did not exist in the Paleozoic include social insects with gut symbionts (lower termites); the array of niches filled by facultative mutualists such as flies and beetles; and external ruminants that maintain mutualistic associations with fungi (macrotermites, siricid wood wasps, and leaf-cutter and fungus-growing ants) (Labandeira 1994; Lavelle et al. 1995).

Nonetheless, the three digestive systems of Lavelle et al. (1995) could have existed in the Late Carboniferous, albeit in different groups than at present. Modern descendants of a Carboniferous detritivore group, the wood roaches, have gut symbionts, which allow them to digest cellulose (Gutherie and Tindall 1968); descendants of another group present during the Carboniferous, modern millipedes, practice facultative mutualism (Hopkin and Read 1992); and ancient coprophagous detritivores were external ruminants.

Although detritivores probably produced most of the coprolites found in permineralized peats, golden coprolites, which contain spores and pollen, could result either from detritivory or spore-pollen predation (Baxendale 1979; Scott and Taylor 1983; Edwards, Selden, and Richardson 1995; Labandeira 1998). Shear and Kukulova-Peck (1990) suggested spores or pollen as the most likely diet of large paleodictyopterids, based on mouth-part morphology and gut contents. Members of this group may have produced the large golden coprolites found in *Medullosa* peats. The size of these coprolites (1.3–5.6 mm minor axis diameter) suggests both large producer organisms and an abundant source of spores or pollen. The pollen organs of *Medullosa* seed ferns (*Whittelseya* and *Dolorotheca* among others) would have provided a concentrated source of pollen, as would the sporophylls of *Botryopteris,* a coenopterid fern and *Cordaites* cones. Labandeira and Phillips (1996) described damage to the medullosan pollen organ, *Dolerotheca,* possibly caused by a pollen predator. In the Williamson No. 3 deposit, the occurrence of large spore and pollen-filled coprolites only in *Medullosa* peat suggests that the producers lived in *Medullosa* communities. Nevertheless, Scott and Taylor (1983) reported golden coprolites with diameters greater than 1 mm from the lycopsid peats of the Lewis Creek Coal, which may be similar to the large golden coprolites of the Williamson No. 3 deposit.

The fate of large spore predators such as paleodictyopterids may have been linked to that of plants producing large concentrations of pollen or spores, such as *Medullosa* seed ferns, *Botryopteris,* arborescent lycopsids, and *Calamites.* Escalation between paleodictyopterids and pollen and spore producers may have resulted in smaller, better protected microsporangia and pollen organs (Shear and Kukulova-Peck 1990). Although paleodictyopterids have a sparse body fossil record, the record of spore and pollen-filled coprolites in permineralized peat and compression–impression assemblages may enable us to investigate this hypothesis.

Small golden coprolites may result from either detritivory or herbivory on the part of Collembola, oribatids, and other groups. Both spore-pollen detritivory and spore-pollen predation occur among modern Collembola (Christiansen 1964; Lawton 1976). Spore-pollen predation also occurs among modern orthopterans (Schuster 1974; Labandeira 1998). In both the Late Silurian–Early Devonian and the Late Carboniferous, some spore-filled coprolites are larger than the largest modern Collembola feces and may derive from millipedes or other groups (Scott and Taylor 1983; Edwards, Selden, and Richardson 1995), although Rolfe (1985) doubted that millipedes produced spore-filled coprolites. In the Williamson No. 3 deposit, the co-occurrence of small golden coprolites and sporangia bearing undispersed spores in the same concretions may support the detritivory hypothesis for the origin of these small golden coprolites. Likewise, the large coprolites (3 mm by 25 mm), composed of seed-fern pollen and cuticle, that occur associated with *Autunia* foliage from Early Permian compression–impression deposits (Meyen 1984; Kerp 1988; Labandeira 1998) probably result from in situ detritivory. It seems unlikely that such large coprolites could withstand transport in water without breaking.

Either oribatid mites or Collembola may have produced the accumulations of golden fecal pellets associated with *Cordaites* cones in *Cordaites* leaf-root peat (figure 10.2A). Three modern species of oribatids eat dispersed pollen (Krantz and Lindquist 1979), although none live in wetlands.

The largest detritivore coprolites of the Williamson No. 3 deposit, found in *Medullosa* stem and leaf-root peats, probably derive from giant millipedes and small arthropleurids. Modern millipedes produce coarsely macerated round to cylindrical coprolites (Scott and Taylor 1983). Although few data exist on the relationship between the size of millipedes and their feces (Scott and Taylor 1983), Rolfe (1985) suggested that the largest coprolites in Late Carboniferous peats (8 mm in diameter) derived from giant millipedes (25 to 40 cm in length) or small arthropleurids. Based on the presence of sclariform tracheids in their guts, Rolfe (1969, 1985) suggested that *Arthropleura* lived in rotting lycopsid logs. Nevertheless, Scott and Taylor (1983) reported *Medullosa* seed-fern pollen (*Monoletes*) from an *Arthropleura* leg. Although *Arthropleura* probably was not a seed-fern pollinator (Shear and Kukulova-Peck 1990), this pollen may indicate that arthropleurids lived near *Medullosa* because the large size of *Medullosa* pollen precludes long-distance transport.

Giant millipedes and large arthropleurids may have required a thick surficial litter layer in which to live. Modern detritivores live in cool, moist crevasses within the litter and upper layers of the soil, which also afford them protection from predators (Kühnelt 1955). Relatively few can tolerate the harsh, dry conditions on the soil surface. The largest modern terrestrial invertebrate detritivores, 2 m long oligochaetes from Africa and 20 cm long beetles, burrow in soil and fallen logs, respectively (Little 1983; Hamilton 1978). The flattened shape of arthropleurid bodies is similar to that of flat-backed millipedes, which move through surface litter rather than burrow in soil (Rolfe 1969, 1985). Although the large size of *Arthropleura* led Rolfe (1985) and Shear and Kukulova-Peck (1990) to suggest that these organisms could not conceal themselves in surface litter, the thick surficial litter layers of Late Carboniferous leaf-root and stem peats could have sheltered considerably larger detritivores than those of modern peats. As rates of decomposition increased with the appearance of new detritivore groups, the thickness of the surficial litter layer may have decreased, eliminating the habitat of giant millipedes and arthropleurids. *Arthropleura* appears confined to the Late Carboniferous (Rolfe 1969).

Within the Williamson No. 3 deposit, *Medullosa* peat poses a paradox. Many characteristics of this peat suggest more decomposition and higher rates of decomposition relative to *Cordaites* peat from the same deposit, including higher matrix frequencies, higher pelletization percentages, larger coprolites, and thinner leaf mats (figure 10.5; table 10.3, 10.7). Yet the low root percentages of *Medullosa* peat compared to *Cordaites* peat (table 10.3) suggests slower rates of decomposition in *Medullosa* communities, and large Late Carboniferous detritivores such as arthropleurids and giant millipedes probably required thick surficial litter layers for shelter, which are more likely to form in peats with slow rates of decomposition. Nor are low root percentages for *Medullosa* unique to the Williamson No. 3 deposit: *Medullosa* has extremely low root percentages (8% and 6%, respectively) in both the Urbandale deposit and the Herrin (No. 6) Coal (Raymond 1988). Pryor (1996) reported low average root percentages (11–24%) for *Medullosa* seed fern and *Psaronius* tree fern peats from the Late Carboniferous (Stephanian) Duquesne Coal of Ohio, and Willard and Phillips (1993) found almost no *Medullosa* roots in the Late Carboniferous (Stephanian) Bristol Hill and Friendsville Coal Members (<1% in the Bristol Hill and 2% in the Friendsville).

A better understanding of the taphonomy of *Medullosa* debris may contribute to the resolution of this paradox. Preferential decay of *Medullosa*, root and stem wood compared to *Medullosa*, stem cortex, and rachises (rachises are the stems of compound leaves) may result in anomalously low root percentages in *Medullosa* peat. Cutlip (1997) used fecal pellet accumulations within plant organs as an indicator of the frequency of detritivore attack in Williamson No. 3 peats. These data suggest that *Medullosa* roots experienced greater amounts of detritivore attack than *Cordaites* stems and roots and

Medullosa stem cortex. Such a pattern of decomposition would lead to a predominance of *Medullosa* stem peat over *Medullosa* root and *Medullosa* leaf-root peat, the pattern observed in the Williamson No. 3 deposit (table 10.3).

Within the Williamson No. 3 deposit, the preponderance of evidence suggests faster rates of decomposition and nutrient cycling in *Medullosa* communities than in *Cordaites* communities. The distribution of large coprolites supports this conclusion. The largest arthropod coprolites in the Williamson No. 3 deposit (those over 4.0 mm minor axis diameter) occur in *Medullosa* peat and most commonly in *Medullosa* stem peat, suggesting that the largest Williamson No. 3 detritivores lived in *Medullosa*-dominated communities.

Leaf Detritivory in the Late Carboniferous and Recent

In Late Carboniferous mires, endophagous oribatids appear to be the principle leaf detritivores. The fecal pellet accumulations associated with *Cordaites* leaves nearly always occur encased in the epidermis of the leaf (Baxendale 1979; this study, figure 10.7D), an observation that also holds for seed-fern and lycopsid leaves (Scott and Taylor 1983; Labandeira, Phillips, and Norton 1997; figure 10.2B). Further, Late Carboniferous *Cordaites-Medullosa* peat provides only equivocal evidence of leaf exophagy. Patterns of epidermal erosion suggest that exophagous microarthropods may have fed on *Cordaites* leaves. Nonetheless, these peats lack direct evidence of exophagous microarthropods and leaf skeletonizers (i.e., fecal pellet accumulations associated with the edges of leaves). Coprolites in *Cordaites* leaf mats may derive from millipedes, roaches, and other detritivores that fed on decaying leaves; yet leaves revealed on the surfaces of broken leaf-root concretions show no evidence of large or small chewing detritivores, such as incomplete margins, or holes (see Zachariae 1963). The distribution of large coarsely-macerated coprolites and debris strings primarily in *Medullosa* stem and leaf-root peats suggests that large chewing detritivores occurred more commonly in *Medullosa* communities than in *Cordaites* communities.

Leaf detritivory in *Cordaites*-dominated communities appears most similar to that of modern conifer forests, in which endophagous oribatids are the primary leaf detritivores, although some Collembola feed on the epidermis of fallen conifer needles (Zachariae 1963; Millar 1974). Within modern angiosperm-dominated ecosystems, microarthropods (oribatid mites and Collembola) skeletonize leaves near the soil or peat surface, and large detritivores (millipedes, and fly and beetle larvae) consume whole, decaying leaves (Eisenbeis and Wichard 1987; Hopkin and Read 1992; Coleman and Crossley 1996).

Whereas most leaf decomposition (including detritivory) occurs near the peat surface in modern wetlands (Clymo 1983), endophagous detritivory of ancient *Cordaites* leaves may have occurred well below the peat surface. The only pelletized *Cordaites* leaves encountered in the seven transects occur in the root peat. Most *Cordaites* leaves (79%, 68 of 86) from the surficial, leaf-mat transects retain both epidermis and mesophyll, and only 2% (2 of 86) lack epidermis (figure 10.8A, table 10.8), suggesting very little endophagous detritivory or skeletonization in the leaf mat. In contrast, within the root-peat transect, less than half of the leaves (38%, 10 of 26) retain both epidermis and mesophyll, and 12% (3 of 26) have disintegrated into scattered vascular bundles and sclerenchyma strands.

The prevalence of endophagous leaf detritivory in Late Carboniferous mires and its occurrence primarily in root peats may contribute to the formation of the thick leaf mats in these ancient mires. Because modern skeletonizers belong to groups (oribatid mites and Collembola) present in the Late Carboniferous (Eisenbeis and Wichard 1987; Rolfe 1985; Shear and Kukulova-Peck 1990), we did not expect to find this difference in ancient and modern leaf detritivory. Nutrient and humidity requirements might explain the prevalence of endophagous leaf detritivory and the preferential attack of buried rather than surficial leaves in Late Carboniferous peats. Most small detritivores (oribatids, Collembola, and isopods) require fungally attacked detritus; some derive all their nutrition from fungal hyphae growing on detritus (Kevan 1962; Luxton 1972). Any factor that led to low rates of fungal attack in Late Carboniferous leaf mats (inefficient or scarce macroinvertebrate detritivores, rarity of skeletonizers, or, possibly, inefficient fungi) might result in surficial leaf mats lacking the necessary nutrients to support small, leaf-mining detritivores. Living inside leaves might have helped ancient detritivores to maintain water balance, control temperature, and escape from predators. Finally, skeletonizers may have an adaptive advantage in modern ecosystems with large leaf-eating detritivores (millipedes, and beetle and fly larvae) that can consume endophagous detritivores along with decaying leaves. Modern conifer forests, in which endophagous leaf detritivory predominates, do not have large, leaf-eating detritivores (Millar 1974).

Conclusions

1. Comparison of the peat characteristics (matrix frequency and root percentage of framework elements) and leaf mat thicknesses in ancient and modern peats suggests that ancient peat experienced less decomposition than modern peats and decomposed at a slower rate.

2. The evolution of efficient detritivores and wood-borers (termites, horntails, beetles, flies, and leaf-cutter ants) probably contributed to increased rates of decomposition in modern terrestrial ecosystems. No evidence suggests that detritus from ancient gymnosperms, taken as a group, decomposed faster than detritus from ancient pteridophytes, taken as a group. Indeed root percentage data from the Herrin (No. 6) Coal (Raymond 1988) and comparisons of *Cordaites* and *Medullosa* peat characteristics suggest that ancient picnoxylic gymnosperms such as *Cordaites* may have decomposed more slowly than ancient lycopsids, tree ferns, and manoxylic gymnosperms such as *Medullosa*. The evolution and diversification of angiosperms may have contributed to increased rates of terrestrial decomposition in the late Mesozoic and Tertiary (Shearer, Moore, and Demchuk 1995); however, the diversification of modern detritivores could also account for late Mesozoic-Tertiary increases in the rate of terrestrial decomposition.

3. Because of the link between nutrient cycling and primary productivity, all Paleozoic terrestrial ecosystems were probably less productive than modern terrestrial ecosystems from equivalent climate zones.

4. *Medullosa* peat has larger coprolites, more pelletization of debris, and more coprophagy than *Cordaites* peat, indicating higher nutrient levels and faster recycling in *Medullosa* communities relative to *Cordaites* communities.

5. Coprolites of the Williamson No. 3 deposit suggest some differentiation between the detritivores and herbivores of *Medullosa* and *Cordaites* communities. The organisms that produced large golden coprolites appear confined to *Medullosa* communities. The evolutionary fate of these organisms, possibly paleodictyopterid spore predators, may have been linked to that of Late Carboniferous plants producing huge accumulations of pollen or spores (*Medullosa* seed ferns, *Botryopteris,* as well as arborescent lycopsids and *Calamites*). *Medullosa* leaf-root and stem peats may have sheltered the largest detritivores in the Williamson No. 3 peat, which were probably giant millipedes and small arthropleurids. As rates of terrestrial decomposition increased, the thickness of leaf and stem litter layers would have decreased, possibly eliminating the habitat of giant litter-dwelling detritivores.

Small golden coprolites occur primarily in *Cordaites* leaf-root peats, invariably in concretions that also contain sporangia having undispersed spores. These coprolites may result either from spore-pollen detritivory by mites, Collembola, millipedes, and other groups, or from spore-pollen predation.

6. Leaf detritivory in Late Carboniferous mires differed from that of the present. Whereas modern microarthropod leaf detritivores both feed endophagously and skeletonize decaying leaves, ancient microarthropod detritivores fed endophagously, leaving the epidermis largely intact. The prevalence of endophagous leaf detritivory and the paucity of leaf skeletonization in Late Carboniferous peat-accumulating ecosystems may contribute to the formation of thick surficial leaf mats.

ACKNOWLEDGMENTS

We owe much to discussions with E. Heise, C. C. Labandeira, S. H. Costanza, W. A. DiMichele, B. Cecil, J. Staub, R. A. Gastaldo, J. Esterle, R. A. Fisher, and T. L. Phillips, and to the SSETI taphonomy team, E. A. Powell, R. Callendar, and G. Staff. We acknowledge the generous financial support of the Bunting Institute of Radcliffe College and the loan of specimens from the Harvard Botanical Museum.

REFERENCES

Anderson, J. A. R. 1983. The tropical peat swamps of western Malesia. In A. P. Gore, ed., *Ecosystems of the World 4B. Mires: Swamp, Bog, Fen, and Moor, Regional Systems,* pp. 181–199. Amsterdam: Elsevier.

Andrews, H. N. 1945. Contributions to our knowledge of American Carboniferous floras: VII. Some pteridosperm stems from Iowa. *Annals of the Missouri Botanical Garden* 32:323–366.

Andrews, H. N. and J. A. Kernen. 1946. Contributions to our knowledge of American Carboniferous floras: VIII. Another *Medullosa* from Iowa. *Annals of the Missouri Botanical Garden* 33:141–146.

Bambach, R. C. 1993. Seafood through time: Changes in biomass, energetics, and productivity in the marine ecosystem. *Paleobiology* 19:372–397.

Baxendale, R. W. 1979. Plant-bearing coprolites from North American Pennsylvanian coal balls. *Palaeontology* 22:537–548.

Beerbower, R. 1985. Early development of continental ecosystems. In B. H. Tiffney, ed., *Geological Factors and the Evolution of Plants,* pp. 47–91. New Haven CT: Yale University Press.

Berbee, M. L. and J. W. Taylor. 1993. Dating the evolutionary radiations of the true fungi. *Canadian Journal of Botany* 71:1114–1131.

Berner, R. A. 1990. Atmospheric carbon dioxide levels over Phanerozoic time. *Science* 249:1382–1385.

Berner, R. A. 1991. A model for atmospheric CO_2 over Phanerozoic time. *American Journal of Science* 291:339–376.

Boelter, D. H. 1969. Physical properties of peat as related to degree of decomposition. *Soil Science Society of America Proceedings* 33:606–609.

Campbell, R. C. 1974. *Statistics for Biologists.* Cambridge MA: Cambridge University Press.

Carpenter, R. M. 1976. Geological history and evolution of insects. *Proceedings of the XV International Congress of Entomology,* Washington DC, pp. 63–70.

Cecil, C. B., R. W. Stanton, F. T. Dulong, and J. J. Renton. 1982. Geologic factors that control mineral matter in coal. In R. H. Filby, B. S. Carpenter, and R. C. Raggini, eds., *Atomic and Nuclear Methods in Fossil Energy Research,* pp. 323–335. New York: Plenum Publishing.

Cecil, C. B., F. T. Dulong, J. C. Cobb, and Supardi. 1993. Allogenic and autogenic controls on sedimentation in the Central Sumatra basin as an analogue for Pennsylvanian coal-bearing strata in the Appalachian basin. In J. C. Cobb, and C. B. Cecil, eds., *Modern and Ancient Coal-Forming Environments,* pp. 3–22. Boulder CO: Geological Society of America.

Christiansen, K. 1964. Bionomics of Collembola. *Annual Review of Entomology* 9:147–178.

Cichan, M. A. and T. N. Taylor. 1982. Wood-borings in *Premnoxylon:* Plant–animal interactions in the Carboniferous. *Palaeogeography, Palaeoclimatology, Palaeoecology* 39:123–127.

Clymo, R. S. 1983. Peat. In A. P. Gore, ed., *Ecosystems of the World 4A. Mires: Swamp, Bog, Fen and Moor,* pp. 159–224. New York: Elsevier.

Clymo, R. S. 1984. The limits to peat bog growth. *Philosophical Transactions of the Royal Society of London B* 303:605–654.

Cohen, A. D. 1968. The petrology of some peats of Southern Florida (with special reference to the origin of coal). Ph.D. dissertation, Pennsylvania State University, University Park PA.

Cohen, A. D. and W. Spackman. 1977. Phytogenic organic sedimentary environments in the Everglades-mangrove complex. *Palaeontographica Abt B* 162:71–114.

Cohen, A. D., W. Spackman and R. Raymond. 1987. Interpreting the characteristics of coal seams from chemical, physical and petrographic studies of peat deposits. In A. C. Scott, ed., *Coal and Coal-Bearing Strata: Recent Advances,* pp. 107–126. Geological Society Special Publication 32. London: Blackwell Scientific Publications.

Coleman, D. C. and D. A. Crossley Jr. 1996. *Fundamentals of Soil Ecology.* San Diego: Academic Press.

Crawford, R. M. M. 1983. Root survival in flooded soils. In A. P. Gore, ed., *Ecosystems of the World 4A. Mires: Swamp, Bog, Fen, and Moor,* pp. 257–283. Amsterdam: Elsevier.

Cridland, A. A. 1964. *Amyelon* in American coal balls. *Palaeontology* 7:186–209.

Crossley, D. A. Jr. 1976. The roles of terrestrial saprophagous arthropods in forest soils: Current status of concepts. In W. J. Mattson, ed., *The Role of Arthropods in Forest Ecosystems,* pp. 49–56. New York: Springer-Verlag.

Cutlip, P. G. 1997. Coprolites and fecal pellets in a Late Carboniferous coal swamp: Morphologic and paleoecologic analysis. M.S. thesis, Texas A&M University, College Station TX.

Davies, D. J., E. N. Powell, and R. J. Stanton Jr. 1989. Relative rates of shell dissolution and net sediment accumulation—a commentary: Can shell beds form by the gradual accumulation of biogenic debris on the sea floor? *Lethaia* 22:207–212.

DiMichele, W. A. and T. L. Phillips. 1994. Paleobotanical and paleoecological constraints on models of peat formation in the Late Carboniferous of Euramerica. *Palaeogeography, Palaeoclimatology, Palaeoecology* 106:39–90.

DiMichele, W. A., J. F. Mahaffy, and T. L. Phillips. 1979. Lycopods of Pennsylvanian age coals: *Polysporia. Canadian Journal of Botany* 57:1740–1753.

Edwards, C. A., D. E. Reichle, and D. A. Crossley Jr. 1970. The role of soil invertebrates in turnover of organic matter and nutrients. In D. E. Reichle, ed., *Temperate Forest Ecosystems*, pp. 147–172. New York: Springer-Verlag.

Edwards, D., P. A. Selden, and J. B. Richardson. 1995. Coprolites as evidence for plant-animal interaction in Siluro-Devonian terrestrial ecosystems. *Nature* 377:329–331.

Eisenbeis, G. and W. Wichard. 1987. *Atlas on the Biology of Soil Arthropods.* Translated by Elizabeth A. Mole. Berlin: Springer-Verlag.

Eriksson, K.-E. L., R. A. Blanchette, and P. Ander. 1990. *Microbial and Enzymatic Degradation of Wood and Wood Components.* Berlin: Springer-Verlag.

Esterle, J. S. 1989. Trends in petrographic and chemical characteristics of tropical domed peats in Indonesia and Malaysia as analogues for coal formation. Ph.D. dissertation, University of Kentucky, Lexington KY.

Gastaldo, R. A. and J. R. Staub. 1996. An explanation for the anomaly of leaf litters preserved in coals: An analogue from the Rajang River Delta, Sarawak, East Malaysia. *GSA Abstracts with Programs* 28(7):A-105.

Guthrie, D. M. and A. R. Tindall. 1968. *The Biology of the Cockroach.* London: Edward Arnold (Publishers) Ltd.

Hamilton, W. D. 1978. Evolution and diversity under the bark. In L. A. Mound and N. Waloff, eds., *Royal Entomological Society of London Symposia No. 9, Diversity of Insect Faunas*, pp. 154–175. Oxford, U.K.: Royal Entomological Society and Blackwell Scientific Publications.

Hasiotis, S. T. and T. M. Demko. 1996. Ant (Hymenoptera: Formicidae) nest ichnofossils, Upper Jurassic Morrison Formation, Colorado Plateau: Evolutionary and ecologic implications. *GSA Abstracts with Programs* 28(7):A-106.

Heal, O. W., P. M Latter, and G. Howson. 1978. A study of the rates of decomposition of organic matter. In O. W. Heal, and D. F. Perkins, eds., *Ecological Studies Vol. 27, Production Ecology of British Moors and Montane Grasslands*, pp. 136–159. Berlin: Springer-Verlag.

Hopkin, S. P. and H. J. Read. 1992. *The Biology of Millipedes.* Oxford, U.K.: Oxford University Press.

Jacot, A. P. 1939. Reduction of spruce and fir litter by minute animals. *Journal of Forestry* 37:858–860.

Jenkinson, D. S. and J. N. Ladd. 1981. Microbial biomass in soil: Measurement and turnover. In J. N. Ladd, and E. A. Paul, eds., *Soil Biochemistry Vol. 5*, pp. 415–517. New York: M. Dekker.

Jordan, C. F. 1985. *Nutrient Cycling in Tropical Forest Ecosystems.* Chichester, U.K.: John Wiley and Sons.

Kerp, J. H. F. 1988. Aspects of Permian palaeobotany and palynology: X. The west- and central European species of the genus *Autunia* Krasser emend. Kerp (Peltaspermaceae) and the form-genus *Rhachiphyllum* Kerp (callipterid foliage). *Review of Palaeobotany and Palynology* 54:249–360.

Kevan, D. K. McE. 1962. *Soil Animals.* New York: Philosophical Library.

Knoll, M. A. and W. C. James. 1987. Effect of the advent and diversification of vascular land plants on mineral weathering through geologic time. *Geology* 15:1099–1102.

Kok, C. J. and G. Van der Velde. 1994. Decomposition and macroinvertebrate colonization of aquatic and terrestrial leaf material in alkaline and acid still water. *Freshwater Biology* 31:65–75.

Koslovskaya, L. S. and A. P. Belous. 1976. Change in the organic fraction of plant waste under the influence of oligochaetes. In N. I. P'yavchenko, ed., *Interrelation of Forest and Bog,* pp. 40–53. Moscow: Nauka Publishers. Translation, Amerind Publishing Co., New Delhi.

Koslovskaya, L. S. and L. M. Zagural'skaya. 1976. Interrelation between insect larvae and soil microflora in forest bogs. In N. I. P'yavchenko, ed., *Interrelation of Forest and Bog,* pp. 88–97. Translated by Amerind Publishing Co., New Delhi. Moscow: Nauka Publishers.

Krantz, G. W. and E. E. Lindquist. 1979. Evolution of phytophagous mites (Acari). *Annual Review of Entomology* 24:121–158.

Kubiena, W. L. 1955. Animal activity in soils as a decisive factor in establishment of humis forms. In D. K. McE. Kevan, ed., *Soil Animals,* pp. 73–82. New York: Philosophical Library.

Kühnelt, W. 1955. A brief introduction to the major groups of soil animals and their biology. In D. K. McE. Kevan, ed., *Soil Animals,* pp. 29–43. New York: Philosophical Library.

Kurcheva, G. F. 1960. The role of invertebrates in the decomposition of oak leaf litter. *Procvovedenie* 4:221–44.

Labandeira, C. C. 1994. A compendium of fossil insect families. *Milwaukee Public Museum Contributions in Biology and Geology* 88.

Labandeira, C. C. 1998. Early history of arthropod and vascular plant associations. *Annual Review of Earth and Planetary Science* 26:329–377.

Labandeira, C. C. and T. L. Phillips. 1996. Insect fluid-feeding on Upper Pennsylvanian tree ferns (Palaeodictyoptera, Marratiales) and the early history of the piercing-and-sucking functional feeding group. *Annals of the Entomological Society of America* 89:157–183.

Labandeira, C. C., A. C Scott, R. Mapes, and G. Mapes. 1994. The biologic degradation of wood through time: New insights from Paleozoic mites and Cretaceous termites. *Geological Society of America Abstracts with Programs* 26:A123.

Labandeira, C. C., T. L Phillips, and R. A. Norton. 1997. Oribatid mites and the decomposition of plant tissues in Paleozoic coal-swamp forests. *Palaois* 12:319–353.

Lavelle, P., E. Blanchart, A. Martin, and S. Martin. 1993. A Hierarchical Model for decomposition in terrestrial ecosystems: Application to soils of the humid tropics. *Biotropica* 25:130–150.

Lavelle, P., C. Lattaud, D. Trigo, and I. Barois. 1995. Mutualism and biodiversity in soils. *Plant and Soil* 170:23–33.

Lawton, J. H. 1976. The structure of the arthropod community on bracken. *Botanical Journal of the Linnean Society* 73:187–216.

Lesnikowska, A. D. and D. A. Willard. 1997. Two new species of *Scolecopteris* (Marattiales), sources of *Torispora securis* Balme and *Thymospora thiessenii* (Kosanke) Wilson and Venkatachala. *Review of Palaeobotany and Palynology* 95:211–225.

Levesque, M. P. and S. P. Mathur. 1979. A comparison of various means of measuring the degree of decomposition of virgin peat materials in the context of their relative biodegradability. *Canadian Journal of Soil Science* 59:397–400.

Little, C. 1983. The Colonization of Land: Origins and Adaptations of Terrestrial Animals. Cambridge MA: Cambridge University Press.

Luxton, M. 1972. Studies of the oribatid mites of a Danish beech wood soil: I. Nutritional biology. *Pedobiologia* 12:161–200.

Martin, M. M. 1984. The role of ingested enzymes in the digestive processes of insects. In J. M. Anderson, A. D. M. Rayner, and D. W. H. Walton, eds., *Invertebrate–Microbial Interactions*, pp. 155–172. Cambridge MA: Cambridge University Press.

Maser, C. and J. R. Sedell. 1994. *From the Forest to the Sea*. Delray Beach FL: St. Lucie Press.

Meyen, S. V. 1984. Is *Thuringia* a gymnosperm synangium or a coprolite? *Z. Geologische Wissenschaften* 12:269–270.

Millar, C. S. 1974. Decomposition of coniferous leaf litter. In C. H. Dickinson, and G. J. F. Pugh, eds., *Biology of Plant Litter Decomposition*, pp. 105–128. London: Academic Press.

Mills, J. T. and R. N. Sinha. 1971. Interactions between a springtail, *Hypogastrura tullergi* and soil-borne fungi. *Journal of Economic Entomology* 64:398–401.

Moore, J. C. 1988. The influence of microarthropods on symbiotic and non-symbiotic mutualism in detrital-based below-ground foodwebs. *Agriculture, Ecosystems and Environment* 24:147–159.

Moore, P. D. 1987. Ecological and hydrological aspects of peat formation. In A. C. Scott, ed., *Coal and Coal-bearing Strata: Recent Advances*, pp. 7–15. Oxford, U.K.: Blackwell Scientific Publications.

Morley, R. J. 1981. Development and vegetation dynamics of a lowland ombrogenous peat swamp in Kalimantan Tengah, Indonesia. *Journal of Biogeography* 8:383–404.

Olin, H. L., R. C. Kinne, and N. H. Hale. 1929. *Analyses of Iowa Coals*. Iowa City IA: Iowa Geological Survey.

Perry, D. A. 1994. *Forest Ecosystems*. Baltimore: The Johns Hopkins University Press.

Phillips, T. L. 1980. Stratigraphic and geographic occurrences of permineralized coal-swamp plants: Upper Carboniferous of North America and Europe. In

D. L. Dilcher and T. N. Taylor, eds., *Biostratigraphy of Fossil Plants*, pp. 25–92. Stroudsburg: Dowden, Hutchinson, and Ross.

Phillips, T. L. and W. A. DiMichele. 1981. Palaeoecology of Middle Pennsylvanian age coal swamps in southern Illinois: Herrin Coal Member at Sahara Mine No. 6. In K. J. Niklas, ed., *Palaeobotany, Palaeoecology, and Evolution*, v. 1, pp. 231–284. New York: Praeger Press.

Phillips, T. L., A. B. Kuntz, and D. J. Mickish. 1977. Paleobotany of permineralized peat (coal balls) from the Herrin (No. 6) Coal Member of the Illinois Basin. In P. H. Givens and A. D. Cohen, eds., *Interdisciplinary Studies of Peat and Coal Origins*, pp. 18–49. Geological Society of America Microform Publication 7, Card 1. Boulder CO: Geological Society of America.

Phillips, T. L., R. A. Peppers, and W. A. DiMichele. 1985. Stratigraphic and interregional changes in Pennsylvanian coal-swamp vegetation: Environmental inferences. *International Journal of Coal Geology* 5:43–109.

Pryor, J. S. 1996. The Upper Pennsylvanian Duquesne Coal of Ohio (USA): Evidence for a dynamic peat-accumulating swamp community. *International Journal of Coal Geology* 29:119–146.

Rader, R. B. 1994. Macroinvertebrates of the Northern Everglades: Species composition and trophic structure. *Florida Scientist* 57:22–33.

Raymond, A. 1987. Interpreting ancient swamp communities: Can we see the forest in the peat? *Review of Palaeobotany and Palynology* 52:217–231.

Raymond, A. 1988. The paleoecology of a coal-ball deposit from the middle Pennsylvanian of Iowa dominated by Cordaitalean gymnosperms. *Review of Palaeobotany and Palynology* 53:233–250.

Raymond, A. 1997. Latitudinal patterns in the diversification of land plants: Climate and the floral break. In R. L. Leary, ed., *Patterns in Paleobotany: Proceedings of a Czech–U.S. Carboniferous Paleobotany Workshop*, pp. 1–18. Illinois State Museum Scientific Papers 26. Springfield: Illinois State Museum.

Raymond, A. and E. A. Heise. 1994. Terrestrial nutrient cycling in the Paleozoic: More stems, less crap. *EOS, Transactions of the American Geophysical Union* 75(44):80.

Raymond, A., E. A. Heise, and P. G. Cutlip. In review. Fungal decomposition of wood in Late Carboniferous permineralized peats: Implications for the rate and processes of terrestrial decomposition through time. *Lethaia*.

Raymond, A. and T. L. Phillips. 1983. Evidence for an Upper Carboniferous mangrove community. In H. Teas, ed., *Biology and Ecology of Mangroves: Tasks for Vegetation Science*, v. 8, pp. 19–30. Dr. W. Junk Pub. Co.

Rayner, A. D. M. and L. Boddy. 1988. *Fungal Decomposition of Wood: Its Biology and Ecology*. New York: John Wiley and Sons.

Robinson, J. M. 1990. Lignin, land plants, and fungi: Biological evolution affecting Phanerozoic oxygen balance. *Geology* 18:607–610.

Robinson, J. M. 1991. Technical comment: Land plants and weathering. *Science* 252:860.

Rolfe, W. D. I. 1969. Arthropleurida. In H. K. Brooks, F. M. Carpenter, M. F. Glaessner, G. Hanhn, R. R. Hesslear, R. L. Hoffman, L. B. Holthuis, R. B. Manning, S. M. Manton, L. McCormick, R. C. Moore, W. A. Newman, A. R. Palmer, W. D. I. Rolfe, P. Tasch, T. H. Withers, and V. A. Zullo, eds., *Part R: Arthropoda 4, Vol. 2.*, pp. R607–R620. In R. C. Moore, series ed., *Treatise on Invertebrate Paleontology.* Lawrence KS: University of Kansas Press and Geological Society of America.

Rolfe, W. D. I. 1985. Aspects of the Carboniferous terrestrial arthropod community. Comptes Rendus. *Ninth International Congress on the Stratigraphy and Geology of the Carboniferous*, v. 5, pp. 303–316.

Rothwell, G. W. 1972. Evidence of pollen tubes in Paleozoic pteridosperms. *Science* 175:772–774.

Schuster, J. C. 1974. Saltatorial Orthoptera as common visitors to tropical flowers. *Biotropica* 6:138–140.

Scott, A. C. and T. N. Taylor. 1983. Plant/animal interactions during the Upper Carboniferous. *The Botanical Review* 49:259–307.

Shear, W. A., and J. Kukalova-Peck. 1990. The ecology of Paleozoic terrestrial arthropods: The fossil evidence. *Canadian Journal of Zoology* 68:1807–1834.

Shearer, J. C., T. A. Moore, and T. D. Demchuk. 1995. Delineation of the distinctive nature of Tertiary coal beds. *International Journal of Coal Geology* 28:71–98.

Spackman, W., A. D. Cohen, P. H. Given, and D. J. Casagrande. 1976. *Environments of Coal Formation, Okefenokee and the Everglades.* College Park PA: Coal Research Station, Pennsylvania State University.

Speight, M. C. D. and R. E. Blackith. 1983. The animals. In A. P. Gore, ed., *Ecosystems of the World 4A. Mires: Swamp, Bog, Fen, and Moor,* pp. 349–365. Amsterdam: Elsevier Scientific Publishing Company.

Subblefield, S. P. and G. W. Rothwell. 1981. Embryogeny and reproductive biology of *Bothrodendrostrobus mundus* (Lycopsida). *American Journal of Botany* 68:625–634.

Swift, M. J., O. W. Heal, and J. M. Anderson. 1979. *Studies in Ecology Vol. 5: Decomposition in Terrestrial Ecosystems.* Berkeley CA: University of California Press.

Taylor, T. N. 1993. The role of Late Paleozoic fungi in understanding the terrestrial paleoecosystem. *Twelfth International Congress on the Stratigraphy and Geology of the Carboniferous and Permian*, v. 2, pp. 147–154.

Taylor, T. N. and J. M Osborn. 1992. The role of wood in understanding saprophytism in the Fossil Record. *Courier Forschungsinstitut Senckenberg* 147:147–153.

Taylor, T. N. and E. L. Taylor. 1993. *The Biology and Evolution of Fossil Plants.* Englewood Cliffs NJ: Prentice Hall.

Thompson, K. and A. C. Hamilton. 1983. Peatlands and swamps of the African continent. In A. P. Gore, ed., *Ecosystems of the World 4B. Mires: Swamp, Bog, Fen, and Moor, Regional Systems,* pp. 331–373. Amsterdam: Elsevier.

Vermeij, G. J. 1995. Economics, volcanoes, and Phanerozoic revolutions. *Paleobiology* 21:125–152.

Webb, D. P. 1976. Regulation of deciduous forest litter, decomposition by soil arthropod feces. In W. J. Mattson, ed., *The Role of Arthropods in Forest Ecosystems,* pp. 57–69. New York: Springer-Verlag.

Willard, D. A. and T. L. Phillips. 1993. Paleobotany and palynology of the Bristol Hill Coal Member (Bond Formation) and Friendsville Coal Member (Mattoon Formation) of the Illinois Basin (Upper Pennsylvanian). *Palaios* 8:574–586.

Witkamp, M. and D. A. Crossley Jr. 1966. The role of arthropods and microflora in breakdown of white oak litter. *Pedobiologia* 6:293–303.

Zachariae, G. 1963. Was leisten Collembolen für den Waldhumus? In J. Doeksen and J. Van Der Drift, eds., *Soil Organisms,* pp. 109–124. Amsterdam: North Holland Publishing Company.

Ziegler, A. M., A. L. Raymond, T. C. Gierlowski, M. A. Horrell, D. B. Rowley, and A. L. Lottes. 1987. Coal, climate and terrestrial productivity: The present and Early Cretaceous compared. In A. C. Scott, ed., *Coal and Coal-Bearing Strata: Recent Advances,* pp. 25–49. Geological Society Special Publication No. 32. London: Blackwell Scientific Publications and Geological Society of London.

11

Ecological Sorting of Vascular Plant Classes During the Paleozoic Evolutionary Radiation

William A. DiMichele, William E. Stein,

and Richard M. Bateman

THE DISTINCTIVE BODY PLANS of vascular plants (lycopsids, ferns, sphenopsids, seed plants), corresponding roughly to traditional Linnean classes, originated in a radiation that began in the late Middle Devonian and ended in the Early Carboniferous. This relatively brief radiation followed a long period in the Silurian and Early Devonian during which morphological complexity accrued slowly and preceded evolutionary diversifications confined within major body-plan themes during the Carboniferous. During the Middle Devonian–Early Carboniferous morphological radiation, the major class-level clades also became differentiated ecologically: Lycopsids were centered in wetlands, seed plants in terra firma environments, sphenopsids in aggradational habitats, and ferns in disturbed environments. The strong congruence of phylogenetic pattern, morphological differentiation, and clade-level ecological distributions characterizes plant ecological and evolutionary dynamics throughout much of the late Paleozoic. In this study, we explore the phylogenetic relationships and realized ecomorphospace of reconstructed whole plants (or composite whole plants), representing each of the major body-plan clades, and examine the degree of overlap of these patterns with each other and with patterns of environmental distribution. We conclude that ecological incumbency was a major factor circumscribing and channeling the course of early diversification events: events that profoundly affected the structure and composition of modern plant communities.

Paleoecological studies of Carboniferous terrestrial environments consistently have revealed distinct ecological centroids for those major architectural groups traditionally described as taxonomic classes. Although the ecological spectra encompassed by the constituent species of these clades overlapped, each predominated in a distinct, broadly construed environmental type, irrespective of whether dominance is assessed by species richness or percentage biomass. In modern landscapes, one of these primordial architectural groups, the seed plants, dominate most habitats, and among seed plants, the angiosperms are now most prominent over most of the Earth's surface. The roots of the modern pattern first appeared at the end of the Paleozoic, when the breadth of ecologically dominant taxa narrowed at the class level, and one class, seed plants, began their rise to prominence in all types of environments. The rise of angiosperms from within the seed plants further narrowed the phylogenetic breadth of ecologically dominant clades. Viewed over geological time, the patterns of clade replacement within major environmental types suggest a self-similar pattern, each new radiation bringing to dominance an even more narrow portion of phylogenetic diversity.

The primordial Carboniferous pattern appears to have become progressively established during the Middle to Late Devonian, when major architectural groups of vascular plants originated from the structurally simple ancestral forms predominant from the Late Silurian to the late Early Devonian (Scott 1980; Gensel and Andrews 1984). This radiation has come under increasing phylogenetic scrutiny (Crane 1990; Kenrick and Crane 1991, 1997; Gensel 1992; Bateman in press). Much still remains to be learned, however, regarding the early evolution of major bauplans (represented by modern classes), which were largely in place by the Early Carboniferous, and of the paleoecological preferences of these groups. The paleoenvironmental distribution of the major clades is well enough known to lead us to conclude that the radiation of architectural types (bauplans) coincided with partitioning of ecological resources, the latter playing an important role in both channeling and constraining the radiation.

The interplay between evolution and ecology, the understanding of which is a primary objective of evolutionary paleoecology, is well illustrated by these mid-Paleozoic events. The evolution of large-scale architectural (and hence taxonomic) discontinuities was made possible in large part by the (evolving) patterns of resource occupation in what was initially an ecologically undersaturated terrestrial world. We believe the basic dynamics of ecological control of plant diversification and morphology are probably general rather than unique to this major radiation (for a parallel pattern in modern seed plants see Lord, Westoby, and Leishman 1996). The architectural results and taxonomic consequences of the mid-Paleozoic radiation, however, are unique, due to both unique environmental opportunities available at the time and the relatively simple morphologies (and by inference low developmental complexity) of the ancestral forms (Stein 1993; DiMichele and Bateman 1996).

The Scenario

Prior to the Middle Devonian, vascular plants were, in structural and developmental terms, relatively simple compared with later forms (e.g., Knoll et al. 1984). Organ, tissue, and cell types were few and such innovations were added piecemeal, gradually building structural complexity in the various lineages (Chaloner and Sheerin 1979). Furthermore, the ecological spectrum encompassed by these plants was limited largely to the wetter parts of lowland environments (Andrews et al. 1977; Edwards 1980; Gensel and Andrews 1984; Beerbower 1985; Edwards and Fanning 1985). The variety of environments colonized through time clearly increased, although relatively slowly, and likely was limited more by constraints produced by primitive vegetative and reproductive morphology (e.g., inadequate root systems, few types of dispersal modes, limited photosynthetic arrays, reproductive phenotypes linked to the need for free water) than by basic physiology, the core aspects of which were probably in place (e.g., photosynthetic pathways, water transport, nutrient use; for discussion of major phases in plant evolution see Bateman 1991). The possible phenotypic disparity (sensu Foote 1994) between ancestor and descendant species was small, although the aggregate spectrum of variation was gradually expanding through time. Clearly, morphological and physiological behavior (capacity) are linked, and as structural complexity increased so did the capacity for energy acquisition and utilization.

During the late Middle and Late Devonian, the vascular plants attained an aggregate "critical mass" of morphological complexity regulated by in creasingly complex developmental systems (Niklas, Tiffney, and Knoll 1980; Rothwell 1987; Wight 1987; Stein 1993; DiMichele and Bateman 1996). This permitted an increase in the maximum ancestor–descendant disparity; phenotypic "experimentation" became greater simply because of greater complexity of the starting forms. The unfilled nature of ecological resource space at this time created a permissive, abiotic selective regime that allowed many of these "hopeful monsters" to locate adequate resources where there was minimal competition from well entrenched incumbent species (Scheckler 1986a;

Bateman 1991; Bateman and DiMichele 1994a). Thus, even though the earliest derivatives would not have been optimally functional, distinctive new plant architectures appeared and, more critically, some established historically persistent ecologically delimited clades.

The radiation was relatively brief for two reasons (DiMichele and Bateman 1996): (1) Nonaquatic plants have a limited range of resource acquisition and exploitation strategies (Niklas 1997), and the number of major resource pools available to vascular plants is rather limited, so the effects of incumbent advantage (Gilinsky and Bambach 1987; Rosenzweig and McCord 1991) in different parts of the ecological landscape developed very rapidly as resources were expropriated; (2) as morphological complexity accrued, the effects of the "epigenetic ratchet" (Levinton 1988) began to limit the size of ancestor-descendant evolutionary disparity–developmental interdependencies progressively limited functional morphological combinations. Moreover, with greater structural and developmental complexity, the evolution of new architectures requires "escape" from the structural organization of complex ancestral forms (Bateman 1996a). This does not mean that speciation rate declined. Rather, the average morphological difference between ancestor and descendant declined, and new species fitted into the existing architectural types (bauplans).

Relative species diversities of class-level clades that evolved during the Middle Devonian radiation, particularly tree forms, appear to have been limited strongly by the resource breadth of the environment into which the clade radiated. Terra firma habitats, the favored territory of seed plants, were the most physically diverse and were thus capable of supporting the most species and the most variation on the basic architectural aspects of the clade. Wetlands, the ecological centroid of the rhizomorphic lycopsids, were much less diverse edaphically and consequently supported fewer architectural types and fewer species. Aggradational and disturbed habitats, the narrowest of all adaptive zones, were occupied by the rhizomatous sphenopsids, which evolved proportionally the fewest variations on their basic tree architecture and also were the group with the lowest species diversity. Early ferns were opportunists that exploited interstitial disturbance in many kinds of environments, permitting them to radiate in significant numbers in ecotonal settings. Tree–fern dominance did not appear until much later, after environmentally induced extinctions created opportunities to exploit previously occupied resources (Pfefferkorn and Thomson 1982; Phillips and Peppers 1984).

We examine this scenario from three perspectives, each of which is usually examined independently as a central attribute of an evolutionary radiation. Only through examination of all three is it possible to evaluate interrelated causal factors for this important period in vascular plant history (Bateman, in press). First is the pattern of phylogenetic diversification that began in the Middle Devonian and substantially terminated by the Early Carboniferous. Second is the nature of the ecophenotypic morphospace that evolved during this radiation and the degree to which it was congruent with the phylogenetic pattern. Third is the ecological preferences of the major lineages and the degree to which such preferences constrained the species diversities of class-level clades.

The Vascular Plant Radiation: Phylogeny

Extant vascular plants can be organized into two major complexes based on their ancestry. The basal groups of the vascular plant phylogenetic tree are represented by the zosterophylls and the trimerophytes, apparently descended from common ancestors among the earliest vascular plants, the rhyniophytes (Banks 1968; Gensel 1992). This fundamental basal dichotomy took place no later than Early Devonian (Banks 1975).

Derivatives of the zosterophylls include at least one and possibly two distinct clades (Kenrick and Crane 1997). The older and more diverse clade is the lycopsids. Within this group are three subclades, likely successively derived from one another in the sequence Lycopodiales, Selaginellales, Isoetales. The other clade is the barinophytes, which may be zosterophylls or derived from a zosterophyll ancestor (Brauer 1981).

Most extant plants are derivatives of the trimerophytes, encompassing several complexes of structurally similar groups. Perhaps least derived are the ferns, first appearing in the Late Devonian (Phillips and Andrews 1968; Rothwell 1996), consisting of several architecturally distinct subgroups, the zygopterids, marattialeans, and filicaleans. The sphenopsids include the equisetophytes and sphenophylls, and may be derived from morphologically intermediate groups in the Middle Devonian that include the iridopterids (Stein, Wight, and Beck 1984) or one or more groups of cladoxylopsids (Skog and Banks 1973; Stein and Hueber 1989). The seed plants and their ancestors, the progymnosperms, form another distinct group. There are two major sublineages of progymnosperms, the archaeopterids and aneurophytes, and there has been considerable debate over which of these groups included the seed-plant ancestor (Rothwell 1982; Meyen 1984; Beck and Wight 1988).

Clearly, classical Linnean taxonomy has not successfully encapsulated this early radiation. Groups with distinctive body plans form a nested hierarchy of relationships, rendering some paraphyletic (e.g., progymnosperms in their possible relationship to seed plants). Crane (1990) and Kenrick and Crane

(1991, 1997) examined this radiation cladistically and compared it with traditional Linnean grouping. Although the nested hierarchy means that some architectural groups are more closely related to one another than to other groups, the basic architectures clearly are distinct, especially when viewed retrospectively in the light of subsequent radiations. Since the Early Carboniferous, speciation largely has taken place within the confines of these existing body plans. Surviving to the present are all three of the lycopsid groups, the filicalean and marattialean ferns, the equisetophyte sphenopsids, and numerous groups of seed plants (although none that were present in the Carboniferous). Here, we view the angiosperms as a subset of the seed plants.

We present a cladistic representation of this early radiation (figure 11.1), noting the points of origin of modern plant body plans. This cladistic phylogeny is an authoritarian composite based on several separate analyses (see figure caption). Placed in the context of geologic time, phylogenetic analysis demonstrates the relative temporal compression of the radiation. It makes no claim for nor does it require unique rates of speciation or evolutionary mechanisms operating only during this time interval. It does emphasize, however, that the outcomes of the evolutionary process appear to have changed in breadth, with ancestor–descendant disparity apparently decreasing in a non-linear fashion through time (Gould 1991; Erwin 1992).

The Angiosperm Problem

The flowering plants are the only group traditionally given high taxonomic rank (i.e., class rank or above) that did not originate during the Middle Devonian–Early Carboniferous radiation. Some mention of them is necessary because the question will arise: Is not the origin and diversification of the angiosperms, the most species-rich groups of vascular plants ever to inhabit the Earth, indeed a radiation as profound structurally as that of the Late Devonian? Angiosperms have been ranked most often between phylum (equivalent to division) and class. This high rank was deemed necessary to accommodate the great species diversity within the group. It reflects a historical accident in plant systematics, where classification systems were initially developed and based upon extant plants, overwhelmingly angiosperms, with the less derived groups subsequently incorporated into the classification in only a quasi-phylogenetic manner. During the past 25 years, it has become traditional to distort this problem even further by treating the angiosperms as a phylum. In attempts to justify this taxonomic strategy, nearly all nonangiospermous seed plant and lower vascular plant orders were inflated to the rank of phylum, simply in order to accommodate the large number of Linnean ranks needed to encapsulate the diversity of the angiosperms (leaving most of these "phyla" encompassing only

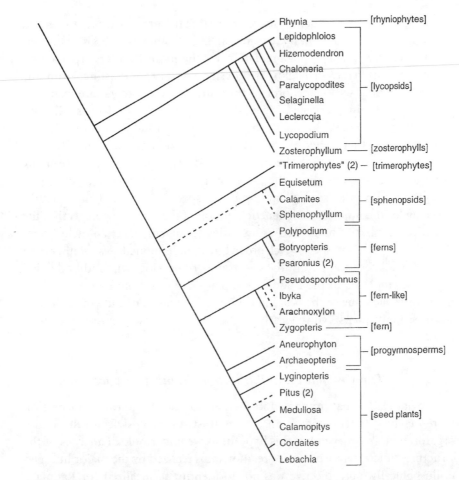

FIGURE 11.1. Tentative phylogeny of representative composite and whole-plant taxa, compiled by synthesizing an amalgam of recent phylogenetic studies. Main framework of the phylogeny is extrapolated from Rothwell (1996, figure 1), also with reference to Crane (1990), Kenrick and Crane (1991, 1997), Pryer, Smith, and Skog (1995), and Stevenson and Loconte (1996). Lycophyte clade (A) follows Bateman (1996a,b, figure 1), also with reference to Kenrick and Crane (1991, 1997) and Bateman, DiMichele, and Willard (1992). Lignophyte clade (B) follows Rothwell and Serbet (1994, figure 1), also with reference to Doyle and Donoghue (1992) and Nixon et al. (1994).

one class and one order). This approach obscures evolutionary relationships among the supposed phyla, the evolutionary significance of morphologies observed within each group, and the morphological disparity among them.

Numerous phylogenetic analyses of seed plants have appeared in recent years (Crane 1985; Doyle and Donoghue 1992; Nixon et al. 1994; Rothwell and

Serbet 1994). All of these demonstrate clearly that the angiosperms are a derived group within the seed-plant body plan that, like all other basic body plans, was a product of the great Devonian radiation. Furthermore, from a morphological perspective, the basics of angiosperm design are no more distinct from any of the traditional seed plant "orders" than any of these "orders" are from one another (cf., Lyginopteridales, Medullosales, Cycadales, Coniferales, Ginkgoales, Peltaspermales, Pentoxylales, Caytoniales, Bennettitales, Gnetales). It cannot be denied that many more variations on the seed-plant architectural theme have evolved among the more derived angiosperms than within any other seed-plant group. Even the most divergent forms, however, are largely confined within the seed plant–progymnosperm bauplan. Consequently, allowing for the implicit phylogenetic outlook of the Linnean perspective, the angiosperms should be ranked as an order (Bateman 1991; DiMichele and Bateman 1996); a very species-rich order, but an order nonetheless. Rather than indicating a later escape from the constraints of development and body plan that were emplaced in the Middle Devonian–Early Carboniferous radiation, they are, in fact, one of the best indications of the inescapability of such constraints. Body plans evolved early and entrained subsequent evolution of form.

The Vascular Plant Radiation: Ecomorphospace

The major clades that originated in the Late Devonian appear by inspection to represent different body plans, or at least different styles of structural organization. In order to test this hypothesis we undertook an analysis of the structure–function morphospace relationships realized by the major lineages. Philosophically, our objective was not to identify an idealized (or Raupian) morphospace, one circumscribed by theoretical limits on plant architectures (e.g., Niklas 1977, 1982), but rather to examine the morphospace as delimited by the plants that actually existed at different times (Foote 1993, 1994). Coded taxa were chosen so that each major lineage (class-level clade) would be represented by several placeholders or (if no single species was fully reconstructed) by composite taxa from different sublineages within each class. In this way, we hope to examine the degree to which each class formed a distinct functional morphological entity that might be termed an ecomorphic group. *Ecomorphic characters* are units of structure–function that will be compared analytically to the results of the ecological analysis; consequently, they were chosen to emphasize shared aspects of functional morphology largely separate from phylogeny, in several instances they are not necessarily homologous (i.e., synapomorphous) between or even within groups. Although characters were

chosen to minimize repetition, we recognize that they are not fully independent. The numerical analyses used do not require that character axes be
orthogonal. Indeed, we were seeking characters that are nonorthogonal due to
covariance, reflecting developmental or ecological associations but not necessarily phylogenetic relatedness. Wherever clear duplication (as opposed to
convergent patterns of expression) of structural or functional attributes was
identified, one of the overlapping characters was excluded from the analysis.
Our intention in constructing the ecomorphospace was to determine the
degree of congruence between the phylogenetic relationships of groups based
on cladistic analysis (figure 11.1) and their similarity as measured by distance
in the multivariate space of structure–function (ecomorphic) characters.

Plants were scored for each of 22 ecomorphic characters listed in table 11.1
and discussed in the following section; a best guess, based on nearest known
relatives or functionally related morphological features, was made for matrix
cells lacking direct observations. The plants and scores of states for particular
ecomorphic characters are listed in table 11.2. Analyses were carried out with
NTSYS, version 1.8 (Applied Biostatistics, Inc.), written by James Rolf. Analyses were exploratory and visual in nature, as no explicit hypotheses of similarity or difference were tested statistically. Techniques utilized included principal
components analysis (PCA) on the correlation matrix of ecomorphic characters, unweighted pair group cluster analysis (UPGMA), and complete linkage
cluster analysis (CLCA), the latter two utilizing the Euclidean distance metric.
Results are shown in figures 11.2 and 11.3.

Ecomorphic Characters

The following aspects of structure–function were used in the analysis. For
convenience of analysis, each ecomorphic character was divided into alternative discrete structure–function "states" by analogy with phenetic or cladistic
characters. In most instances, more than two states have been identified for
each ecomorphic character; if a linear sequence was hypothesized, the end-
point states were considered to be more distinct from each other than either
was from any of the intermediate states. In other instances, where linear chains
of states were not inferred (ecomorphic characters 2–3, 11–12, 14–16, 19–20),
multistate ecomorphic characters were converted to two-state or multistate,
allowing for several possible proximity relationships among states.

Ecomorphic character 1 expresses the growth capacity of the axis/shoot
apex. The growth dynamics of both the root and shoot system in vascular
plants are regulated by the meristems, which are places where cell division
takes place and from which much of the internal and external architecture is

TABLE 11.1. Features of Morphology used in the Analysis of Ecomorphospace[a]

1 - Growth capacity of the apex - shoot apex:
  0 = only more-or-less isodichotomous apex [rhyniophytes]
  1 = distinctly anisodichotomous apex organization present [in addition to isodi-chotomous; trimerophytes]
  2 = large central apex producing much smaller lateral appendages/leaves [herbaceous lycopsids]
  3 = relatively small central apex producing large lateral appendages [seed plants, ferns]
2 - Capacity for light reception by appendicular laminar surfaces:
  0 = low: cylindrical axes only responsible for light reception
  1 = medium: capacity enhanced by small scales/leaves providing laminar photosynthetic surfaces
  2 = high: photosynthetic capacity mostly by laminar photosynthetic surfaces
3 - Capacity for light reception by branch ramification:
  0 = none: dichotomous axes [rhyniophytes]
  1 = significant ramification but not filling of all the space [conifers]
  2 = space-filling ramifications
4 - Degree of apical dominance over lateral branch systems:
  0 = none or very little
  1 = moderate, anisodichotomous
  2 = strong apical dominance [conifers, lycopsids, *Equisetum*]
5 - Main shoot (apex) geotropism:
  0 = ascendant
  1 = main axis prostrate with ascendent lateral axes
  2 = fully upright main axis
6 - Support mechanisms of main axis:
  0 = minimal supporting structures
  1 = "semi–self supporting" (Rowe et al. 1993)
  2 = fully self-supporting
7 - Reproduction - life cycle:
  0 = homosporous, free-sporing
  1 = heterosporous, free-sporing
  2 = heterosporous, retained megaspores
  3 = heterosporous, single functional megaspores, retained
8 - Fructification display:
  0 = solitary or paired sporangia
  1 = small clusters of terminal or lateral sporangia, borne on lateral branch systems
  2 = "cones" (= masses of sporangia), either as separate structures or periodically as part of main or lateral shoot development
9 - Growth habit - rooting:
  0 = adventitious roots/rootlets
  1 = functional central root
10 - Cortical - ground tissue specialization for air flow:
  0 = none present
  1 = well developed aerenchyma or schizogeneous air channels
11 - Stelar architecture:
  0 = solid protostele

(*Continued on next page*)

TABLE 11.1. (*continued*)

       1 = solid but distinctly ribbed protosteles

       2 = dissected primary vascular system

12 - Pith or ground parenchyma - both stem and petiole/rachis:

       0 = center of stem with well developed primary xylem tracheids

       1 = pith or ground parenchyma present

13 - Secondary growth:

       0 = absent

       1 = present, unifacial or otherwise limited in extent (developmental capacity)

       2 = present, unlimited developmental capacity and potential extent

14 - Support tissues - internal physiology:

       0 = distributed or undifferentiated support

       1 = cortical support / limited internal physiology

       2 = peripheral bark support of unlimited extent / limited internal physiology

15 - Root mantle support:

       0 = distributed or undifferentiated support

       1 = root mantle support including peripheral vascular support

16 - Support tissues - external physiology:

       0 = distributed or undifferentiated support

       1 = wood support / external physiology

17 - Separation of sexes in sporophyte population:

       0 = plants showing no separation or homosporous

       1 = plants monoecious, but with spatial or temporal separation of megasporangiate and microsporangiate structures

18 - Capacity for continued vegetative growth following reproduction:

       0 = determinate sporangial structures terminate axial/lateral branch shoots of crown

       1 = sporangial structures periodic; vegetative growth continues on main axis of lateral shoots of the crown

19 - Cortical or ground tissue specialization:

       0 = none present

       1 = massive, not in discrete bundles

       2 = discrete bundles at periphery of cortex ['sparganum', 'dictyoxylon', etc.]

20 - Cortical specialization involving a periderm:

       0 = none present

       1 = 'secondary modification' of cortex by means of continued cell division, but not organized in tissue systems [*Triloboxylon* (Stein, Wight, and Beck 1983)]

       2 = periderm - all kinds - involving discrete zones of cell proliferation

       3 = massive permanent covering periderm [rhizomorphic lycopsids]

21 - Secondary xylem architecture:

       0 = no secondary xylem

       1 = manoxylic or intermediate secondary xylem

       2 = distinctly pycnoxylic secondary xylem [conifers]

22 - Propagule size:

       0 = small

       1 = medium/small

       2 = medium/large

       3 = large

[a] Defined states of ecomorphic characters used in the ecomorphospace analysis. In some instances, exemplar taxa are included in brackets.

TABLE 11.2. Taxa and States of Features Used in Analysis of Ecomorphospace

TAXON	CODE	ECOMORPHIC CHARACTERS																					
		1	2	3	4	5	6	7	8	9	10	11	12	13	14	15	16	17	18	19	20	21	22
Aneurophytes	ANE	1	0	1	1	1	1	0	1	0	0	0	0	2	0	0	1	0	0	2	1	1	0
Aneurophytes	ANE	1	0	1	2	1	1	0	1	0	0	1	0	2	0	0	0	0	0	2	1	1	0
Aneurophytes	ANE	1	0	1	1	2	1	0	1	0	0	1	0	2	0	0	1	0	0	2	1	1	0
Aneurophytes	ANE	1	0	1	2	2	1	0	1	0	0	1	1	2	0	0	1	0	0	2	1	2	0
Archaeopterids	ARC	3	2	1	2	2	2	1	2	1	0	2	1	2	0	0	1	0	0	2	2	2	1
Botryopteris	BOT	3	1	1	2	0	0	0	1	0	0	0	0	0	0	0	0	0	0	0	0	0	0
Calamites	CAL	3	2	2	2	2	2	0	2	0	0	2	0	2	0	0	1	0	0	0	0	1	0
Calamopitys	CPS	3	2	0	2	1	1	3	2	1	0	2	1	2	0	0	0	2	0	2	1	1	2
Chaloneria	CHA	2	2	0	2	2	2	1	2	1	0	0	0	1	1	0	0	1	0	3	1	1	1
Cladoxylopsids	CLA	1	0	2	2	2	2	0	1	0	0	2	0	0	1	0	0	0	0	0	0	0	0
Coniferales	CON	3	2	1	2	2	2	3	2	1	0	2	1	2	0	0	1	1	0	2	2	2	2
Cordaitales	COR	3	2	1	2	2	2	3	2	1	0	2	1	2	0	0	1	1	0	2	2	2	3
Equisetum	EQU	2	0	2	1	1	1	0	2	0	1	2	0	0	2	0	0	0	0	0	1	0	0
Filicales	FIL	3	2	0	1	1	1	0	2	0	0	2	0	0	2	0	0	0	0	0	0	0	0
herbaceous lycopsids	HLY	2	1	0	0	0	1	0	2	0	0	0	0	0	0	0	0	0	0	0	0	0	0
herbaceous lycopsids	HLY	2	1	0	1	0	1	0	2	0	0	0	0	0	0	0	0	0	0	0	0	0	0
Hizemodendron	HIZ	2	1	1	2	2	2	3	2	1	0	0	1	1	0	0	2	2	2	0	0	1	2
Iridopteridales	IRI	1	0	1	1	1	1	0	1	0	1	1	0	0	0	0	0	0	0	0	0	0	0
Lepidophloios	LEP	2	2	0	2	2	2	3	2	1	0	0	1	1	2	0	0	1	1	3	3	1	3
Lyginopteris	LYG	3	1	1	1	0	1	3	1	1	0	2	1	0	0	0	2	2	0	2	2	1	2
Marattiales	MAR	3	2	0	2	1	2	0	2	0	1	2	1	0	0	1	0	0	0	0	0	0	0
Marattiales	MAR	3	2	0	2	2	2	2	2	0	1	2	1	0	0	0	0	0	0	0	0	0	0
Medullosa	MED	3	2	0	2	2	1	3	2	1	0	2	1	1	1	0	2	2	0	1	1	2	3
Paralycopodites	PAR	2	1	1	2	2	2	2	2	1	1	0	0	1	2	0	0	0	0	0	3	1	1
Pitus	PIT	3	2	0	2	2	2	3	2	1	0	2	1	2	0	1	1	1	0	0	2	2	2

(Continued on next page)

TABLE 11.2. (continued)

TAXON	CODE	ECOMORPHIC CHARACTERS																					
		1	2	3	4	5	6	7	8	9	10	11	12	13	14	15	16	17	18	19	20	21	22
Pittus	PIT	3	2	0	2	2	2	3	2	1	0	2	1	2	0	0	1	2	0	2	2	2	2
Rhyniophytes	RHY	0	0	0	0	0	0	0	0	0	0	0	0	0	0	0	0	0	0	0	0	0	0
Selaginella	SEL	2	1	0	1	0	0	1	2	0	1	0	0	0	0	0	0	0	0	0	0	0	0
Sphenophyllum	SPH	2	1	1	1	0	1	0	2	0	0	0	0	1	0	0	1	0	0	0	2	1	0
Trimerophytes	TRI	1	0	1	1	1	2	0	1	0	0	0	0	0	0	0	0	0	0	1	0	0	0
Trimerophytes	TRI	1	0	1	1	1	2	0	1	0	0	0	0	0	0	0	0	0	0	1	0	0	0
Zosterophylls	ZOS	0	1	0	0	0	0	0	2	0	0	0	0	0	0	0	0	0	0	0	0	0	0
Zygopteris	ZYG	3	1	1	1	0	0	0	2	0	0	0	0	0	0	0	0	0	0	0	0	0	0

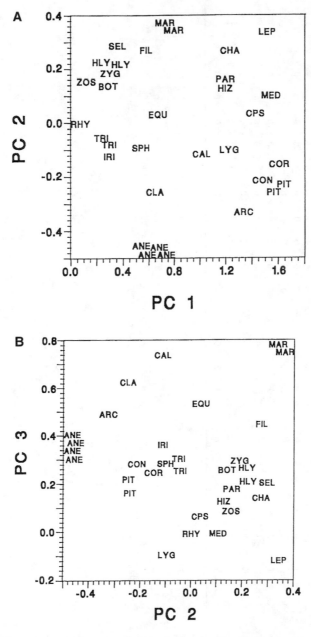

FIGURE 11.2. Principal components analysis of representative composite and whole-plant taxa. Analysis based on data matrix presented in table 11.2. See table 11.2 for key to acronyms: (A) Axis 1 vs. Axis 2; (B) Axis 2 vs. Axis 3. See text for details.

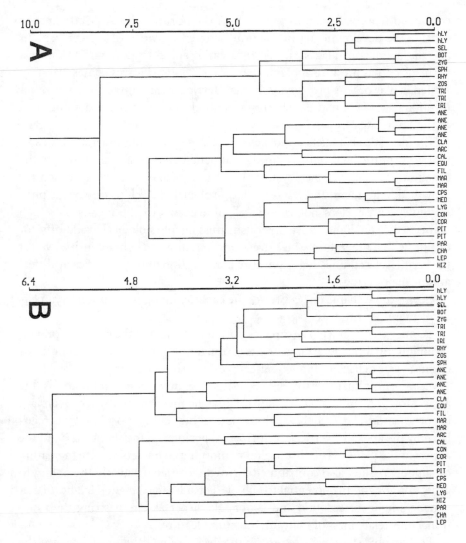

FIGURE 11.3. Cluster analyses of representative composite and whole-plant taxa. Analysis based on data matrix presented in table 11.2. See table 11.2 for key to codes. (A) Unweighted Pair Group Method of Analysis (UPGAC); (B) Complete Linkage Cluster Analysis (CLCA). See text for details.

controlled (e.g., Bateman, in press). This ecomorphic character attempts to capture the developmental interaction between the apex of the central axis and the lateral appendages. Simple, bifurcating apices (states 0 and 1) contrast with apices that produce appendicular organs, notably leaves (states 2 and 3). Appendicular organs may have minimal developmental impact on axis organization (state 2) or may have a strong feedback effect on further development of the axis (state 3).

Ecomorphic characters 2 and 3 express the potential of the aerial shoot system to capture light, either through appendicular organs or through the organization of the branch system. Light is a primary resource, needed for growth, development, and reproduction in all plants. Architecture of the photosynthetic array does not capture all dimensions of the means by which plants sequester light, most obviously missing physiological trade-offs for dealing with temperature, water stress, and variations in light intensity. Even at the basic level examined here, however, construction of the array clearly differs among major clades.

Ecomorphic character 2 expresses the capacity for light reception by appendicular laminar surfaces.

Ecomorphic character 3 expresses the enhancement of the capacity for light reception by branch arrangement through the support of appendicular organs, or by density of branching, increasing axis surface area.

Ecomorphic characters 4, 5, and 6 describe, in various combinations, the aerial growth form of a plant. They include the mechanisms by which a plant branches and the degree to which the lateral branching morphology is regulated by the apex, the vertical or horizontal position of the shoot axis, and the ability of the plant to support itself. Ecomorphic characters 13–16, the nature of support tissues, also contribute to growth form, particularly the ability to grow upright and the potential to achieve considerable height. Height further enhances light interception and propagule dispersal, emphasizing the functional interrelatedness of many architectural features.

Ecomorphic character 4 expresses the degree of apical dominance over the lateral branching system. This character identifies the extent to which the plant developed a main axis and either wholly suppressed lateral branches or relegated them to a subordinate, lateral growth position.

Ecomorphic character 5 expresses main-shoot (apex) geotropism. This character seeks to identify the developmental basis of growth habit. In ascendant growth forms (state 0) the apex is elevated dynamically as the axis "unrolls" along a substrate. Other shoot systems of clonal organisms differentiate into strictly prostrate and strictly vertical stems (state 1). A fully upright plant (state 2) may be centrally rooted or may be supported by adventitious roots, but is fundamentally unitary in construction and vertical in orientation.

Ecomorphic character 6 describes the support mechanism of main axis. It describes the degree to which the main axis was capable of supporting itself in an upright (vertical) position. Some plants with upright structure are not capable of standing vertically without leaning against or climbing on other plants (i.e., "semi–self supporting" *sensu* Rowe, Speck, and Galtier 1993; Speck 1994).

Ecomorphic character 7 describes reproduction and life cycle. This character focuses on the ability of a plant to disseminate its reproductive organs and gain control over the vagaries of environmental conditions. Each of the basic life histories in plants places different constraints on the likelihood of reproductive success and dispersal to suitable habitats. Homosporous plants, for example, often have great dispersal abilities and can produce a new population from a single propagule. They must, however, accommodate two independent life history phases (gametophyte and sporophyte) during evolution of habitat tolerance (Bateman and DiMichele 1994b). Seed plants, in contrast, greatly compress the life cycle, almost into one life history phase, the sporophyte, but must contend with the necessity for pollination and special mechanisms to disperse seeds. Using the major life cycles as short-hand, this character differentiates homosporous, heterosporous, and seed-bearing plants, considering degree of heterospory to represent a structure–function morphocline. "Seeds" are considered to be any integumented megasporangium; consequently, the "aquacarps" of the rhizomorphic lycopsids (Phillips and DiMichele 1992) are coded as functional seeds, even though, in phylogenetic terms, they were not borne by seed plants.

Ecomorphic character 8 expresses fructification display. Reproductive effort, both in any single reproductive event and over the entire lifetime of the individual, is an important ecological characteristic related to energy allocation patterns. Reproductive effort can be measured in various ways for fossil plants, although, unfortunately, not by the precise dry weight measurements preferentially made for extant plants. Here we used the degree of aggregation of sporangia as a measure of reproductive effort put into any one, short-term event. Note that fern fronds bearing abundant sporangial clusters (sori) are coded as "cones," reflecting shared high reproductive output per reproductive event (state 3).

Ecomorphic character 9 expresses growth habit and rooting. This ecomorphic character is intended to separate centrally rooted plants from those in which adventitious roots are the main fluid absorbing and/or support organs. Plants with central rooting have limited potential to spread from the point of rooting. The root system and shoot system in such plants must remain continually connected by live tissue. Adventitious roots provide flexibility in point of growth through time if combined with a prostrate growth habit, but also limit the ability of the plant to undergo extended vertical growth.

Ecomorphic character 10 describes ground tissue specialization for gaseous flow within the plant body. The presence of aerenchyma tissues is most frequently associated with growth under conditions of periodic flooding or standing water. In some plants (e.g., some species of *Psaronius* tree ferns), aerenchymatous tissues provide structural strength with limited carbon input and may have permitted attainment of tree habit with minimal energetic outlay.

Ecomorphic characters 11 and 12 differentiate the primary vascular architecture of the plant. Vascular architecture influences both support of the primary body and water conducting capacity. This is especially important in herbaceous plants with little or no secondary vascular tissues, but is also important in many groups that achieve tree habit using support mechanisms other than wood. The combination of ecomorphic characters 11 and 12 permits specialized types of nonlobed steles to be identified (e.g., siphonosteles are coded as protosteles in ecomorphic character 11 and as having a pith in ecomorphic character 12).

Ecomorphic character 11 describes the stelar architecture of main axis. States are differentiated by the degree of stelar surface area created by xylary lobing: smooth, ribbed, or dissected.

Ecomorphic character 12 describes the pith or ground parenchyma in main axis or leaves. Pith parenchyma has been shown to be a mechanism of water storage and a means to reduce the energetic cost of vascular support of an axis (a cylinder is a more effective means of support than a solid rod).

Ecomorphic character 13 expresses presence and mode of secondary vascular growth. It reflects the potential of the vascular cambium (if present) to produce a cylinder of secondary xylem and phloem. Plants with unifacial cambia have limited growth potential (Cichan and Taylor 1984), as do some plants with bifacial cambia.

Ecomorphic characters 14, 15, and 16 differentiate contrasting means of support of the shoot system and the effects these means of support have on location of the major physiological processes of the plant. Plants with states 1 or 2 of ecomorphic character 14 have cortical or peripheral bark support. Such plants retain much of their physiology within the support tissues, most notably water and nutrient transport. Plants with state 1 of ecomorphic character 15 have peripheral support from mantles of adventitious roots. Such root mantles are functionally analogous to wood, transporting water and nutrients to active leaves and meristems. In such systems, water transport, although external, is compartmentalized within the individual roots. Most other stem-based physiology resides within the support tissue. Ecomorphic character 16 identifies those plants supported by secondary xylem. In such cases, vascular tissue development confines all the active physiology outside of or marginal to the main locus of support of the plant: the periphery of the vascular cylinder.

Ecomorphic character 17 describes separation of the sexes in the sporophyte population. The distribution of male and female sex organs in plant populations is a complex phenomenon that would benefit from greater detail than can be obtained from most fossil taxa. For example, many homosporous plants (state 0) with potentially bisexual gametophytes regulate sex organ production through complex chemical regulation, creating distinct male and female plants; in other cases, male and female sex organs function on the same gametophyte. Many heterosporous plants produce both microspores ("male") and megaspores ("female") in the same cone, but when dispersed, the mature sex organs from the same plant, borne on different gametophytes, may not be in close proximity. Unfortunately, this character can be identified for few taxa; in fossils its level of refinement is far below that possible for extant plants. In our matrix, where the state in the fossil could not be determined, scoring was based on comparison with the nearest living relative.

Ecomorphic character 18 describes the capacity of the plant to continue vegetative growth following sexual reproduction. This character separates those plants that are monocarpic (semelparous) from those that are polycarpic (iteroparous). In the case of rhizomatous and other clonal plants (e.g., *Equisetum*), reproductive capacity was considered for the whole clone rather than just the "individual" upright shoots.

Ecomorphic character 19 describes cortical ground tissue specialization. Some plants have specialized sclerenchyma bundles or regions of the cortex that can serve as support tissues. Such tissues offer flexible support in young stems of plants with limited secondary support tissues in all or part of the plant, or in vine-like stems.

Ecomorphic character 20 describes cortical specialization involving a periderm. Periderms are secondary nonvascular tissues developing at the periphery of the stem. In certain lycopsids, the periderm was a permanent, largely supportive tissue (state 2). In most woody plants periderm is strictly protective, given the peripheral position of the active physiological attributes of the plant. In such plants new layers of periderm form during each increment of growth, reorganized within the secondary phloem of the previous increment (state 3).

Ecomorphic character 21 describes secondary xylem architecture. It is intended to separate plants that produce dense, pycnoxylic wood from those that have more parenchymatous wood or produce larger diameter, thinner walled tracheids.

Ecomorphic character 22 expresses propagule size. It distinguishes plants by the size of their female disseminules (see Bateman and DiMichele 1994b). Homosporous plants are routinely small, generally less than 160–180 μm (state 0). Heterosporous plants are apportioned between the two medium states (1 and 2), generally between 160–180 μm and 1 mm, or 1 mm to 5 mm. State 3 is for propagules greater than 5 mm in diameter. Seeds may fall in states 1, 2, or 3, although all the seeds scored for Paleozoic plants were of medium-large or large size.

Numerous potential ecomorphic characters could not be evaluated satisfactorily in the suite of fossils under consideration. Examples include epiphytic habit, parasitism or saprophytism, the nature of the relationships of root systems with mycorrhizal fungi, and of course numerous physiological features. In many cases there are grounds to speculate on certain of these characters for a few, exceptionally well-known species, but provide no basis for assessing the vast majority of species. At this juncture, therefore, we present this list of ecomorphic characters as a preliminary examination of the ecomorphospace created by the Middle Devonian–Early Carboniferous evolutionary radiation.

Coded Taxa

The following vascular land plants were selected for the analysis. Our objective was to include as many growth forms as possible in the analysis, including different variants within the major groups. Where unavoidable, composite taxa were constructed to permit combination of anatomical, morphological, and reproductive features, generally not all known from a single "species" (it is possible that different fossil species actually may represent the same plant preserved differently).

Rhyniophytes

Rhyniophytes are the basal plexus from which the two major Early Devonian clades are thought to have evolved (Banks 1968). The rhyniophytes are represented by a generalized composite plant similar in character to *Rhynia gwynne-vaughnii* (Kidston and Lang 1917; D. S. Edwards 1980). This is the most phylogenetically primitive plant in the analysis.

Trimerophyte Lineage

TRIMEROPHYTES. Two types of generalized trimerophytes are represented, based on *Psilophyton dawsonii* (Banks, Leclercq, and Hueber 1975) and *Pertica* spp. (Kaspar and Andrews 1972; Granoff, Gensel, and Andrews 1976). They differ according to the form of growth of the axis (ecomorphic character 5), either ascendant or with inferred upright axes borne on a prostrate axis system.

IRIDOPTERIDS. These are represented by the genera *Ibyka* (Skog and Banks 1973: gross morphology) and *Arachnoxylon* (Stein 1981; Stein, Wight, and Beck 1983: anatomy).

CLADOXYLOPSIDS. These are represented by the genus *Pseudosporochnus* (Leclercq and Banks 1962; Stein and Hueber 1989; Berry and Fairon-Demaret 1997).

SPHENOPSIDS. Herbaceous equisetophytes are represented by the modern genus, *Equisetum*. Woody tree-equisetophytes are represented by a generalized, homosporous calamitean; all members of this group shared common basic architectural features (Andrews and Agashe 1965; Barthel 1980). Sphenophylls are represented by the prostrate, but woody, *Sphenophyllum plurifoliatum* (Williamson and Scott 1894).

FERNS. Three major variants of fern architecture are included in the analysis. The Filicales include two taxa, a generalized filicalean based on the modern polypodiaceous ferns and *Botryopteris antiqua* from the Late Carboniferous (Phillips 1974). The Zygopteridales are represented by *Zygopteris illinoensis* (Dennis 1974), a form very similar to the filicalean types but differing phylogenetically rather than in gross architecture; both botryopterids and zygopterids are known from the Early Carboniferous. The Marattiales are represented by two generalized forms of Late Carboniferous *Psaronius*. The more primitive is of scrambling habit and lacks a well developed root mantle; the more derived is arborescent with a large supportive root mantle (Lesnikowska 1989).

PROGYMNOSPERMS. There are two widely recognized types of progymnosperms, the aneurophytes and archaeopterids. Four alternative interpretations of aneurophyte morphology are presented, based on different interpretations and combinations of ecomorphic characters 4 and 5 (degree of apical dominance and main shoot geotropism), both contributing to general growth habit. The concept of aneurophytes is based on *Rellimia* (Bonamo 1977), *Triloboxylon* (Matten and Banks 1966; Scheckler 1976), and *Tetraxylopteris* (Bonamo and Banks 1967). Archaeopterids are represented as a composite, combining information from petrified stem remains and compressions of foliage and reproductive organs (Beck and Wight 1988; Trivett 1993).

SEED PLANTS. Included in the analysis are several kinds of seed plants, which can be grouped broadly into two lineages. "Cycadophytic" seed plants share radially symmetrical seeds, manoxylic wood, and limited secondary vascular development, and include *Calamopitys* (Rowe and Galtier 1988; Galtier and Meyer-Berthaud 1989), *Medullosa primaeva* (Delevoryas 1955; Stidd 1980), and *Lyginopteris oldhamia* (Oliver and Scott 1904). "Coniferophytic" seed plants share more dense, pycnoxylic wood with unlimited potential for secondary growth, bilaterally symmetrical seeds, and generally small leaves; included are two variations of *Pitus* that differ on uncertainty in ecomorphic character 17 (dioecy or monoecy), a generalized cordaitalean (Costanza 1985; Trivett and Rothwell 1985), a generalized primitive conifer (Clement-Westerhoff 1988).

Zosterophyll Lineage

ZOSTEROPHYLL. Generalized zosterophyll are based on *Rebuchia, Serrulacaulis,* and *Gosslingia* (Niklas and Banks 1990; Lyon and Edwards 1991; Hueber 1992).

LYCOPODIOPSIDS. These are represented by an herbaceous lycopod similar in form to extant *Huperzia* and by the Middle Devonian *Leclercqia* (Banks, Bonamo, and Grierson 1972). The two types of lycopodioid morphology included here differ in degree of inferred apical dominance (ecomorphic character 4).

SELAGINELLIDS. Most evidence for the presence of selaginellids in the Devonian and Early Carboniferous is equivocal (Thomas 1992; although see Rowe 1988). However, the phylogeny of the lycopsids (Bateman 1996b) suggests that selaginellids evolved prior to the isoetoids, which have an excellent Late Devonian and Carboniferous record. Consequently, we scored a modern *Selaginella* of the *S. kraussiana* type.

ISOETALEANS. The isoetalean lycopsids were a far more diverse group in the Paleozoic than they are today. Many fossil forms are known in exceptional detail, and whole plants have been reconstructed. We used as exemplars the best known extinct tree forms, *Lepidophloios hallii* (the most derived species) and *Paralycopodites brevifolius* (the most primitive species), the pseudoherb *Hizemodendron serratum,* and the more typically isoetalean *Chaloneria* (Bateman, DiMichele, and Willard 1992).

Analysis Results

Principal Components Analysis

The principal components analysis (figure 11.2) reveals a pattern that is strongly, but not perfectly, congruent with the phylogeny. In other words, the clusters of taxa based on ecomorphic characters, and the relative proximity of these clusters, are similar to the grouping of taxa in the phylogenetic analysis. Ninety percent of the variance is accounted for by the first eight axes and 66% by the first three, indicating (perhaps expected) high dimensionality of the morphospace. The basic pattern shows clearly when the coded taxa are plotted on the first two axes (figure 11.2A). The seed plants and their pteridophytic ancestors, the progymnosperms, are largely distinct from the rest of the lower vascular plants. There is some overlap among the cycadophytic seed plants and the woody equisetophyte and cladoxylopsid lower vascular plants. Differentiated along the first axis are forms with centralized root systems and upright growth habits (wood or bark supported) versus those with trailing habit. This latter gradient in habit is anchored at one end by the phylogenetically basal rhyniophytes and at the other by plants with tree habit, divided into two major groups: the lycopsids plus ferns and the progymnosperms plus seed plants. Various other groups, mostly pteridophytes, occupy the middle ground. Most differentiated by a third axis (figure 11.2B) are the marattialean tree ferns, the calamites (tree sphenopsids), and, to a lesser extent, the cladoxylopsids, indicating that these groups represent different ways to build a tree and may share some similarities in detail, despite significant, including phylogenetic, differences.

The following aspects of this analysis are noteworthy:

1. Using rhyniophytes as a basis for comparison, isoetoid lycopsids are the most ecologically-structurally divergent members of the zosterophyll lineage. They form a fairly tight cluster in the PCA. Members of this group have a distinctive basic body plan. They are bark-supported, have strong apical dominance, specialized root systems, and heterosporous reproduction. Branches play little or no role in the construction of their photosynthetic array, and leaves and branches are based on different developmental programs, resulting in few architectural similarities such as those found in leaves and shoots of groups descended from trimerophytes. Bark support permits them to separate support and water transport functions in different specialized tissues; Cichan (1986), in model studies, found lycopsid wood to have high efficiency in water transport. The structural and developmental similarity of the rhizomorphic rootlets to the microphyllous leaves, and the lack of identifiable secondary phloem to permit transport of photosynthate from the shoot to the root systems, suggest that root systems may have been photosynthetically self-supporting (Phillips and DiMichele 1992). Thus, these plants may come closer than any other vascular plant trees to a colonial growth habit, similar to that seen in metazoans such as bryozoans.

2. The seed plants represent the evolutionary pinnacle of the trimero-phyte lineage. Although diverse, they form four distinct clusters on the PCA plot, clearly separated from all pteridophytes except the archaeopterid pro-gymnosperms, which are virtually identical to seed plants in vegetative architecture. Of course, except for the archaeopterids, the seed plants are united by the presence of their specialized reproductive apparatus, which permits them to escape free-water constraints during critical phases of the life cycle. All woody members of this group have distinctly bifacial vascular cambia, although many have limited wood development. Leaves are primi-tively large and share many developmental and structural characteristics with branches. The stems tend to be centrally rooted. Most of the plants included here are trees, except for the scrambling or semi–self supporting *Calamopitys* and *Lyginopteris,* which plot away from the other seed plants. The medullosans and lyginopterids converge with the lycopsids in some design aspects, particularly in their nonwoody peripheral support and con-sequent high flow capacity in the specialized secondary xylem (Cichan 1986).

3. The aneurophyte progymnosperms do not cluster with the archae-opterid progymnosperm-seed plant group. Aneurophytes are similar to other Middle Devonian plants on PCA axis 1. They occupy an intermediate position, possibly indicating the general progression of many lineages toward larger size and greater phenotypic complexity than Early Devonian ancestors, while being less divergent than Late Devonian forms. PCA axis 2 indicates that aneurophytes, although the most divergent, are nevertheless most like their cousins, the archaeopterids and (to a lesser extent) the early seed plants. The overall isolation of the aneurophytes probably signifies a divergent ecological role for these plants in the Middle Devonian, perhaps reflecting movement into the better drained habitats more fully exploited later by archaeopterids and seed plants.

4. The sphenopsids and their potential ancestors, iridopterids, spheno-phylls, and herbaceous and woody equisetophytes (Stein, Wight, and Beck 1984), form a loose cluster that is closer than any other lineage to the ances-tral trimerophytes. The relationship is closest between the trimerophytes and iridopterids, with the sphenophylls somewhat more distant. All share trailing or rhizomatous growth habits. *Equisetum* and the calamites are considerably less similar in the ecomorphospace than would be expected, given their clear phylogenetic and structural similarities, including shared nodal whorls of leaves, branches, and reproductive organs, and unique attributes of the reproductive organs (notably sporangia borne on "sporan-giophores" of uncertain homology to reproductive organs in other plants).

Perhaps the relatively small, nonwoody *Equisetum* is not a particularly close ecological analogue of the ancient woody, arborescent calamites.

5. Filicalean and zygopterid ferns plus lycopsids, all of generalized trailing morphology, form a distinct group that is more phylogenetically heterogeneous than the other clusters. In contrast, ground cover plants with sprawling, thicket-forming, or climbing habits (e.g., *Hizemodendron* among the rhizomorphic lycopsids, *Sphenophyllum* among the sphenopsids, and *Lyginopteris* among the primitive seed plants) appear closer to the tree forms of the respective taxonomic groups. The cluster of ferns and primitive lycopsids may reflect phylogenetically retained (i.e., plesiomorphic) structural simplicity, whereas the other groups with sprawling habit have converged by secondary morphological simplification from relatively derived architectures (Bateman 1994, 1996a). Thus, specialization followed by simplification actually may have created more ways to exploit the resources available to ground cover than were available to the more primitive ancestral forms.

Cluster Analyses

Several ecomorphic groups appear in both the CLCA and UPGMA cluster analyses (Figure 11.3). Because CLCA uses the most distant relationship of taxa in constructing clusters, this method tends to suggest clusters of greater compactness, but lower similarity, than UPGMA. The latter assesses relationships of unlinked taxa with the arithmetic means of clusters that have already been formed. Both methods use the full dimensionality of the data in calculating distances between taxa. Thus, the methods serve as a useful comparison with the more incomplete, but highly suggestive, patterns observed in PCA.

1. As with the PCA, there remains a distinction between ground cover and the more structurally complex trees. The first dichotomy in the complete linkage dendrogram, and the second in the UPGMA dendrogram, separate ground cover from tree forms. Forming one group are primitive rhyniophytes, trimerophytes, zosterophylls, lycopodiopsids, sellaginellopsids, filicalean and zygopterid ferns, iridopterids, and *Sphenophyllum*. Missing from this group are the ground cover rhizomorphic lycopsids, the ground cover pteridosperms, and, interestingly, both the trailing filicalean and the marattialean ferns.

2. Another feature also seen in the PCA is evident in the dendrograms. The tree-sized groups show strong concordance with phylogeny. The following architectural groups are distinct in both CLCA and UPGMA analyses:

the woody seed-plants, the bark-supported isoetalean lycopsids, and the relatively primitive aneurophytes plus cladoxylopsids. More difficult to interpret are two consistent associations: (1) the marattialeans (both tree and trailing habits) plus *Equisetum* plus the advanced filicaleans, and (2) the archaeopterids plus calamites.

3. In UPGMA, which, compared to CLCA, de-emphasizes highly divergent taxa, the first dichotomy separates woody taxa from those that generate little if any woody tissue. Unsurprisingly, most of the nonwoody forms are also nontrees. Of the woody forms, the rhizomorphic lycopsids are the only group not supported primarily by wood. In contrast, the CLCA separates most ground cover, including clonal mats, from tree and shrub habits.

Patterns of Ecological Distribution of Major Vascular Plant Clades

Clade Distributions

The major vascular plant classes appear to occupy distinct ecological centroids soon after their appearance, their divergence continuing through the Late Devonian and Early Carboniferous. This assertion is borne out not only by the previous ecomorphospace analysis but by numerous paleoecological studies that relate fossil plant species to sedimentary environments and thereby reconstruct ancient plant communities. In the most generalized terms the clades differentiate as follows.

Isoetalean lycopsid trees became dominant elements in the wettest parts of the lowlands. Beginning as important elements in minerotrophic (clastic substrate) swamps (Scheckler 1986a,b), they quickly became the dominant elements in peat-forming habitats as well (Daber 1959; Phillips and Peppers 1984; Scott, Galtier, and Clayton 1985). Scott (1979) documented their occurrence in a broad spectrum of wetland habitats well into the Late Carboniferous, including swamps, mires, point bars, and the wetter parts of flood plains.

Calamitean tree sphenopsids appeared in stream and lakeside settings as early as the Late Devonian (Scheckler 1986a) and continued in these types of aggradational environments throughout the Carboniferous and into the Permian (Teichmüller 1962; Scott 1978, 1979; Bateman 1991; Gastaldo 1992). Their success in habitats with high clastic influx reflects their clonal, rhizomatous growth habit and the ability it confers to recover from burial (Potonié 1909; Gastaldo 1992). The literature suggests that calamites became relatively widespread in wetlands but were only truly abundant in environments marginal to water bodies, a narrow "adaptive zone."

The progenitors of the seed plants, the progymnosperms, appear to have become tolerant of soil moisture deficits early in their evolutionary history. Beck (1964) and Retallack (1985) suggested that archaeopterid progymnosperms exploited a wide range of habitats in the Late Devonian, including better drained interfluves. Although such environments may not have been "dry" by later standards, they appear to have been among the drier settings colonized by Middle and Late Devonian vascular plants. Substrate-penetrating centralized root systems may have been a key morphological feature permitting this ecology (Bateman and DiMichele 1994b; Algeo et al. 1995; Retallack 1997; Elick, Driese, and Mora 1998; Driese et al. 1997), along with the moderate reproductive diapause offered by heterospory, where spores of some species can persist in a pregermination state without water.

Seed plants appear to have arisen in wetland settings, possibly in habitats similar to those occupied by heterosporous progymnosperms (Scheckler 1986a,b). Of course, early seed habit may have been very similar in its functional attributes to heterospory and have worked most effectively in habitats with regularly available free water (Bateman and DiMichele 1994b). In the Early Carboniferous, however, seed plants radiated in terra firma habitats (Bridge, Van Veen, and Matten 1980; Matten, Tanner, and Lacey 1984; Retallack and Dilcher 1988), a resource zone made available to them by their reproductive biology, basically through reproductive preadaptation (or exaptation in the terminology of Gould and Vrba 1982) to survive moisture stress. Terra firma settings offer a wider array of environmental variations and a vastly greater physical space than that colonized by either the isoetalean lycopsids or the sphenopsids.

Late Devonian ferns and fernlike plants are not particularly well understood ecologically. *Rhacophyton,* a fernlike possible zygopterid, was a dominant element in organic-rich swamps (Scheckler 1986a,b). Certainly, by the Early Carboniferous there is excellent documentation of ferns occupying a variety of habitats, especially those subject to significant disturbance. Scott and Galtier (1985) described volcanigenic landscapes in which ferns are common elements, often preserved as fusain (mineralized charcoal), suggesting growth in habitats frequently swept by ground fires. In the Late Carboniferous, both ground-cover and larger marattialean ferns occurred in a wide range of habitats, except those that appear to have been flooded for long periods of time (Mickle 1980; Lesnikowska 1989; Rothwell 1996). The ground cover elements often are found in highly diverse assemblages within coal-ball deposits from coal seams (Phillips and DiMichele 1981; DiMichele and Phillips 1988), suggesting colonization of short-lived areas of disturbance within mire and swamp forests. Marattialean tree ferns, on the other hand, appear to have arisen in terra firma settings and from there began a penetration of forests dominated by seed plants, perhaps as weedy elements in more disturbed parts of landscapes. Following extinctions of tree lycopsids in the Late Pennsylvanian of the western tropical belt (not in China, however), tree ferns became dominants in many lowland, wetland habitats throughout the later part of the Late Carboniferous and into the Permian.

Timing of Origination of the Patterns

The most comprehensive analyses of Paleozoic plant ecology focus on the Carboniferous. Certainly, for the Late Carboniferous, dominance of class-level groups has been documented in distinctive sedimentological settings in the tropics, reflecting original resource partitioning. There is considerable literature, extending back into the early part of the twentieth century, that documents the occurrence of plants in coal measures environments in particular. Scott (1977, 1978, 1979, 1980, 1984) summarized this literature, particularly as it relates to compression–impression (adpression) floras from the coal measures, and provided extensive quantitative data documenting plant distribution by sedimentological setting. Eble and Grady (1990) and DiMichele and Phillips (1994) also summarized general patterns of distribution, focusing mostly on floristic associations within peat swamps. Many studies (e.g., Chaloner 1958; Cridland and Morris 1963; Havlena 1970; Pfefferkorn 1980; Lyons and Darrah 1989) documented the coeval existence of lowland–wetland and extrabasinal, more xeric floras in the Late Carboniferous tropics. Little is known of the extrabasinal floras until the last stage of the Late Carboniferous (Stephanian). At that time periodic climatic oscillations created intermittent drier conditions in the lowlands, which permitted immigration of extrabasinal elements (e.g., Winston 1983; Mapes and Gastaldo 1984; Rothwell and Mapes 1988; Mamay and Mapes 1992), providing the earliest unequivocal evidence of conifers and other more derived seed plant groups.

The distributional patterns of Late Carboniferous plants raised our awareness of the broadly distinct ecological centroids of the major clades (e.g., Scott 1980; DiMichele and Bateman 1996; DiMichele and Phillips 1996). Because the pattern already existed and was well differentiated in the Late Carboniferous, it became clear that its origin lay deeper in time.

Early Carboniferous floras have received extensive study recently, largely by Scott, Galtier, and colleagues (Scott and Galtier 1985; Scott, Galtier, and Clayton 1985; Scott et al. 1986; Rex 1986; Rex and Scott 1987; Bateman and Scott 1990; Scott 1990; Bateman 1991), who documented sedimentary environments and revised the systematics of the plants. During the Early Carboniferous, Europe and most of North America were outside the tropical rainy belt (Raymond, Parker, and Parrish 1985). Consequently, the paleoclimate and spectrum of habitats represented in the fossil record differs from that found in the Late Carboniferous. Yet, the basic patterns of clade-by-habitat distribution revealed by these recent studies appear fundamentally similar to later times. All major body plans, except marattialean ferns, are recorded in the Early Carboniferous. It is in rocks of this age that the early ecological role of ferns as interstitial opportunists is most evident (Scott and Galtier 1985). By the early Namurian, floral assemblages similar to those of the later Carboniferous coal measures began to appear intermittently in wetland habitats (Jennings 1984, 1986; Raymond 1996).

The Carboniferous studies point squarely back to the Devonian as the time of origin of the ecological patterns that would persist directly for the next 30 million years and beyond through influencing patterns of clade replacement through time. Although there are relatively few paleoecological studies of Late Devonian plants, some are quite comprehensive. For example, Scheckler (1986a) provided a thorough examination of the major lineages and their ecological distributions. The basic patterns of partitioning can be detected quite readily in these floras. However, the spectrum of environments occupied and the details of distribution vary from the more prominent patterns that would follow. It thus appears that this was a time of sorting out. The seed plants and their early occurrences in swampy habitats, the dominance of fern-like plants in organic-rich swamps, the occurrence of isoetalean lycopsids in peri-swamp wetlands, and the commonness and numerical dominance of progymnosperms in many floras all are patterns that are close, but not identical, to those that would be more firmly established by the Early Carboniferous. Ecological differentiation can be recognized much earlier still, among the more primitive plants of the Early and Middle Devonian (e.g., Matten 1974; Stein, Wight, and Beck 1983; Edwards and Fanning 1985; Hotton et al. in press). These patterns, however, have yet to be linked clearly to those found in younger vegetation.

Demise of the Primeval Ecosystem

The ecological patterns that evolved in the Late Devonian and earliest Early Carboniferous persisted until the Westphalian, the middle of the Late Carboniferous, throughout most of the world. During the early to middle part of the Late Carboniferous extinctions began, evidently driven by global climatic changes related to the dynamics of polar glaciations (Frakes, Francis, and Sytka 1992). These floristic changes were globally asynchronous but everywhere had the effect of breaking up the primitive clade-by-habitat patterns of resource partitioning. Seed plants rose to prominence as resources were vacated, and long-term patterns of clade incumbency ended. Climatic changes took place earlier and to a greater extent in the northern and southern temperate zones, driving extinctions and permitting the rise of seed plants in these parts of the world (Meyen 1982; Cúneo 1996). The pattern of disassembly of the tropical wetland biome began later. Its ultimate replacement by a seed-plant dominated biome was hierarchical, beginning first within the wetlands, followed later by replacement of the whole wetland biome on a larger scale (Fredericksen 1972; Broutin et al. 1990; DiMichele and Aronson 1992). In parts of China, Westphalian-type floras persisted well into the Permian (Guo 1990) but ultimately disappeared. Within the subsequent Mesozoic ecosystems there is evidence that subgroups of seed plants and ferns partitioned ecospace along phylogenetic lines and that such patterns of partitioning can be found among angiosperm groups in modern ecosystems (Lord, Westoby, and Leishman 1996).

Diversity Patterns: Did Breadth of Resource Space Constrain the Species Diversity of the Major Clades?

The major clades differ considerably in species and generic diversity. Although species from different clades co-occurred and their aggregate ecological amplitudes overlapped, each clade nonetheless had a distinct ecological centroid. We are led to speculate, then, whether differences in overall species diversity may have been controlled in part by the ecological opportunities available within the core environment colonized by each clade. Do these opportunities reflect not only the physical variability of the habitat but also the simple area available for colonization, which Rosenzweig (1995) suggests as the most important regulator of diversity patterns?

Quantification of past diversity is a problem, however. Two approaches have been used: global species diversity and average floristic diversity. The only comprehensive global database was compiled by Niklas, Tiffney, and Knoll (Knoll, Niklas, and Tiffney 1979; Niklas, Tiffney, and Knoll 1980, 1985), which indicates considerable differences in the diversity of the major clades through time. In the Niklas et al. (1985) compilation, Late Carboniferous arborescent lycopsids are the most diverse group, with over 100 species reported, followed by approximately 50 species of ferns, 40 seed plants, and 20 sphenopsids. In contrast, Knoll (1986) reported diversity of major groups as average floristic diversity. In this latter compilation, during the Early and Middle Pennsylvanian, seed plants were the most diverse group, accounting for approximately 55% of the species; ferns and sphenopsids account for somewhat less than 20% each; and lycopsids account for just 7%.

We determined to follow the method of Knoll (1986) because his results paralleled our direct experiences with Carboniferous tropical floras at local and regional scales, which suggest that seed plants are most diverse, followed in order by ferns, lycopsids, and sphenopsids. Raymond (1996) also reported patterns similar to those of Knoll (1986) on the basis of a regional stage-level compilation. After examination of numerous "local" Late Carboniferous floras (narrow time intervals and uniform depositional environments) and several large monographic treatments of regional floras (longer time intervals and mixed depositional settings), we selected eight examples from tropical Euramerica, confining our analysis to floras Namurian and Westphalian in age. The objective was to examine taxonomic diversity of each major clade in several different kinds of environments or taphonomic settings, representing time windows of different durations. Confining analyses to monographic treatments of floras enabled us to minimize problems of form taxonomy (different names for parts of the same whole plant) and the vagaries of taxonomic use. Finally, by working only with Namurian and Westphalian floras, the confounding effects of major extinctions and ecosystem reorganizations in the later Late Carboniferous were avoided. No attempt was made to combine the floras into a single database because of the problems of inconsistent taxonomic use. Following Knoll (1986), our intent is to seek diversity averages across a representative sampling of locales and environments.

We also examined modern environments as "actualistic" analogues to the Carboniferous, to investigate how patterns of species diversity were constrained by broad environmental types; specifically, wetlands versus terra firma. If our assertions are correct, basic patterns of species packing should vary among these two broad environments. Even though modern taxonomic composition is very different from that of the past, relative diversity patterns should reflect underlying, taxonomically independent controls. The existing numbers for modern environments were obtained from many different sources; studies had a variety of objectives and thus used many different sampling strategies and methods of reporting data. Hence, our compilation is of necessity neither complete nor rigorously statistical. Campbell (1993) provided an excellent summary comparing flooded and terra firma forest diversity in the Amazon basin, which we discovered after undertaking our compilation.

Diversity Patterns in Late Carboniferous Floras

The floras examined can be divided broadly into two landscape types within the wetland biome: peat-forming mires and floodplains. Within each of these there are clearly differentiable subenvironments that have recurrent, distinctive species composition; however, we wish to examine only relative diversities of the major clades and so have chosen to average spatially across subenvironments

TABLE 11.3. Species Diversity of Selected Pennsylvanian Floras[a]

	Average spp. #	Leary & Pfef 1977 Compression Flora Spencer Farm Flora Single Site	Scott 1977 Compression Flora Annbank-Coal Roof Single Site	DiMichele et al. 1991 Coal-Ball Flora Secor Coal Single Site
Lycopsids	8.8	1	1	7 (74.4)
Ferns	15.8	3	2	10 (5.9)
Sphenopsids	7.1	4	5	2 (7.4)
Pteridosperm	22.25	9	9	10 (3.9)
Cordaites	2	5	1	3 (8.3)
Unident. Seed Plant	0.4	0	0	1 (0.1)
Total Seed Plant	24.6	14	10	14 (12.3)

[a] Numbers in parentheses are percent abundance values based on quantitative analyses of respective floras. For each flora the citation, compressed versus coal ball, locality, and single or multisite origin are noted. If all groups of seed plants are combined, average species number is 26.4.

within each of the landscape types. The patterns are clear. Without exception, seed plants are the most species-rich group in all floras, generally two to three times as diverse as lycopsids (table 11.3). This is the case even in those habitats where lycopsids dominate in terms of biomass. Fern diversity rarely exceeds that of seed plants but generally is slightly lower. Sphenopsids are the least diverse major clade. Diversity pattern should not be confused with ecological dominance (Wing, Hickey, and Swisher 1993). Lycopsids have been shown clearly to dominate many Late Carboniferous swamps and mires (Phillips, Peppers, and DiMichele 1985), whereas pteridosperms and sphenopsids dominate most Westphalian and Namurian compression floras from floodplain habitats (Pfefferkorn and Thomson 1982).

Similar patterns prevail in both the Southern Hemisphere Gondwana floras (Cúneo 1996) and the Northern Hemisphere Angara floras (Meyen 1982). Although lycopsids are major elements of the landscape and dominate many environments, in overall diversity at either the generic or species level, seed plants are two to three times more diverse than lycopsids.

Modern Diversity Patterns

Our goal was to contrast the species diversity of modern wetlands with that of terra firma environments, expecting that species diversity would be lower in wetlands. Obtaining robust data proved to be as difficult as for fossils but for different reasons. Few regional studies break floras down by habitat, and diver-

TABLE 11.3. (*continued*)

DiMichele et al. 1991 Compression Flora Secor Coal Roof Single Site	Pfefferkorn 1979 Compression Flora Mazon Creek Multisite	Willard et al. 1995 Compression Flora Springfield Coal Split Single Site	Josten 1991 Compression Flora NW German Coal Measure Multisite	DiMichele and Phillips 1996 Coal Ball Flora Eastern USA Coal Measures Multisite
1 (0.1)	15	5	32	10
7 (28.2)	37	3	48	16
4 (14.0)	9	3	25	5
11 (57.8)	33	6	82	18
0 (0)	2	1	3	1
0 (0)	0	0	0	2
11 (57.8)	35	7	85	21

sity patterns remain poorly known in many parts of the world. Furthermore, where diversity has been reported by habitat type, the size of the area sampled varies greatly. Without the original data it is not possible to normalize for differences in sampling area. So, once again, a general impression must be gained by inspection rather than by rigorous statistical analysis.

The data we compiled are summarized in table 11.4. Clearly, well-drained, terra firma habitats are more diverse than wetlands, both in total species numbers and in range of variation (Peters et al. 1989; Campbell 1993). In Brazil, for example, a 0.5 ha area of seasonally flooded varzéa forest contains 37 tree species, whereas the same area of nearby terra firma forest contains 165 tree species (Prance 1994). Compilation of species by habitat type in Costa Rican forest preserves (Hartshorn and Poveda 1983) demonstrates much lower diversity in swamps and riparian habitats than in terra firma environments. Organic-rich wetlands (peat swamps, bogs, and fens) are lower in species diversity than the surrounding areas, both in temperate and tropical environments (Best 1984; Kartawinata 1990; Westoby 1993; Wheeler 1993). Even in arctic habitats, where species diversity is low in general, the wettest habitats are the least diverse (Muc and Bliss 1977).

The closest approximation to a global summary is given by Gore (1983) in *Ecosystems of the World 4B.* An appendix summarizes all families and genera referenced in the aforementioned volume, a total of 489 genera from 161 families of vascular plants. These are probably gross underestimates, given the

TABLE 11.4. Species Diversity of Selected Extant Floras, Divided by Geographic Region and Terra Firma or Wetland Environment within Region

Sampling Region	Environment or Flora	Sample	Diversity	Geographic Location	Citation
Southeast Asian Floras					
	TERRA FIRMA	10.5 h	406	E. Kalimantan lowland forest	Kartawinata 1990
		1.6 h	239	E. Kalimantan lowland forest	Kartawinata 1990
		25 'plots'	332	N. Sumatra lowland forest	Kartawinata 1990
		1 h	62	lowland Indonesian forest	Kartawinata 1990
		1 h	70	lowland Indonesian forest	Kartawinata 1990
		1 h	95	lowland Indonesian forest	Kartawinata 1990
		1 h	100	lowland Indonesian forest	Kartawinata 1990
		1 h	108	lowland Indonesian forest	Kartawinata 1990
		1 h	110	lowland Indonesian forest	Kartawinata 1990
		1 h	120	lowland Indonesian forest	Kartawinata 1990
		1 h	175	lowland Indonesian forest	Kartawinata 1990
		1 h	208	lowland Indonesian forest	Kartawinata 1990
	WETLANDS	0.5 h	20	Semboja kerangas (swamp) forest	Kartawinata 1990
		0.5 h	10	Semboja kerangas (swamp) forest	Kartawinata 1990
		1 h	70	Indonesian freshwater swamp	Kartawinata 1990
		0.2 h	29	Indonesian peat swamp	Kartawinata 1990
		0.2 h	35	Indonesian peat swamp	Kartawinata 1990
		0.2 h	40	Indonesian peat swamp	Kartawinata 1990
		0.2 h	45	Indonesian peat swamp	Kartawinata 1990
		0.2 h	50	Indonesian peat swamp	Kartawinata 1990
	FLORAS	flora	14500	Malaysia—angiosperms	
		flora	650	Malaysia—ferns	

Neotropical Floras

	Description	Value	Measure	Reference
	Peninsular Malaysia—angiosp. trees	2398	flora	
	Peninsulat Malaysia—ferns	500	flora	
TERRA FIRMA	Neotropics	33–276	69–0.1 h plots	Prance 1994
	Brazilian terra firma forest	165 spp/ 39 families	0.5 h	Hartshorn 1983
	Costa Rica terra firma forest	44	4 h	Hartshorn 1983
	Costa Rica terra firma forest	68	4 h	Hartshorn 1983
	Costa Rica terra firma forest	88	4 h	Hartshorn 1983
	Costa Rica terra firma forest	112	4 h	Hartshorn 1983
	Costa Rica terra firma forest	108	2 h	Hartshorn 1983
	Costa Rica terra firma forest	108	1 h	Hartshorn 1983
	Costa Rica terra firma forest	102	0.75 h	Hartshorn 1983
	Costa Rica terra firma forest	61	0.2 h	Hartshorn 1983
	Costa Rica terra firma forest	36	0.2 h	Hartshorn 1983
WETLANDS	Brazilian varzea forest	37 spp/ 17 families	0.5 h	Prance 1994
	Costa Rica natural levee	24	0.1 h	Hartshorn 1983
	Costa Rica swamp	115	2 h	Hartshorn 1983
FLORAS	La Selva, Costa Rica, tree species	253	flora	Hartshorn & Poveda 1983
	swamp diversity	27		Hartshorn & Poveda 1983
	wet floodplain diversity	94		Hartshorn & Poveda 1983
	Costa Rica, trees of 7 perserves	863	flora	Hartshorn & Poveda 1983
	species occur in swamps	58		Hartshorn & Poveda 1983
	species exclusively swamp	4		Hartshorn & Poveda 1983
	riparian species	43		Hartshorn & Poveda 1983
	exclusively riparian species	6		Hartshorn & Poveda 1983

(*Continued on next page*)

TABLE 11.4. (continued)

Sampling Region	Environment or Flora	Sample	Diversity	Geographic Location	Citation
Southeast USA Floras					
	FLORAS				
		flora	9	Okeefenokee swamp trees	Best 1984
			18	Cypress domes—trees	Best 1984
			27	Bayhead swamps—trees	Best 1984
			30	Mixed hardwoods wetland—trees	Best 1984
			71	Mesic mixed hardwood forest—trees	Best 1984
		flora	101	Okeefenokee swamp	Lowe & Jones 1984
			65	forb and grass spp.	Lowe & Jones 1984
			6	vines	Lowe & Jones 1984
			21	shrubs	Lowe & Jones 1984
			9	trees	Lowe & Jones 1984
		flora	33 tree / 100 total	River floodplain	Lugo 1984
			13 tree / 99 total	Old fields	Lugo 1984
			5 tree / 179 total	Pine barrens	Lugo 1984
			7 tree / 23 total	Sand scrub	Lugo 1984
Floras of Great Britain					
	FLORAS				
		flora	3354	Total vascular plant species	Kent 1992
		flora	464	Total fen species	Wheeler 1993
		flora	109	Total bog species	Wheeler 1993

Floras of Australia

FLORAS

Angiosperm species	approx. 25,000	flora	Westoby 1993
Tropical mangrove forests	37	flora	Gore 1983
Tidal salt marshes	55	flora	Gore 1983

Floras of the Arctic

TERRA FIRMA

Average spp. diversity of 10 sites	15.2	flora	Muc & Bliss 1977

WETLAND

Wet sedge meadow	12	flora	Muc & Bliss 1977
Tidal salt marsh	4	flora	Muc & Bliss 1977

Global Patterns

TOTAL DIVERSITY

Angiosperm genera	15,000	Takhtajan 1997
Angiosperm familiea	383	Takhtajan 1997
Nonangiosperm genera	127	Tryon and Tryon 1982
Nonangiosperm families	41	Tryon and Tryon 1982

WETLAND DIVERSITY

Angiosperm genera	450	Gore 1983
Angiosperm families	147	Gore 1983
Nonangiosperm genera	39	Gore 1983
Nonangiosperm families	24	Gore 1983

poor state of our knowledge of diversity in many parts of the tropics. Even so, they pale by comparison with global estimates of total vascular plant diversity of more than 15,000 genera and 424 families (Takhtajan 1997; Tryon and Tryon 1982), a gap that is unlikely to bridged by greater sampling (which should also add proportionally more terra firma taxa).

Wetlands account for between 5,570,000 and 8,558,000 km^2 of a total world land area (approximately 6%: Mitsch and Gosselink 1993). In the United States and Canada, for example, total land area is 18,616,960 km^2 (Hofstetter 1983; Zoltai and Pollett 1983; Ian Davidson, personal communication, 1997). Of this area, wetlands account for between 1,970,000 and 2,370,000 km^2 (10–13%), depending on the chosen definition of *wetland*. Peatlands in Southeast Asia account for only 50,000 km^2 of nearly 2,250,000 km^2 of total land area (about 2%). In 29 temperate countries, peatlands account for approximately 5% of total land area (Gore 1983). Clearly, the possibility that habitable area is an important constraint on wetland diversity must be considered seriously (Rosenzweig 1995). This should be qualified by the realization that not only is significantly more of the earth's surface terra firma than wetland, but the variation in edaphic conditions in that terra firma area is much greater than in the wetlands.

Overview

The "tessarae model" of Valentine (1980) provides a framework for the origin of "higher taxa" that may be the best single descriptor for the dynamics of the origin of vascular plant classes and, more generally, provides a clear link between ecology and the evolutionary process. In brief, the model suggests that unexploited or underexploited resource space is highly permissive of large morphological discontinuities early in an evolutionary radiation. Resource space is visualized in three dimensions as a field of tessarae similar to a checkerboard, with time along the vertical axis and niche space along the horizontal axes. Space filling begins at the bottom and, through time, the niche space (squares on the checkerboard) is progressively filled. Valentine considered body plans that differ significantly from ancestral forms to be less well integrated developmentally than the ancestral forms, and thus in need of a large, low-competition resource space in which to stabilize. Forms that were small modifications of ancestral body plans were assumed to be able to survive in resource pools of narrower breadth. As space fills, the likelihood of successful establishment of major morphological innovations declines because resources become increasingly scarce and thus more difficult for highly divergent forms to locate, especially if large targets of opportunity are needed. In effect, this aspect of selection acts as a filter that becomes finer and thus more restrictive as the radiation proceeds. Yet forms only slightly divergent from the ancestors will continue to be able to locate resources because they are more readily accommodated within existing resource limitations.

Using the categorization of Erwin (1992), the spectrum of modern vascular-plant architectures originated in a "novelty radiation," where the limits of the morphological envelope were described early and filling of the adaptive space with a broad spectrum of increasingly specialized species followed later. Incumbent or "home-field" advantage (Gilinsky and Bambach 1987; Pimm 1991; Rosenzweig and McCord 1991) provides ecological bounds to such radiations; in theory, once resource space is occupied, incumbents impede invasions of new species (DiMichele and Bateman 1996).

Whether or not such equilibrium models accurately describe patterns in the short term, studies of the origin of plant morphological features during the Devonian (Chaloner and Sheerin 1979; Knoll et al. 1984; Bateman in press) suggest that complexity initially accrued gradually and that the appearance of major architectures was concentrated in a relatively narrow time interval conforming fundamentally to equilibrium models (e.g., Valentine 1980). The paleobotanical studies of diversity patterns were not linked explicitly to either ecological or phylogenetic patterns, but they mesh well with subsequent data.

Although all aspects of this model still need considerable refinement, the general patterns seem fairly clear. High level phylogeny and morphospace conform well. Patterns of ecological distribution map with a high level of consistency onto the pattern of phylogenetic relationships. The Middle Devonian–Early Carboniferous radiation rapidly became highly channeled ecologically, resulting in an exceptional degree of high-level phylogenetic partitioning of ecological resources that continued to influence ecological dynamics through the middle of the Late Carboniferous (DiMichele and Phillips 1996). It was not until this system began to break down under the influence of major changes in global climate that seed plants began their rise to prominence in the late Paleozoic. The seed plant rise appears to have been globally asynchronous (Knoll 1984) but driven by similar dynamics in temperate as well as tropical regions. These linked patterns suggest a strong role for incumbent advantage in mediating major evolutionary replacements. As cautioned by Valentine (1980), however, radiations following breakdown of the primordial system are less likely to generate new body plans. Rather, they are fueled by innovations from ever more restricted (i.e., less inclusive) parts of the phylogenetic tree, representing modifications of existing forms rather than radically new solutions to the age-old problems of evolutionary opportunity.

ACKNOWLEDGMENTS

We thank Ian Davidson of Wetlands International, Ottawa, Canada, for providing access to their estimates of wetland areas. William Mitsch, Ohio State University, provided helpful advice on relevant wetlands literature. Jerrold Davis, Cornell University, kindly provided data on global diversity of angiosperms. Anne Raymond, Hermann Pfefferkorn, Warren Allmon, and the late Jack Sepkoski provided helpful comments on the manuscript. This research was partially supported by the Evolution of Terrestrial Ecosystems Program of the National Museum of Natural History and represents ETE Contribution Number 56. The Royal Botanic Garden Edinburgh is supported by the Scottish Office, Agriculture Environment and Fisheries Department.

REFERENCES

Algeo, T. J., R. A. Berner, J. B. Maynard, and S. E. Scheckler. 1995. Late Devonian oceanic anoxic events and biotic crises: "Rooted" in the evolution of vascular plants? *GSA Today* 5(45)64–66.

Andrews, H. N. and S. N. Agashe. 1965. Some exceptionally large calamite stems. *Phytomorphology* 15:103–108.

Andrews, H.N., A. E. Kasper, W. H. Forbes, P. G. Gensel, and W. G. Chaloner. 1977. Early Devonian flora of the Trout Valley Formation of northern Maine. *Review of Palaeobotany and Palynology* 23:255–285.

Banks, H. P. 1968. Early history of land plants. In E. T. Drake, ed., *Evolution and Environment, a Symposium Presented on the Occasion of the Hundredth Anniversary of the Peabody Museum*, pp. 73–107. New Haven CT: Yale University Press.

Banks, H. P. 1975. Reclassification of Psilophyta. *Taxon* 24:401–413.

Banks, H. P., P. M. Bonamo, and J. D. Grierson. 1972. *Leclercqia complexa* gen. et sp. nov., a new lycopod from the late Middle Devonian of eastern New York. *Review of Palaeobotany and Palynology* 14:19–40.

Banks, H. P., S. Leclercq, and F. M. Hueber. 1975. Anatomy and morphology of *Psilophyton dawsonii* sp. n. from the late Lower Devonian of Quebec (Gaspé), and Ontario, Canada. *Paleontographica Americana* 8:77–127.

Barthel, M. 1980. Calamiten aus dem Oberkarbon und Rotliegenden des Thüringer Waldes. *Festschrift 100 Jahre Arboretum (1879–1979), Berlin*, pp. 237–258.

Bateman, R. M. 1991. Palaeoecology. In C. J. Cleal, ed., *Plant Fossils in Geological Investigation: The Palaeozoic*, pp. 34–116. Chichester, U.K.: Ellis Horwood.

Bateman, R. M. 1994. Evolutionary-developmental change in the growth architecture of fossil rhizomorphic lycopsids: Scenarios constructed on cladistic foundations. *Biological Reviews* 69:527–597.

Bateman, R. M. 1996a. Nonfloral homoplasy and evolutionary scenarios in living and fossil land plants. In M. J. Sanderson and L. Hufford, eds., *Homoplasy: The Recurrence of Similarity in Evolution*, pp. 91–130. New York: Academic Press.

Bateman, R. M. 1996b. An overview of lycophyte phylogeny. In J. M. Camus, M. Gibby, and R. J. Johns, eds., *Pteridology in Perspective*, pp. 405–415. Kew, U.K.: Royal Botanic Gardens.

Bateman, R. M. In press. Architectural radiations cannot be optimally interpreted without morphological and molecular phylogenies. In M. H. Kurmann and A. R. Hemsley, eds., *The Evolution of Plant Architecture*. Kew, U.K.: Royal Botanic Gardens.

Bateman, R. M. and A. C. Scott. 1990. A reappraisal of the Dinantian floras at Oxroad Bay, East Lothian, Scotland: II. Volcanicity, palaeoenvironments and palaeoecology. *Transactions of the Royal Society of Edinburgh* 81:161–194.

Bateman, R. M. and W. A. DiMichele. 1994a. Heterospory: The most iterative key innovation in the evolutionary history of the plant kingdom. *Biological Reviews* 69:345–417.

Bateman, R. M. and W. A. DiMichele. 1994b. Saltational evolution of form in vascular plants: A neoGoldschmidtian synthesis. In D. S. Ingram and A. Hudson, eds., Shape and Form in Plants and Fungi. *Linnean Society of London, Symposium Series* 16:61–100.

Bateman, R.M., W. A. DiMichele, and D. A. Willard. 1992. Experimental cladistic analyses of anatomically preserved arborescent lycopsids from the Carboniferous of Euramerica: An essay in paleobotanical phylogenetics. *Annals of the Missouri Botanical Garden* 79:500–559.

Bateman, R.M., P. R. Crane, W. A. DiMichele, P. Kenrick, N. P. Rowe, T. Speck, and W. E. Stein. 1998. Early evolution of land plants: Phylogeny, physiology, and ecology of the primary terrestrial radiation. *Annual Review of Ecology and Systematics* 29:263–292.

Beck, C. B. 1964. Predominance of *Archaeopteris* in Upper Devonian flora of western Catskills and adjacent Pennsylvania. *Botanical Gazette* 125:126–128.

Beck, C. B. and D. C. Wight. 1988. Progymnosperms. In C. B. Beck, ed., *Origin and Evolution of Gymnosperms*, pp. 1–84. New York: Columbia University Press.

Beerbower, J. R. 1985. Early development of continental ecosystems. In B. Tiffney, ed., *Geological Factors and the Evolution of Plants*, pp. 47–92. New Haven CT: Yale University Press.

Berry, C. M. and M. Fairon-Demaret. 1997. A reinvestigation of the cladoxylopsid *Pseudosporochnus nodosus:* Leclercq et Banks from the Middle Devonian of Goé, Belgium. *International Journal of Plant Sciences* 158:350–372.

Best, G. R. 1984. An old-growth cypress stand in Okefenokee Swamp. In A. D. Cohen, D. J. Casagrande, M. J. Andrejko, and C. R. Best, eds., *The Okefenokee Swamp: Its Natural History, Geology, and Geochemistry*, pp. 132–143. Los Alamos NM: Wetland Surveys.

Bonamo, P. M. 1977. *Rellima thompsonii* (Progymnospermopsida) from the Middle Devonian of New York State. *American Journal of Botany* 64:1272–1285.

Bonamo, P. M. and H. P. Banks. 1967. *Tetraxylopteris schmidtii:* Its fertile parts and its relationship with the Aneurophytales. *American Journal of Botany* 54:755–768.

Brauer, D. F. 1981. Heterosporous barinophytacean plants from the Upper Devonian of North America, and a discussion of the possible affinities of the Barinophytaceae. *Review of Palaeobotany and Palynology* 33:347–362.

Bridge, J. S., P. M. Van Veen, and L. C. Matten. 1980. Aspects of sedimentology, palynology and palaeobotany of the Upper Devonian of southern Kerry Head, Co. Kerry, Ireland. *Geological Journal* 15:143–170.

Broutin, J., J. Doubinger, G. Farjanel, P. Freytet, H. Kerp, J. Langiaux, M.-L. Lebreton, S. Sebban, and S. Satta. 1990. Le renouvellment des floras au passage Carbonifère Permien: Approches stratigraphique, biologique, sédimentologique. *Compte Rendu Academie Sciences des Paris* 311:1563–1569.

Campbell, D. G. 1993. Scale and patterns of community structure in Amazonian forests. In P. J. Edwards, R. M. May, and N. R. Webb, eds., *Large Scale Ecology and Conservation Biology*, pp. 179–197. Oxford, U.K.: Blackwell Science.

Chaloner, W. G. 1958. The Carboniferous upland flora. *Geological Magazine* 95:261–262.

Chaloner, W. G. and A. Sheerin. 1979. Devonian macrofloras. In M. R. House, C. T. Scrutton, and M. G. Bassett, eds., *The Devonian System*, pp. 145–161. Palaeontological Association Special Paper 23.

Cichan, M. A. 1986. Conductance in the wood of selected Carboniferous plants. *Paleobiology* 12:302–310.

Cichan, M. A. and T. N. Taylor. 1984. A method for determining tracheid lengths in petrified wood by analysis of cross-sections. *Annals of Botany* 53:219–226.

Clement-Westerhof, J. A. 1988. Morphology and phylogeny of Paleozoic conifers. In C. B. Beck, ed., *Origin and Evolution of Gymnosperms*, pp. 298–337. New York: Columbia University Press.

Costanza, S. H. 1985. *Pennsylvanioxylon* of Middle and Upper Pennsylvanian coals from the Illinois Basin and its comparison with *Mesoxylon*. *Palaeontographica* 197B:81–121.

Crane, P. R. 1985. Phylogenetic relationships of seed plants. *Cladistics* 1:329–348.

Crane, P. R. 1990. The phylogenetic context of microsporogenesis. In S. Blackmore and S. B. Knox, eds., *Microspores: Evolution and Ontogeny*, pp. 11–41. London: Academic Press.

Cridland, A. A. and J. E. Morris. 1963. *Taeniopteris, Walchia,* and *Dichophyllum* in the Pennsylvanian System of Kansas. *University of Kansas Science Bulletin* 44:71–85.

Cúneo, N. R. 1996. Permian phytogeography in Gondwana. *Palaeogeography, Palaeoclimatology, Palaeoecology* 125:75–104.

Daber, R. 1959. Die Mittle-Vise-Flora der Tiefbohrungen von Doberlug-Kirchhain. *Geologie* 26:1–83.

Delevoryas, T. 1955. The Medullosaceae: Structure and relationships. *Palaeontographica* 97B:114–167.

Dennis, R. L. 1974. Studies of Paleozoic ferns: *Zygopteris* from the Middle and Upper Pennsylvanian of the United States. *Palaeontographica* 148B:95–136.

DiMichele, W. A. and T. L. Phillips. 1988. Paleoecology of the Middle Pennsylvanian-age Herrin coal swamp (Illinois) near a contemporaneous river system, the Walshville paleochannel. *Review of Palaeobotany and Palynology* 56:151–176.

DiMichele, W. A. and R. B. Aronson. 1992. The Pennsylvanian-Permian vegetational transition: A terrestrial analogue to the onshore-offshore hypothesis. *Evolution* 46:807–824.

DiMichele, W. A. and T. L. Phillips. 1994. Paleobotanical and paleoecological constraints on models of peat formation in the Late Carboniferous of Euramerica. *Palaeogeography, Palaeoclimatology, Palaeoecology* 106:30–90.

DiMichele, W. A. and T. L. Phillips. 1996. Clades, ecological amplitudes and ecomorphs: Phylogenetic effects and persistence of primitive plant communities in the Pennsylvanian-age lowland tropics. *Palaeogeography, Palaeoclimatology, Palaeoecology* 127:83–105.

DiMichele, W. A. and R. M. Bateman. 1996. Plant paleoecology and evolutionary inference: Two examples from the Paleozoic. *Review of Palaeobotany and Palynology* 90:223–247.

DiMichele, W. A., T. L. Phillips, and G. E. McBrinn. 1991. Quantitative analysis and paleoecology of the Secor Coal and roof-shale floras (Middle Pennsylvanian, Oklahoma). *Palaios* 6:390–409.

Doyle, J. A. and M. J. Donoghue. 1992. Fossils and seed plant phylogeny reanalyzed. *Brittonia* 44:89–106.

Driese, S. G., C. I. Mora, and J. M. Elick. 1997. Morphology and taphonomy of root and stump casts of the earliest trees (Middle to Late Devonian), Pennsylvania and New York, U.S.A. *Palaios* 12:524–537.

Eble, C. F. and W. C. Grady. 1990. Paleoecological interpretation of a Middle Pennsylvanian coal bed in the Appalachian basin. *International Journal of Coal Geology* 16:255–286.

Edwards, David S. 1980. Evidence for the sporophytic status of the Lower Devonian plant *Rhynia gwynne- vaughanii* Kidston and Lang. *Review of Palaeobotany and Palynology* 29:177–188.

Edwards, Diane. 1980. Early land floras. In A. L. Panchen, ed., *The Terrestrial Environment and the Origin of Land Vertebrates*, pp. 55–85. Systematics Association Special Volume 15. London: Systematics Association.

Edwards, D. and U. Fanning. 1985. Evolution and environment in the Late Silurian–Early Devonian: The rise of the pteridophytes. *Philosophical Transactions of the Royal Society of London* 309B:147–165.

Elick, J. M., S. G. Driese, and C. I. Mora. 1998. Very large plant and root traces from the Early to Middle Devonian: Implications for early terrestrial ecosystems and atmospheric $p(CO_2)$. *Geology* 26:97–192.

Erwin, D. H. 1992. A preliminary classification of evolutionary radiations. *Historical Biology* 6:133–147.

Foote, M. J. 1993. Discordance and concordance between morphological and taxonomic disparity. *Paleobiology* 19:185–204.

Foote, M. J. 1994. Morphological disparity in Ordovician–Devonian crinoids and the early saturation of morphological space. *Paleobiology* 20:320–344.

Frakes, L. E., J. E. Francis, and J. I. Sykta. 1992. *Climate Modes of the Phanerozoic.* Cambridge MA: Cambridge University Press.

Frederiksen, N. O. 1972. The rise of the Mesophytic flora. *Geoscience and Man* 4:17–28.

Galtier, J. and B. Meyer-Berthaud. 1989. Studies of Early Carboniferous pteridosperm *Calamopitys:* A redescription of the type material from Saalfeld (GDR). *Palaeontographica* 213B:1–36.

Gastaldo, R. A. 1992. Regenerative growth in fossil horsetails following burial by alluvium. *Historical Biology* 6:203–219.

Gensel, P. G. 1992. Phylogenetic relationships of the zosterophylls and lycopsids: Evidence from morphology, paleoecology, and cladistic methods of inference. *Annals of the Missouri Botanical Garden* 79:450–473.

Gensel, P. G. and H. N. Andrews. 1984. *Plant Life in the Devonian.* New York: Praeger.

Gilinsky, N. L. and R. K. Bambach. 1987. Asymmetrical patterns of origination and extinction in higher taxa. *Paleobiology* 13:427–445.

Gore, A. J. P. 1983. *Ecosystems of the World 4B: Swamp, Bog, Fen and Moor.* Amsterdam: Elsevier Science.

Gould, S. J. 1991. The disparity of the Burgess Shale arthropod fauna and the limits of cladistic analysis: Why we must strive to quantify morphospace. *Paleobiology* 17:411–423.

Gould, S. J. and E. S. Vrba. 1982. Exaptation: A missing term in the science of form. *Paleobiology* 8:4–15.

Granoff, J. A., P. G. Gensel, and H. N. Andrews. 1976. A new species of *Pertica* from the Devonian of eastern Canada. *Palaeontographica* 155B:119–128.

Guo Yingting. 1990. Paleoecology of flora from coal measures of Upper Permian in western Guizhou. *Journal of the China Coal Society* 15:48–54.

Hartshorn, G. S. 1983. Plants: Introduction. In D. H. Janzen, ed., *Costa Rican Natural History*, pp. 118–157. Chicago: University of Chicago Press.

Hartshorn, G. S. and L. J. Poveda. 1983. Checklist of trees. In D. H. Janzen, ed., *Costa Rican Natural History,* pp. 158–183. Chicago: University of Chicago Press.

Havlena, V. 1970. Einige Bemerkungen zur Phytogeographie und Geobotanik des Karbons und Perms. C. R. *Sixth International Congress of Carboniferous Stratigraphy and Geology,* Sheffield, U.K., 1967, v. 3, pp. 901–912.

Hofstetter, R. H. 1983. Wetlands in the United States. In A. J. P. Gore, ed., *Ecosystems of the World 4B: Swamp, Bog, Fen, and Moor,* pp. 201–244. New York: Elsevier.

Hotton, C. L., F. M. Hueber, D. H. Griffing, and J. S. Bridge. 1996. Paleoenvironments of early land plants: An example from the Early Devonian of Gaspé, Canada. *Fifth Conference of the International Organization of Palaeobotany,* June 30–July 5, 1996, Abstracts, p. 47.

Hotton, C. L., D. H. Griffing, J. S. Bridge and F. M. Hueber. In press. Early terrestrial plant environments: An example from the Emsian of Gaspe, Quebec, Canada. In P. G. Gensel and D. Edwards, eds., *Early Land Plants and Their Environments.* New York: Columbia University Press.

Hueber, F. M. 1992. Thoughts on the early lycopsids and zosterophylls. *Annals of the Missouri Botanical Garden* 79:474–499.

Jennings, J. R. 1984. Distribution of fossil plant taxa in the Upper Mississippian and Lower Pennsylvanian of the Illinois Basin. *C. R. Ninth International Congress of Carboniferous Stratigraphy and Geology*, 1979, Urbana IL, v. 2, pp. 310–312.

Jennings, J. R. 1986. A review of some fossil plant compressions associated with Mississippian and Pennsylvanian coal deposits in the central Appalachians, Illinois Basin, and elsewhere in the United States. *International Journal of Coal Geology* 6:303–325.

Josten, K.-H. 1991. Die Steinkohlen-Floren Nordwestdeutschlands. *Fortschritte in der Geologie von Rheinland und Westfalen* 36(Textband):1–436.

Kartawinata, K. 1990. A review of natural vegetation studies in Malesia, with special reference to Indonesia. In P. Baas, K. Kalkman, and R. Geesink, eds., *The Plant Diversity of Malesia*, pp. 121–132. Dordrecht, Netherlands: Kluwer Academic Publishers.

Kaspar, A. and H. N. Andrews. 1972. *Pertica*, a new genus of Devonian plants from northern Maine. *American Journal of Botany* 59:897–911.

Kenrick, P. and P. R. Crane. 1991. Water-conducting cells in early fossil land plants: Implications for the early evolution of trachaeophytes. *Botanical Gazette* 152:335–356.

Kenrick, P. and P. R. Crane. 1997. *The Origin and Early Diversification of Land Plants: A Cladistic Study*. Studies in Comparative Evolutionary Biology 4. Washington DC: Smithsonian Institution Press.

Kent, D. H. 1992. *List of Vascular Plants of the British Isles*. London: Botanical Society of the British Isles.

Kidston, R. and W. H. Lang. 1917. On Old Red Sandstone plants showing structure, from the Rhynie Chert Bed, Aberdeenshire. I. *Rhynia gwynne-vaughanii* Kidston and Lang. *Transactions of the Royal Society of Edinburgh* 51:761–784.

Knoll, A. H. 1984. Patterns of extinction in the fossil record of vascular plants. In M. H. Nitecki, ed., *Extinctions*, pp. 21–68. Chicago: University of Chicago Press.

Knoll, A. H. 1986. Patterns of change in plant communities through geological time. In J. Diamond and T. Case, eds., *Community Ecology*, pp. 126–141. New York: Harper and Row.

Knoll, A. H., K. J. Niklas, and B. H. Tiffney. 1979. Phanerozoic land plant diversity in North America. *Science* 206:1400–1402.

Knoll, A. H., K. J. Niklas, P. G. Gensel, and B. H. Tiffney. 1984. Character diversification and patterns of evolution in early vascular plants. *Paleobiology* 10:34–47.

Leary, R. L. and H. W. Pfefferkorn. 1977. An Early Pennsylvanian flora with *Megalopteris* and Noeggerathiales from west-central Illinois. *Illinois State Geological Survey Circular* 500:1–77.

Leclercq, S. and H. P. Banks. 1962. *Pseudosporochnus nodosus* sp. nov., a Middle Devonian plant with cladoxylalean affinities. *Palaeontographica* 110B:1–34.

Lesnikowska, A. D. 1989. Anatomically preserved marattiales from coal swamps of the Desmoinesian and Missourian of the midcontinent United States: Systematics, ecology, and evolution. Ph.D. thesis, University of Illinois at Urbana-Champaign, Urbana IL.

Levinton, J. 1988. *Genetics, Paleontology and Macroevolution.* Cambridge MA: Cambridge University Press.

Lord, J., M. Westoby, and M. Leishman. 1996. Seed size and phylogeny in six temperate floras: Constraints, niche conservatism, and adaptation. *American Naturalist* 146:349–364.

Lowe, J. A. and S. B. Jones Jr. 1984. Checklist of the vascular plants of the Okefenokee Swamp. In A. D. Cohen, D. J. Casagrande, M. J. Andrejko, and C. R. Best, eds., *The Okefenokee Swamp: Its natural History, Geology, and Geochemistry*, pp. 702–709. Los Alamos NM: Wetland Surveys.

Lugo, A. E. 1984. A review of early literature on forested wetlands in the United States. In K. T. Ewell and H. T. Odum, eds., *Cypress Swamps*, pp. 7–15. Gainesville: University of Florida Press.

Lyon, A. G. and D. Edwards. 1991. The first zosterophyll from the Lower Devonian Rhynie Chert, Aberdeenshire. *Transactions of the Royal Society of Edinburgh, Earth Sciences* 83:323–332.

Lyons, P. C. and W. C. Darrah. 1989. Earliest conifers in North America: Upland and/or paleoclimatic indicators. *Palaios* 4:480–486.

Mamay, S. H. and G. Mapes. 1992. Early Virgillian megafossils from the Kinney Brick Company quarry, Manzanita Mountains, New Mexico. *New Mexico Bureau of Mines and Mineral Resources Bulletin* 138:61–85.

Mapes, G. and R. A. Gastaldo. 1984. Late Paleozoic non-peat accumulating floras. In T. Broadhead, ed., *Plants: Notes for a Short Course*, pp. 115–127. University of Tennessee Department of Geological Sciences, Studies in Geology 15. Knoxville: University of Tennessee, Dept. of Geological Sciences.

Matten, L. C. 1974. The Givetian flora from Cairo, New York: *Rhacophyton, Triloboxylon*, and *Cladoxylon. Botanical Journal of the Linnean Society* 68:303–318.

Matten, L. C. and H. P. Banks. 1966. *Triloboxylon ashlandicum* gen. et sp. n. from the Upper Devonian of New York. *American Journal of Botany* 53:1020–1028.

Matten, L. C., W. R. Tanner, and W. S. Lacey. 1984. Additions to the silicified Upper Devonian/Lower Carboniferous flora from Ballyheigue, Ireland. *Review of Palaeobotany and Palynology* 43:303–320.

Meyen, S. V. 1982. The Carboniferous and Permian floras of Angaraland (a synthesis). *Biological Memoirs* 7:1–109.

Meyen, S. V. 1984. Basic features of gymnosperm systematics and phylogeny as evidenced by the fossil record. *Botanical Review* 50:1–112.

Mickle, J. E. 1980. *Ankyropteris* from the Pennsylvanian of eastern Kentucky. *Botanical Gazette* 141:230–243.

Mitsch, W. J. and J. G. Gosselink. 1993. *Wetlands,* 2nd ed. New York: Van Nostrand Reinhold.

Muc, M. and L. C. Bliss. 1977. Plant communities of Truelove lowland. In L. C. Bliss, ed., *Truelove Lowland, Devon Island, Canada: A High Arctic Ecosystem*, pp. 143–154. Edmonton: University of Alberta Press.

Niklas, K. J. 1977. Branching patterns and mechanical design in Paleozoic plants: A theoretic assessment. *Annals of Botany* 42:33–39.

Niklas, K. J. 1982. Computer simulations of early land plant branching morphologies: Canalization of patterns during evolution? *Paleobiology* 8:196–210.

Niklas, K. J. 1997. *Evolutionary Biology of Plants*. Chicago: University of Chicago Press.

Niklas, K. J. and H. P. Banks. 1990. A reevaluation of the Zosterophyllophytina with comments on the origin of the lycopods. *American Journal of Botany* 77:274–283.

Niklas, K. J., B. H. Tiffney, and A. H. Knoll. 1980. Apparent changes in the diversity of fossil plants. In M. K. Hecht, W. C. Steere, and B. Wallace, eds., *Evolutionary Biology*, v. 12, pp. 1–89. New York: Plenum Publishing.

Niklas, K. J., B. H. Tiffney, and A. H. Knoll. 1985. Patterns in vascular land plant diversification: An analysis at the species level. In J. W. Valentine, ed., *Phanerozoic Diversity Patterns: Profiles in Macroevolution*, pp. 97–128. Princeton NJ: Princeton University Press.

Nixon, K. C., W. L. Crepet, D. Stevenson, and E. M. Friis. 1994. A reevaluation of seed plant phylogeny. *Annals of the Missouri Botanical Garden* 81:484–533.

Oliver, F. W. and D. H. Scott. 1904. On the structure of the Palaeozoic seed *Lagenostoma lomaxii*, with a statement of the evidence upon which it is referred to *Lyginodendron*. *Philosophical Transactions of the Royal Society of London* 197B:193–247.

Peters, C. M., M. J. Balik, F. Kahn, and A. B. Anderson. 1989. Oligarchic forests of economic plants in Amazonia: Utilization and conservation of an important tropical resource. *Conservation Biology* 3:341–349.

Pfefferkorn, H. W. 1979. High diversity and stratigraphic age of the Mazon Creek flora. In M. H. Nitecki, ed., *The Mazon Creek Fossils*, pp. 129–142. New York: Academic Press.

Pfefferkorn, H. W. 1980. A note on the term "upland flora." *Review of Palaeobotany and Palynology* 30:157–158.

Pfefferkorn, H. W. and M. C. Thomson. 1982. Changes in dominance patterns in upper Carboniferous plant-fossil assemblages. *Geology* 10:641–644.

Phillips, T. L. 1974. Evolution of vegetative morphology in coenopterid ferns. *Annals of the Missouri Botanical Garden* 61:427–461.

Phillips, T. L. and H. N. Andrews. 1968. *Rhacophyton* from the Upper Devonian of West Virginia. *Journal of the Linnean Society of London* 61:37–64.

Phillips, T. L. and W. A. DiMichele. 1981. Paleoecology of Middle Pennsylvanian age coal swamps in southern Illinois: Herrin Coal Member at Sahara Mine No. 6. In K. J. Niklas ed., *Paleobotany, Paleoecology, and Evolution*, v. 1, pp. 231–284. New York: Praeger Press.

Phillips, T. L. and W. A. DiMichele. 1992. Comparative ecology and life-history biology of arborescent lycopsids in Late Carboniferous swamps of Euramerica. *Annals of the Missouri Botanical Garden* 79:560–588.

Phillips, T. L. and R. A. Peppers. 1984. Changing patterns of Pennsylvanian coal-swamp vegetation and implications of climatic control on coal occurrence. *International Journal of Coal Geology* 3:205–255.

Phillips, T. L., R. A. Peppers, and W. A. DiMichele. 1985. Stratigraphic and interregional changes in Pennsylvanian coal-swamp vegetation: Environmental inferences. *International Journal of Coal Geology* 5:43–109.

Pimm, S. L. 1991. *The Balance of Nature?* Chicago: University of Chicago Press.

Potonié, H. 1909. Die Tropen-Sumpfflachmoor-Natur der Moore des Produktiven Carbons. *Jahrbuch der Königl. Preufs. Geologischen Landesanstalt* 30:389–443.

Prance, G. T. 1994. A comparison of the efficacy of higher taxa and species numbers in the assessment of biodiversity in the tropics. *Philosophical Transactions of the Royal Society of London* 345B:89–99.

Pryer, K. M., A. R. Smith, and J. E. Skog. 1995. Phylogenetic relationships of extant ferns based on evidence from morphology and rbcL sequences. *American Fern Journal* 85:205–282.

Raymond, A. 1996. Latitudinal patterns in the diversification of mid-Carboniferous land plants: Climate and the floral break. *Illinois State Museum Scientific Papers* 26:1–18.

Raymond, A., W. C. Parker, and J. T. Parrish. 1985. Phytogeography and paleoclimate of the Early Carboniferous. In B. H. Tiffney, ed., *Geologic Factors and the Evolution of Plants*, pp. 169–222. New Haven: Yale University Press.

Retallack, G. J. 1985. Fossil soils as grounds for interpreting the advent of large plants and animals on land. *Philosophical Transactions of the Royal Society of London* 309B:108–142.

Retallack, G. J. 1997. Early Devonian forest soils and their role in Devonian global change. *Science* 276:583–585.

Retallack, G. J. and D. L. Dilcher. 1988. Reconstructions of selected seed ferns. *Annals of the Missouri Botanical Garden* 75:1010–1057.

Rex, G. M. 1986. The preservation and palaeoecology of the Lower Carboniferous plant deposits at Esnost, near Autun, France. *Géobios* 19:773–780.

Rex, G. M. and A. C. Scott. 1987. The sedimentology, palaeoecology and preservation of the Lower Carboniferous plant deposits at Pettycur, Fife, Scotland. *Geological Magazine* 124:43–66.

Rosenzweig, M. L. 1995. *Species Diversity in Time and Space.* Cambridge: Cambridge University Press.

Rosenzweig, M. L. and R. McCord. 1991. Incumbent replacements: Evidence for long-term evolutionary progress. *Paleobiology* 17:202–213.

Rothwell, G. W. 1982. New interpretations of the earliest conifers. *Review of Palaeobotany and Palynology* 37:7–28.

Rothwell, G. W. 1987. The role of development in plant phylogeny: A paleobotanical perspective. *Review of Palaeobotany and Palynology* 50:97–114.

Rothwell, G. W. 1996. Phylogenetic relationships of ferns: A palaeobotanical perspective. In J. M. Camus, M. Gibby, and R. J. Johns, eds., *Pteridology in Perspective*, pp. 395–404. Kew, U.K.: Royal Botanic Gardens.

Rothwell, G. W. and G. Mapes. 1988. Vegetation of a Paleozoic conifer community. In G. Mapes and R. H. Mapes, eds., *Regional Geology and Paleoecology of Upper Paleozoic Hamilton Quarry Area in Southeastern Kansas,* pp. 213–223. Kansas Geological Survey Guidebook 6. Lawrence KS: Kansas Geological Survey.

Rothwell, G. W. and R. Serbet. 1994. Lignophyte phylogeny and the evolution of spermatophytes: A numerical cladistic analysis. *Systematic Botany* 19:443–482.

Rowe, N. P. 1988. A herbaceous lycophyte from the Lower Devonian Drybrook Sandstone of the Forest of Dean, Gloucestershire. *Palaeontology* 31:69–83.

Rowe, N. P. and J. Galtier. 1988. A large calamopityacean stem compression yielding anatomy from the Lower Carboniferous of France. *Géobios* 21:109–115.

Rowe, N. P., T. Speck, and J. Galtier. 1993. Biomechanical analysis of a Palaeozoic gymnosperm stem. *Proceedings of the Royal Society of London* 252B:19–28.

Scheckler, S. E. 1976. Ontogeny of progymnosperms: I. Shoots of Upper Devonian Aneurophytales. *Canadian Journal of Botany* 54:202–219.

Scheckler, S. E. 1986a. Floras of the Devonian-Mississippian transition. In T. Broadhead, ed., *Plants: Notes for a Short Course,* pp. 81–96. University of Tennessee Department of Geological Sciences, Studies in Geology 15. Knoxville: University of Tennessee, Dept. of Geological Sciences.

Scheckler, S. E. 1986b. Geology, floristics and paleoecology of Late Devonian coal swamps from Appalachian Laurentia (U.S.A.). *Annales Societé Geologique Belgique* 109:209–222.

Scott, A. C. 1977. A review of the ecology of Upper Carboniferous plant assemblages, with new data from Strathclyde. *Palaeontology* 20:447–473.

Scott, A. C. 1978. Sedimentological and ecological control of Westphalian B plant assemblages from west Yorkshire. *Proceedings of the Yorkshire Geological Society* 41:461–508.

Scott, A. C. 1979. The ecology of coal measures floras from northern Britain. *Proceedings of the Geologists Association* 90:97–116.

Scott, A. C. 1980. The ecology of some Upper Palaeozoic floras. In A. L. Panchen, ed., *The Terrestrial Environment and the Origin of Land Vertebrates,* pp. 87–115. New York: Academic Press.

Scott, A. C. 1984. Studies on the sedimentology, palaeontology and palaeoecology of the Middle Coal Measures (Westphalian B, Upper Carboniferous) at Swillington, Yorkshire: I. Introduction. *Transactions of the Leeds Geological Association* 10:1–16.

Scott, A. C. 1990. Preservation, evolution and extinction of plants in Lower Carboniferous volcanic sequences in Scotland. In M. G. Lockley and A. Rice, eds., *Volcanism and Fossil Biotas,* pp. 25–38. Geological Society of America Special Paper 244. Boulder CO: Geological Society of America.

Scott, A. C. and J. Galtier. 1985. Distribution and ecology of early ferns. *Proceedings of the Royal Society of Edinburgh* 86B:141–149.

Scott, A. C., J. Galtier, and G. Clayton. 1985. Distribution of anatomically preserved floras in the Lower Carboniferous of western Europe. *Transactions of the Royal Society of Edinburgh, Earth Sciences* 75:311–340.

Scott, A. C., B. Meyer-Berthaud, J. Galtier, G. M. Rex, S. A. Brindley, and G. Clayton. 1986. Studies on a new Lower Carboniferous flora from Kingswood near Pettycur, Scotland: I. Preliminary report. *Review of Palaeobotany and Palynology* 48:161–180.

Skog, J. E. and H. P. Banks. 1973. *Ibyka amphikoma*, gen. et sp. n., a new protoarticulate precursor from the late Middle Devonian of New York State. *American Journal of Botany* 60:366–380.

Speck, T. 1994. A biomechanical method to distinguish between self-supporting and non self-supporting fossil plants. *Review of Palaeobotany and Palynology* 81:65–82.

Stein, W. E. 1981. Reinvestigation of *Arachnoxylon kopfii* from the Middle Devonian of New York State, USA. *Palaeontographica* 177B:90–117.

Stein, W. E. 1993. Modeling the evolution of stelar architecture in vascular plants. *International Journal of Plant Science* 154:229–263.

Stein, W. E. and F. M. Hueber. 1989. The anatomy of *Pseudosporochnus: P. hueberi* from the Devonian of New York. *Review of Palaeobotany and Palynology* 60:311–359.

Stein, W. E., D. C. Wight, and C. B. Beck. 1983. *Arachnoxylon* from the Middle Devonian of southwestern Virginia. *Canadian Journal of Botany* 61:1283–1299.

Stein, W. E., D. C. Wight, and C. B. Beck. 1984. Possible alternatives for the origin of the Sphenopsida. *Systematic Botany* 9:102–118.

Stevenson, D. W. and H. Loconte. 1996. Ordinal and familial relationships of pteridophyte genera. In J. M. Camus, M. Gibby, and R. J. Johns, eds., *Pteridology in Perspective*, pp. 435–467. Kew, U.K.: Royal Botanical Gardens.

Stidd, B. M. 1980. The current status of medullosan seed ferns. *Review of Palaeobotany and Palynology* 32:63–101.

Takhtajan, A. 1997. *Diversity and Classification of Flowering Plants.* New York: Columbia University Press.

Teichmüller, M. 1962. Die genese der Kohle. *C. R. Fourth International Congress of Carboniferous Stratigraphy and Geology*, v. 3, pp. 699–722.

Thomas, B. A. 1992. Paleozoic herbaceous lycopsids and the beginnings of extant *Lycopodium* sens. lat. and *Selaginella* sens. lat. *Annals of the Missouri Botanical Garden* 79:623–631.

Trivett, M. L. 1993. An architectural analysis of *Archaeopteris*, a fossil tree with pseudomonopodial and opportunistic adventitous growth. *Botanical Journal of the Linnean Society* 111:301–329.

Trivett, M. L. and G. W. Rothwell. 1985. Morphology, systematics, and paleoecology of Paleozoic fossil plants: *Mesoxylon priapi* sp. nov. (Cordaitales). *Systematic Botany* 10:205–223.

Tryon, R. M. and A. F. Tryon. 1982. *Ferns and Allied Plants with Special Reference to Tropical America.* New York: Springer-Verlag.

Valentine, J. W. 1980. Determinants of diversity in higher taxonomic categories. *Paleobiology* 6:444–450.

Westoby, M. 1993. Biodiversity in Australia compared with other continents. In R. Ricklefs and D. Schluter, eds., *Species Diversity in Ecological Communities*, pp. 170–177. Chicago: University of Chicago Press.

Wheeler, B. D. 1993. Botanical diversity of British mires. *Biodiversity and Conservation* 2:490–512.

Wight, D. C. 1987. Non-adaptive change in early land plant evolution. *Paleobiology* 13:208–214.

Willard, D. A., W. A. DiMichele, D. L. Eggert, J. C. Hower, C. B. Rexroad, and A. C. Scott. 1995. Paleoecology of the Springfield Coal Member (Desmoinesian, Illinois Basin) near the Leslie Cemetery paleochannel, southwestern Indiana. *International Journal of Coal Geology* 27:59–98.

Williamson, W. C. and D. H. Scott. 1894. Further observations on the organisation of fossil plants from the Coal Measures: I. *Calamites, Calamostachys* and *Sphenophyllum. Philosophical Transactions of the Royal Society of London* 185B:863–959.

Wing, S. L., L. J. Hickey, and C. C. Swisher. 1993. Implications of an exceptional fossil flora for Late Cretaceous vegetation. *Nature* 363:342–363.

Winston, R. B. 1983. A Late Pennsylvanian upland flora in Kansas: Systematics and environmental interpretations. *Review of Palaeobotany and Palynology* 40:5–31.

Zoltai, S. C. and F. C. Pollett. 1983. Wetlands in Canada: Their classification, distribution, and use. In A. J. P. Gore, ed., *Ecosystems of the World 4B: Mires: Swamp, Bog, Fen and Moor*, pp. 245–268. New York: Elsevier.

Author Index

Subject Index